Boundary Elements and other Mesh Reduction Methods XXXIII

WIT*PRESS*

WIT Press publishes leading books in Science and Technology.
Visit our website for the current list of titles.
www.witpress.com

WIT*eLibrary*

Home of the Transactions of the Wessex Institute.
Papers presented at BEM/MRM XXXIII are archived in the WIT elibrary in volume 52 of WIT Transactions on Modelling and Simulation (ISSN 1743-355X).
The WIT electronic-library provides the international scientific community with immediate and permanent access to individual papers presented at WIT conferences.
http://library.witpress.com

Thirty-Third International Conference on Boundary Elements and Other Mesh Reduction Methods

BEM/MRM 2011

Conference Chairmen

C.A. Brebbia
Wessex Institute of Technology, UK

V. Popov
Wessex Institute of Technology, UK

International Scientific Advisory Committee

Organised by
Wessex Institute of Technology, UK

Sponsored by
International Journal of Engineering Analysis with Boundary Elements (EABE)

WIT Transactions

Transactions Editor

Carlos Brebbia
Wessex Institute of Technology
Ashurst Lodge, Ashurst
Southampton SO40 7AA, UK
Email: carlos@wessex.ac.uk

Editorial Board

Boundary Elements and other Mesh Reduction Methods XXXIII

Editors

C.A. Brebbia
Wessex Institute of Technology, UK

V. Popov
Wessex Institute of Technology, UK

WIT*PRESS* Southampton, Boston

Editor:

C.A. Brebbia
Wessex Institute of Technology, UK

V. Popov
Wessex Institute of Technology, UK

Published by

WIT Press
Ashurst Lodge, Ashurst, Southampton, SO40 7AA, UK
Tel: 44 (0) 238 029 3223; Fax: 44 (0) 238 029 2853
E-Mail: witpress@witpress.com
http://www.witpress.com

For USA, Canada and Mexico

Computational Mechanics Inc
25 Bridge Street, Billerica, MA 01821, USA
Tel: 978 667 5841; Fax: 978 667 7582
E-Mail: infousa@witpress.com
http://www.witpress.com

British Library Cataloguing-in-Publication Data

A Catalogue record for this book is available
from the British Library

ISBN: 978-1-84564-542-7
ISSN: (print) 1746-4064
ISSN: (on-line) 1743-355X

The texts of the papers in this volume were set individually by the authors or under their supervision. Only minor corrections to the text may have been carried out by the publisher.

Printed in Great Britain by Quay Digital, Bristol.

Preface

This book contains edited versions of many of the papers presented at the 33rd International Conference on Boundary Element and other Mesh Reduction Methods, which was held in our New Forest Wessex Institute Campus in 2011.

Having been instrumental in originating the Method through the combined efforts of my associates at Southampton University, I do not cease to be surprised by its vitality. Since 1978 when the first conference took place and the first book bearing the BEM name was published, the method has constantly evolved.

It is most unusual for a scientific conference dealing with a particular methodology to survive for as long as this Conference has. I attribute this capacity to the wider remit of this Meeting, ie finding ways of reducing, or even better, totally suppressing the need for a mesh.

In the 1970s we started to question the need to discretise the problem in order to find a numerical solution. It was the right question at the wrong time, as finite element was emerging as the dominant computational technique. This was the moment when our research on Boundary Elements started. It is difficult nowadays to envisage the hostile reception that BEM received at that time. It took another 10 to 15 years to demonstrate its advantages and by then we were already looking at ways of BEM evolving beyond its classical formulation.

This led to the development of meshless techniques – research which has continued until today. A substantial part of the current work however seems to concentrate on finding simple solutions which may not always lead to the best results.

It seems appropriate to put more emphasis on achieving robustness and accuracy in particular when solving large engineering problems. We need to develop general purpose computer codes equivalent to those produced with finite elements.

This requires to progress from the realm of mathematical possibilities into that of engineering certainty.

The BEM/MRM Conference is the ideal forum to discuss these and other issues which are crucial to the development of the Method. Since 1978, this series of conferences have produced more than 60 volumes recording all major advances in BEM/MRM.

Judging by the papers published in this Volume, the topic continues to develop, validating the need to hold these meetings on a regular basis. The next conference, due to take place in Split, Croatia from 25–27 June 2012, will ensure that.

I must express my sincere gratitude to all contributors for the quality of the papers and in particular to the members of the International Scientific Advisory Committee and other colleagues who helped to select them.

Carlos A. Brebbia
New Forest, UK, 2011

Contents

Section 1
Advances in solid mechanics

Probabilistic fatigue crack growth using BEM and reliability algorithms

E. D. Leonel & W. S. Venturini
University of São Paulo. School of Engineering of São Carlos, Department of Structural Engineering, Brazil

Abstract

Fatigue and crack propagation are phenomena affected by high uncertainties, where deterministic methods fail to predict accurately the structural life. This paper aims at coupling reliability analysis with boundary element method (BEM) in modeling probabilistic fatigue crack growth. BEM has been recognized as an accurate and efficient numerical technique in modeling crack growth problems. The dual BEM approach was adopted to model crack growth. The couple BEM and reliability algorithms allows us to consider uncertainties during the crack propagation process. In addition, it calculates the probability of fatigue failure for complex structural geometry and loading. Two coupling procedures are considered: direct coupling of reliability and mechanical solver and indirect coupling by the response surface method. Numerical applications show the performance of the proposed models in lifetime assessment under uncertainties, where the direct method has shown faster convergence than response surface method.
Keywords: fatigue crack growth, BEM, structural reliability.

1 Introduction

Fatigue and crack propagation problems have been widely studied by the scientific community in recent years, because crack growth can explain the failure of structures. The accurate modelling of complex engineering structures, including complex geometries and boundary conditions, requires numerical techniques. Therefore, to model crack propagation problems, numerical models are required because the structural geometry and, consequently, the boundary conditions in these problems change at each crack length increment.

WIT Transactions on Modelling and Simulation, Vol 52,
www.witpress.com, ISSN 1743-355X (on-line)
doi:10.2495/BE110011

Particularly, the boundary element method (BEM) has already been recognized as an accurate and efficient numerical technique to deal properly with many problems in engineering, especially crack growth problems. Compared to finite element and other domain mesh methods, BEM is recommended to analyze crack propagation problems due to its efficiency of dimensionality reduction, as only the boundary is discretized. The BEM is even more efficient for mixed mode propagation, as the remeshing problems are avoided, whatever the structural complexity. For these reasons, BEM has been widely applied to deal with fatigue and crack propagation problems [1–4].

Fatigue crack growth is a slow process that includes large uncertainties and requires appropriate inspection plan in order to prevent the risk of failure. The fatigue crack growth is governed by several parameters as structural geometry, initial crack size and configuration, material properties and loading history. All these conditions are random and appropriate analysis method is required on the basis of probabilistic models. An appropriate analysis of fatigue phenomena is performed by considering the problem in a probabilistic manner. Numerous studies have been carried out on probability-based fatigue assessment as [5–8]. Although these works, and many others [9–11], have developed probability-based fatigue assessment approaches and applications, the model assumptions are often very restrictive, leading to inappropriate application to practical engineering structures.

This paper aims at developing a general procedure allowing to deal efficiently random fatigue crack growth for complex structures. This goal is achieved by coupling a reliability model with the mechanical model based on BEM, which one simulates accurately fatigue and arbitrary crack propagation. The BEM model is based on the dual BEM formulation in which singular (displacement integral equation) and hyper-singular (traction integral equation) integral equations are adopted. Displacement integral equations are used for collocation points along the crack surface, whereas traction integral equations are used for collocation points along the opposing crack surface. This technique avoids singularities in the resulting algebraic system of equations, despite the fact that two of the collocation points defined on the opposite crack surfaces have the same coordinates.

Two reliability procedures are considered: the direct method (DM) based on implicit limit state function and the response surface method (RSM) based on polynomial approximation of the mechanical behavior. The DM is based on the direct application of first order reliability method (FORM), where the derivatives of the mechanical response are calculated by finite difference technique applied to BEM model. In the RSM, the implicit mechanical response, given by BEM, is approximated by a polynomial function, known as response surface. In this approach, the implicit mechanical response is converted into an explicit one. The reliability procedure is then applied to this response surface instead of the BEM itself. Finally, the failure probability regarding fatigue can be defined and the most probable crack path determined.

The developed procedures are applied to random fatigue problems, where the DM has shown to give faster convergence, with respect to RSM.

2 BEM formulation

The BEM has shown to be an attractive approach in various engineering fields, such as contact problems, fatigue and fracture mechanics, due to its precision and robustness in modelling structures with high stress concentration. In two-dimensional elasticity, the boundary element formulation is obtained considering a homogeneous domain Ω, with a boundary Γ. The equilibrium equation can be written in terms of displacements as:

$$u_{i,jj} + \frac{1}{1-2\upsilon} u_{j,ji} + \frac{b_i}{\mu} = 0 \qquad (1)$$

where μ is the shear modulus, υ is Poisson's ratio, u_i are the displacement components and b_i are the body forces. In this equation i=1,2 because it is 2D case. Using Betti's theorem, the singular integral for displacements is given, without body forces, by:

$$c_{il}(f,c)u_l(f) + ⨍_\Gamma P_{il}^*(f,c)u_l(c)\,d\Gamma = \int_\Gamma P_l(c)u_{il}^*(f,c)\,d\Gamma \qquad (2)$$

where P_l and u_l are respectively the tractions and displacements at the boundary, $⨍$ is the integral of Cauchy principal value and the term c_{il} is equal to $\delta_{il}/2$ for smooth contours. P_{il}^* *and* u_{il}^* are respectively the fundamental solutions, which ones are shown in [12].

Linear elastic two dimensional domains can be analyzed by evaluating Eqn. (2) on the elements located at the body's boundary. However, for solids with cracks, using this equation for the discretization of all boundaries leads to singularities, because both crack surfaces are located on the same geometrical position. To deal with crack problems using BEM, many formulations have been proposed. The dual boundary element formulation stands out, as it applies to analysis of arbitrary crack growth. In this formulation, four algebraic relationships are required at each node along the crack path. To avoid redundant relationships, these four relationships are obtained from two different integral equations, which are written in terms of displacements, Eqn. (2), and tractions.

The hyper-singular integral representation at the boundary, in terms of tractions, can be obtained from Eqn. (2), which must be differentiated to obtain the integral representation in terms of strains. Then, Hooke's law is applied to obtain the integral representation in terms of stress. Finally, multiplication by the director cosines of the normal to crack surfaces at the collocation point leads to the traction representation, as follow:

$$\frac{1}{2}P_j(f) + \eta_k ⨎_\Gamma S_{kj}(f,c)u_k(c)\,d\Gamma = \eta_k ⨍_\Gamma D_{kj}(f,c)P_k(c)\,d\Gamma \qquad (3)$$

where $⨎$ is the integral of Hadamard finite part, the terms S_{kj} and D_{kj} contain the derivatives of P_{ij}^* and u_{ij}^* respectively, as indicated in [12]. In this paper, only linear boundary elements were used. High order elements could be used. However, this simple boundary element allow us calculate accurately the boundary values with low computational memory requirements. The singular integrals are evaluated in numerical forms, using sub-element procedure, whereas the hyper-singular integrals are calculated by analytical expressions. This procedure has shown to be accurate and efficient enough in arbitrary crack growth analyses.

3 Fracture mechanics model

The fatigue life prediction is a challenging problem in engineering design, inspection and maintenance. It is highly important to give accurate estimation of the life distribution of mechanical and structural components, in terms of the number of load cycles. For various kinds of materials, the crack growth rate can be modeled using the Paris' law [13]:

$$\frac{da}{dN} = C\Delta K^n \tag{4}$$

where a is the crack length, N is the number of loading cycles, C and n are material constants, and ΔK is the stress intensity factor range.

Stress intensity factors depend on the crack and structural geometry as well as on the stress field at the crack tip. For complex structures, these parameters can only be properly calculated by numerical methods. In this paper, stress intensity factors are evaluated through the displacement correlation technique using BEM model. For plane structures, stress intensity factors for modes I (opening) and II (sliding) are given throughout the following expressions:

$$K_I = \sqrt{\frac{2\cdot\pi}{r}}\frac{\xi}{(\kappa+1)}u_o \tag{5}$$

$$K_{II} = \sqrt{\frac{2\cdot\pi}{r}}\frac{\xi}{(\kappa+1)}u_s \tag{6}$$

where K_I and K_{II} are respectively stress intensity factors for modes I and II, u_o is the crack opening displacement, u_s is the crack sliding displacement, r is the distance between the crack tip and the computation point (i.e. mesh node) and ξ and κ are material properties. These variables are evaluated for six couples of mesh nodes near to the crack tip. Then, stress intensity factors at the crack tip are obtained by a local extrapolation process. This process has shown accurate according our results.

In mixed mode propagation, it is necessary to calculate the equivalent stress intensity factor K, and the crack propagation (or bifurcation) angle θ_p. For this purpose, the maximum circumferential stress criterion is adopted, which yields:

www.witpress.com, ISSN 1743-355X (on-line)

$$K = K_I \cos^3\left(\frac{\theta_p}{2}\right) - 3K_{II}\cos^2\left(\frac{\theta_p}{2}\right)\sin\left(\frac{\theta_p}{2}\right) \tag{7}$$

$$\tan\left(\frac{\theta_p}{2}\right) = \frac{1}{4}\left[\frac{K_I}{K_{II}} \pm \sqrt{\left(\frac{K_I}{K_{II}}\right)^2 + 8}\right] \tag{8}$$

For each load cycle, the equivalent stress intensity factor is evaluated at the minimum and maximum load levels, namely $K_{\min}$ and $K_{\max}$ respectively. If the equivalent stress intensity factor value is bigger than the material toughness, brittle failure is considered and the analysis is stopped; in this case, we talk about failure. When $K_{\max} < K_{Ic}$, the stress intensity factor range ΔK is computed and compared to the threshold ΔK_{th}. If $\Delta K \leq \Delta K_{th}$, no crack propagation is considered. Otherwise, the crack growth rate is computed using the Paris' law. The fatigue failure is defined by large crack propagation. As recommended by [14], when the crack growth rate da/dN is bigger than 0.1 mm/cycle, the analysis is stopped and structural failure takes place (the remaining life is negligible).

4 Reliability analysis

The reliability analysis aims at computing the failure probability Pf regarding a specific failure scenario, known as the limit state (note that reliability and failure probability are complementary).

The leading step in the reliability assessment is to identify the basic set of random variables $X = [x_1, x_2, ..., x_n]^T$ for which uncertainties have to be considered. For all these variables, probability distributions are attributed to model randomness. These probability distributions can be defined by physical observations, statistical studies, laboratory analysis and expert opinion. The number of random variables is an important parameter to determine the computing time consumed during the reliability analysis. In order to reduce the size of the random variable space, it is strongly recommended to consider as deterministic all variables whose uncertainties lead to minor effects on the value of the failure probability.

The second step consists in defining a number of potentially critical failure modes. For each of them, a limit state function G(X) separates the space into two regions: the safe domain, where G(X)>0, and the failure domain where G(X)<0. The boundary between these two domains is defined by G(X)=0, known as the limit state itself. It is worth to mention that an explicit expression of the limit state function is not possible, and only desired points can be computed by running the BEM analysis.

The failure probability is evaluated by the integral, [15]:

$$P_f = \int_{G\leq 0} f_X(x_1, x_2, \ldots, x_n)\ dx_1, dx_2, \cdots dx_n \tag{9}$$

where $f_X(x_1, x_2, \ldots, x_n)$ is the joint density function of the variables X. As the evaluation of the above integration is impossible in practice, alternative procedures have been developed on the basis of the concept of reliability index β, [16]. This variable is defined by the distance between the median point and the failure domain in the normalized space of random variables. The reliability index allows us to compute the failure probability, using the first order reliability method (FORM), as: $P_f = \Phi(-\beta)$, where $\Phi(\cdot)$ is the standard Gaussian cumulated distribution function.

When numerical mechanical methods are involved, the structural reliability analysis can be performed by one of two approaches: direct application of the reliability procedure using the mechanical analysis tool, or the use of the response surface method as an explicit representation of the structural behavior, in order to perform the reliability analysis. These two approaches will be discussed below.

5 Coupled BEM-reliability procedures

In order to consider random fatigue crack propagation, it is required to couple the reliability procedures with the BEM model. As mentioned above, this coupling can be performed by either the direct method or the response surface method.

5.1 Direct method (DM)

The basic procedure consists in directly coupling the reliability model with the mechanical model. As described in the previous section, the limit state function defines the safety and failure domains. For fatigue crack growth, this limit state function can be written in terms of number of load cycles:

$$G(X) = N_{\text{Resistant}}(X) - N_{\text{Applied}}(X) \tag{10}$$

where N*Resistant*(*X*) is the number of cycles corresponding to structural failure and N*Applied*(*X*) is the applied number of cycles during the service life. In order to give invariance measure of safety, the random variables, defined in the physical space, are transformed into independent standard Gaussian variables [16], by using appropriate probabilistic transformation. Figure 1 illustrates this transformation showing that the performance function G(*X*) in the physical space is transformed to H(*U*) in the standard normalized space, where $U = [u_1, u_2, \ldots, u_n]^T$ denote the standard Gaussian variables.

In this standard space, the reliability index β is given by the minimum distance between the failure domain and the origin of the standard space, therefore evaluated by solving the constrained optimization problem:

$$\begin{gathered}\text{find: } U^* \\ \text{which minimizes: } \beta = \sqrt{U^T \cdot U} \\ \text{subject to: } \mathrm{H}(U) = 0\end{gathered} \tag{11}$$

The solution of this problem converges to the failure point nearest to the space origin, known as the design point or the most probable failure point X*. In the standard space, the distance between this point and the origin is the reliability index. The reliability index β can be found by applying the Rackwitz and Fiessler algorithm [17], directly to the mechanical model. As the number of cycles to failure is known point-by-point, the resistance N*Resistant*(*X*) is implicit, and therefore the derivatives of the limit state function can only be computed by the finite difference technique. In our case, the forward finite difference scheme was chosen because of its low computation cost. The numerical error due to finite difference may affect the convergence of the coupled procedure, as well as the precision of the solution, especially for nonlinear phenomena. However, for the problems studied in this paper, it was verified that this coupling procedure gives accurate results and stable convergence rate, with a reasonable number of mechanical analyses.

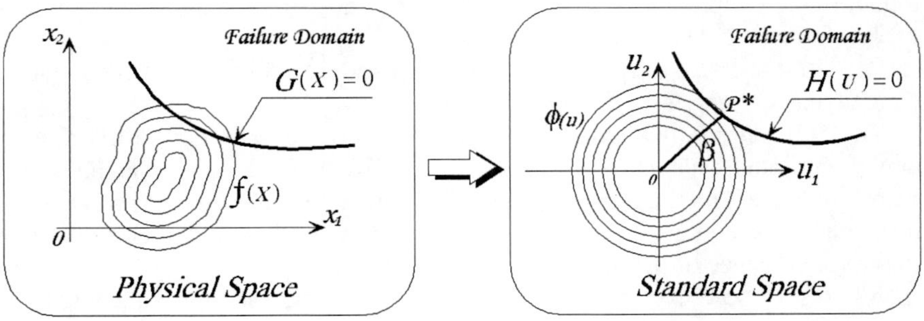

Figure 1: Probabilistic transformation from physical to standard space.

5.2 Response surface method (RSM)

The response surface method (RSM) is an efficient method for solving optimization problems, such one presented in Eqn. (11). The RSM allows us to replace complex models by approximate analytical functions based on the response values at various points in the design space. For reliability applications, the RSM is used to approximate the structural response at the vicinity of the most probable failure point, in terms of input variables related to geometrical data, material properties and boundary conditions. Naturally, the variables to be considered are those undergoing randomness within the reliability analysis. The set of realizations of these variables is named as experiment design (ED), which defines a set of structural responses from which a surface may be fitted using least square regression. Any shape surface, named response surface, can be adopted to represent the structural response. In this paper, the complete quadratic polynomial, defined in n-dimensional space, has been chosen to approximate the structural response.

The general procedure for evaluating the failure probability using response surface methods is divided into three steps:

1. In the first step, different sets of points are chosen according to the experiment design procedure. Each set constitutes the input values for which the mechanical response has to be computed. For the problem considered in this paper, the BEM model is performed to compute the mixed mode crack growth and the resisting number of load cycles corresponding to structural failure, for each set. During the iterative procedure, the points to be used in the ED are given in a hyper-cube centered at the current search point. The hyper-cube dimensions are given as a multiple of the variables standard-deviations. The mean values of the random variables are usually assumed as the first trial of the search point. After computing the mechanical responses for the selected points, the response surface can be approximated by polynomials identified by regression techniques.

2. The second step is defined by rewriting the limit state function in the standard normalized space using probabilistic transformations. Then, the minimum distance between the limit state function and the coordinate origin is calculated using an appropriate optimization procedure. This distance is the reliability index, β, as defined by Hasofer and Lind [16], and the design point as well as the direction cosines can be defined.

3. The third step is the estimation of the failure probability, which can be computed according to FORM approximation.

The procedure is iterative and it is continued by re-defining the ED at each new design point calculated. The center of the ED in the iteration *k* is the design point calculated in the iteration *k-1*. The convergence is given by the error measured between the reliability indexes of two successive iterations, in addition to the convergence of the design point coordinates.

In this procedure, the polynomial coefficients, *a*, are calculated by minimizing the quadratic error, ζ, between the exact responses, given by the BEM number of load cycles at failure $N_{\text{Resistant}}(w_k)$ at each ED point w_k, and the approximate response surface $RS(w_k)$ evaluated at the same points. The polynomial coefficients are obtained by minimizing:

$$\text{find: } a_c \textit{ and } a_{ij}$$

$$\text{which minimizes: } \zeta = \sum_k \left[RS(w_k) - N_{\text{Resistant}}(w_k) \right]^2 \tag{12}$$

$$\text{where } RS(w_k) = a_c + \sum_{i=1}^{n} a_{ii} x_i^2 + \sum_{i=1}^{n} \sum_{\substack{j=1 \\ j \neq i}}^{n} a_{ij} x_i x_j$$

It is to be noted that each numerical experiment to be used in the ED is obtained by complete fatigue crack growth analysis, in order to define the corresponding number of cycles. Therefore, each mechanical call represents a significant computing cost and experiments should be optimally designed, in order to reduce the numerical effort. In our case, the response surface represents a local approximation of the structural lifetime and is computed by:

$$G(X) = RS(X) - N_{\text{Applied}}(X) \tag{13}$$

Using the response surface, the design point search can be carried out by Rackwitz and Fiessler algorithm [17] which gives good results in this situation.

Beside its accuracy, the RSM is known to be robust for reliability analysis of complex structural systems, especially when high nonlinearities and bifurcations are involved. However, from the numerical point of view, this method becomes expensive when the number of random variables increases. For this reason, various experiment design schemes have been developed to reduce the number of the required points.

6 Application

A perforated panel is fixed at the bottom and subjected to uniform and cyclic tensile load at the top edge. An initial crack is located as shown in Fig. 2. The random variables considered in this analysis are the tensile load *P*, the hole diameter *D*, the location of the hole center with respect to the panel bottom, D_f, and the applied number of cycles N_{applied} (see Table 1).

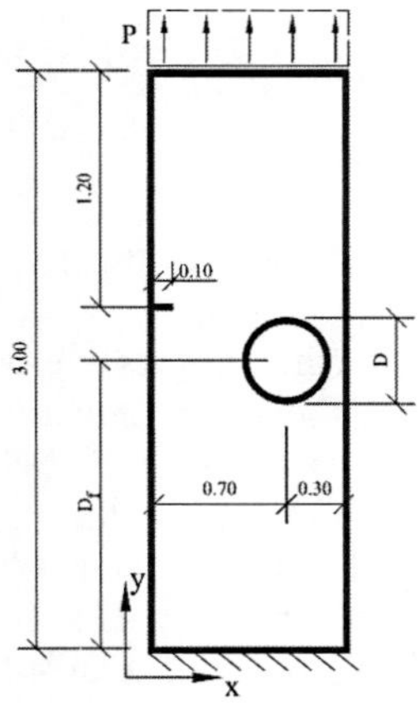

Figure 2: Perforated plate with initial crack. (dimensions in meter).

Table 1: Deterministic and random variables for this example.

Variable	Distribution	Mean	Standard Deviation
Young's modulus E *(MPa)*	deterministic	30 000	
Poisson's ratio ν	deterministic	0.20	
Paris'coefficient$C$ $(m^{1-1.5n}/cycle/kN^n)$	deterministic	2.0×10^{-10}	
Paris' Law parameter n	deterministic	2.7	
Toughness K_c $(MN/m^{1.5})$	deterministic	104	
Hole diameter D *(m)*	normal	0.4	0.025
Hole location D_f*(m)*	normal	1.5	0.15
Applied load P*(kN/m)*	normal	5.00	0.80
Applied cycles N_{Cycles}^{Load} (cycles)	normal	5×10^6	10^5

The RSM, with various experiments design, and DM were applied to analyze this structure considering the four random variables described in Table 1. The convergence curves for two representative random variables, number of load cycles and reliability index, are shown in Fig. 3. Regarding these results, we observe that the convergence is achieved, for DM and RSM progressive resizing, with maximum 12 iterations, while the RSM stepped resizing needs not less than 21 iterations to the convergence. In this example, the DM has shown faster convergence than RSM approaches. The reliability index obtained is 1.912 that corresponds to a failure probability of 0.0279.

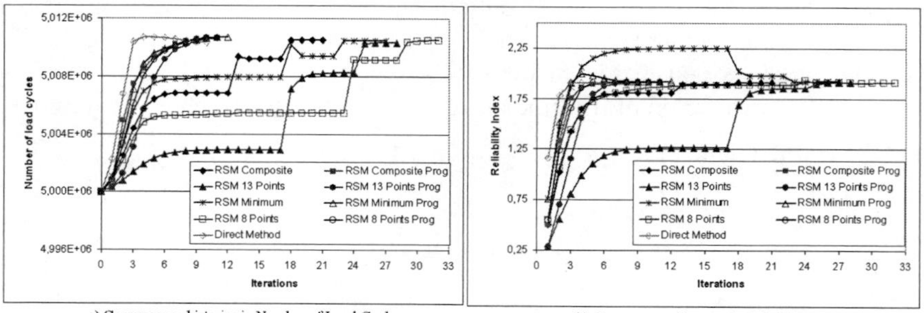

a) Convergence histories in Number of Load Cycles

b) Convergence histories in Reliability Index

Figure 3: Convergence history for load cycles and reliability index.

In spite of DM and RSM with progressive resizing needed maximum 12 iterations to the convergence, the computing time associated is considerable different. Figure 4 shows the number of BEM runs for different reliability methods used. Concerning RSM analysis, the maximum and minimum BEM runs were observed when 13 Points and Minimum ED were adopted, respectively. Using the first one, it required 728 BEM runs, while with Minimum ED only 120 mechanical runs were demanded. Considering DM only 40 mechanical calls were required to achieve the convergence. Thus, the good performance of DM is confirmed, as it requires a low number of BEM runs when compared with RSM procedures.

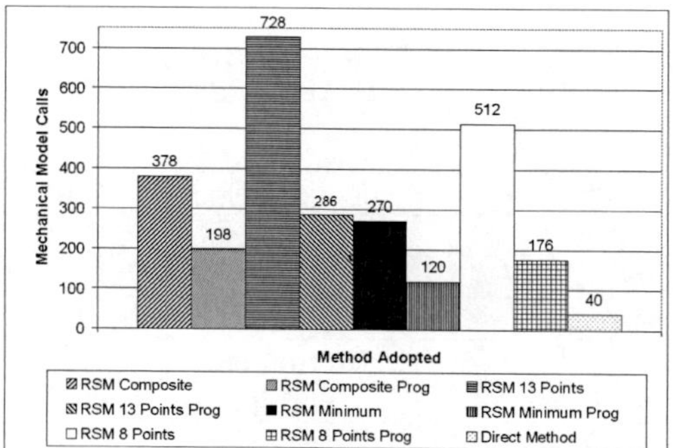

Figure 4: Number of BEM calls for the reliability analysis.

The crack growth path changes according the vertical position of hole, D_f. If the center hole position is aligned, or almost aligned, to the initial crack position, the crack grows up to the hole. However, below a certain position, the hole changes the stress field distribution in the structures as well as the crack growth path.

To better understand the role of the hole position, a parametric analysis has been performed by varying the mean hole position D_f from 1.00 m to 2.50 m (the other random and deterministic variables remains the same as described in Table 1. The reliability index and the most probable crack paths are shown in Fig. 5. We observe high sensitivity and dependence between reliability index and hole position. It can be seen that the minimum reliability index is achieved when the mean location is equal to 1.75 m. In this position, the reliability index calculated is 1.52.

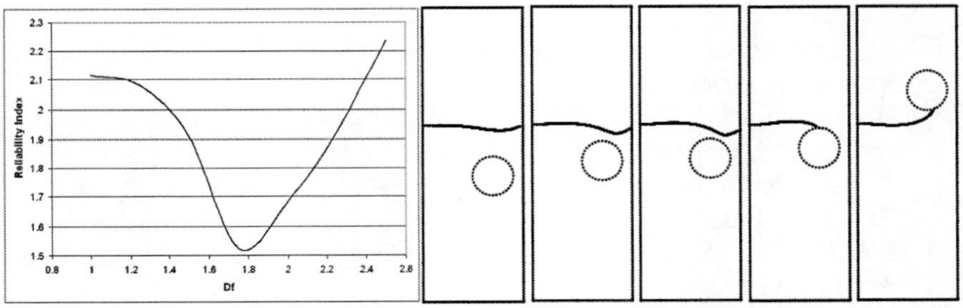

Figure 5: Influence of the hole position on the reliability index.

7 Conclusions

In this paper, a couple reliability and BEM model has been proposed for analysis of mixed mode crack propagation in structures subjected to fatigue. The BEM is an accurate approach to model random crack growth, especially in the framework of reliability analysis where many mechanical model runs are required. The DM and RSM have been applied to solve the reliability problem. The numerical application has shown good agreement between these two approaches. However, DM has demonstrated to be stable and more efficient than RSM.

The coupled model proposed is an interesting tool for probabilistic fatigue life assessment. Based on the reliability index results, inspections and maintenance plans can be developed and its costs, as well as the failure cost, can be considerably reduced.

Extension of the BEM code to consider multi materials and heterogeneities can be done in the future. In this case, structural systems can be simulated and a structural system reliability index can be achieved.

Acknowledgement

Sponsorship of this research project by the São Paulo State Foundation for Research – FAPESP – is greatly appreciated.

References

[1] Cisilino, A.P., Aliabadi, M.H. Three-dimensional boundary element analysis of fatigue crack growth in linear and non-linear fracture problems, Engineering Fracture Mechanics, 63, (1999), 713-733.
[2] Dell'erba, D.N., Aliabadi, M.H. Three-dimensional thermo-mechanical fatigue crack growth using BEM, International Journal of Fatigue, 22, (2000), 261-273.
[3] Leonel, E.D., Venturini, W.S. Dual boundary element formulation applied to analysis of multi-fractured domains, Engineering Analysis with Boundary Elements, 34, (2010), 1092-1099.
[4] Leonel, E.D., Venturini, W.S. Non-linear boundary element formulation with tangent operator to analyse crack propagation in quasi-brittle materials, Engineering Analysis with Boundary Elements, 34, (2010), 122-129.
[5] Bogdanoff, J.L., Kozin, F. Probabilistic models of cumulative damage. New York: Wiley. (1985).
[6] Wirsching, P.H., Torng, T.Y., Martin, W.S. Advanced fatigue reliability analysis. International Journal of Fatigue. 13, (1991), 389-483.
[7] Zhao, Y.X. A methodology for strain-based fatigue reliability analysis. Reliability Engineering and System Safety, 70, (2000), 205-213.
[8] Sobczyk, K., Spencer, B.F. Random Fatigue: From Data to Theory. Academic Press. (1992).
[9] Yu, L., Purnendu, K.D. Zheng, Y. A response surface approach to fatigue reliability of ship structures, Ships and Offshore Structures, 4, (2009), 253-259.
[10] Liu, W.K., Chen, Y., Belytschko, T., Lua, Y.J. Three Reliability methods for fatigue crack growth, Engineering Fracture Mechanics, 53, (1996), 733-752.
[11] Lua, Y.J., Liu, W.K., Belytschko, T. A stochastic damage model for the rupture prediction of a multi-phase solid. Part I1: statistical approach, Int. J. Fracture Mech. 55, (1992), 341-361.
[12] Hong, H.K., Chen, J.T. Derivations of Integral Equations of Elasticity, Journal of Engineering Mechanics, V.114, 6:1028-1044, 1988.
[13] Paris P.C., Erdogan F. A critical analysis of crack propagation laws. Journal of Basic Engineering, 85, (1963),528-534,.
[14] Pimenta, P.M., Ricardo, L.C.H. Camarao, A.F. Modern trends on fatigue, SAE Brasil International Conference on Fatigue, São Paulo, Brazil, (2001).
[15] Ditlevsen O., Madsen H.O. Structural reliability method. New York: John Wiley and Sons, (1996).
[16] Hasofer, A.M., Lind, N.C. Exact and invariant second moment code format, Journal of the Engineering Mechanics Division, ASCE. 100 (EM1), (1974), 111-121.
[17] Rackwitz, R., Fiessler, B. Structural reliability under combined random load sequences, Computers & Structures, 9, (1978), 489-494.

Modification of Gauss-Chebyshev quadrature for modelling of crack growth in the field of residual stresses

A. N. Galybin[1,2] & S. M. Aizikovich[3]
[1]*Wessex Institute of Technology, Southampton, UK*
[2]*Institute of Physics of the Earth, RAS, Moscow, Russia*
[3]*Rostov-on-Don Technical University, Russia*

Abstract

A modification of the Gauss-Chebyshev quadrature for solving singular integral equations appearing in plane fracture problems is proposed. This modification is aimed at accurate modelling of the crack growth in non-homogeneous stress fields, which is the case that presents certain difficulties because the positions of collocation points on the crack vary with the crack length. In the proposed modification they are fixed, which extends the application of the well-established technique for oscillating or piecewise loads acting on the crack surfaces. In the latter case, despite lowering the degree of approximation, the proposed method provides better accuracy in calculations as confirmed by comparisons of numerical and analytic results for some examples.
Keywords: singular integral equations, quadrature, crack, residual stresses.

1 Introduction

Some natural and artificial materials exhibit considerable level of residual stresses. The latter can significantly change fracture characteristics of the material and affect crack propagation. In practice the residual stresses are rarely known as continues functions of spatial coordinates but they can routinely be measured at discrete points. It is, therefore, important to possess accurate numerical methods for the prediction of crack behaviour based on experimental data. This paper is aimed to adapt the Gauss-Chebyshev quadrature method for studying crack propagation in elastic media under combined load of external and internal non-homogeneous stresses.

WIT Transactions on Modelling and Simulation, Vol 52, © 2011 WIT Press
www.witpress.com, ISSN 1743-355X (on-line)
doi:10.2495/BE110021

Quadrature of the Gauss type is often applied for solving singular integral equations, SIEs, appearing in plane elastic problem especially for bodies with cracks. It is believed that this approach had been originally developed by Multhopp in 1938 for the wing equation (see references in [1]) and became known in fracture mechanics after the paper by Erdogan and Gupta [2] published in 1972. The approach attracts attention of many researches due to simple programming and high accuracy in calculations of fracture characteristics. However there are some restrictions in the application of the method, for instance, discontinuities in applied loads or irregular crack paths cannot be directly handled by this method. For instance, Savruk [3] applied a modified scheme for the case of concentrated loads, otherwise the solution will be trivial because of the zero right hand side of the linear system obtained after discretization of SIE.

Modelling of the crack growth in non-homogeneous stress fields also presents certain difficulties because the positions of collocation points on the crack vary with the crack length. The essence of the method is that the choice of collocation points depends on the nodes of quadrature formulae, which provides highest degree of approximation. The positions of the nodes are also dependent on the contour. Therefore, after every finite step in quasistatic crack growth the positions of the collocation points are different from those at the previous steps. As a result due the right hand sides of a linear system algebraic equations obtained after the SIE discretisation (i.e., the loads acting at the crack surfaces) are always calculated at different collocation points. This is unimportant if the distributions of internal stresses are described by polynomial functions of spatial coordinates (because of convergence). However the original method produce large errors for discontinuous loads. For the discrete data (on residual stress measurements) the arrays of the right hand sides will be strongly dependant on interpolation used to calculate the loads at the collocation points. This may lead to the loss of accuracy in calculation of stress intensity factors and crack paths especially for curvilinear cracks.

The paper reports a modification of the Gauss-Chebyshev quadrature for solving SIEs. The SIE for curvilinear crack in the form presented in [3] is used in the analysis, although the results can be easily applied to any other forms of SIE or systems of SIEs appearing in fracture mechanics problem see survey [4].

2 Singular integral equation for curvilinear cracks

The SIE of the problem for curvilinear cracks in an elastic plane has the form [3]

$$\frac{1}{2\pi}\int_{\Gamma}\left\{\left[\frac{2}{t-t'}+k_1(t,t')\right]Q(t)\,dt+k_2(t,t')\overline{Q(t)\,dt}\right\}=N(t')+iT(t'),\quad t'\in\Gamma \qquad (1)$$

Here Γ is the contour that represents the union of all K cracks Γ_k (k=1,…K) and other boundaries; the sought function $Q(t)$ is proportional to the density of the jump of the complex displacement vector, $u(t)+iv(t)$, across Γ, this function is unbounded at the crack ends; $N(t')+iT(t')$ is the complex stress vector acting on

Γ. For the case of cracks in the infinite plane the regular kernels in (1) are given as follows

$$k_1(t,t') = \frac{d}{dt'} \ln \frac{t-t'}{\bar{t}-\bar{t}'}, \quad k_2(t,t') = -\frac{d}{dt'} \frac{t-t'}{\bar{t}-\bar{t}'} \tag{2}$$

For other configurations, e.g. cracks in a half plane or in a plane with a circular hole these kernels have more complex expressions, see [3] for details.

In order to provide a unique solution of the SIE the following condition of single valuedness of displacements should take place (for non-intersecting cracks)

$$\int_{\Gamma_k} Q(t)\, dt = 0, \quad k = 1, \ldots K \tag{3}$$

If some of the cracks intersect of bifurcate then the condition (3) should be modified by assuming single-valuedness of displacements after complete traverse of the contours formed by intersecting cracks.

Let us further consider the case of a single curvilinear crack in the complex plane, i.e. $\Gamma=\Gamma_1$. This case is sufficient to address the goal of this paper; more complex cases do not introduce essential difficulties.

It is convenient to introduce the parameterisation of the crack in the following form

$$t = t(s) = \int_0^s e^{i\theta(r)} dr, \quad a \le s \le b \tag{4}$$

where $\theta=\theta(s)$ is the angle between the tangent to the crack path and the positive direction of the *x*-axis, *a* and *b* are parameters associated with the crack length, $2L=b-a$. Substitution of (4) into (1) leads to the following SIE

$$\frac{1}{\pi} \int_a^b \left\{ \left[\frac{1}{s-r} + k_1(s,r) \right] \tilde{Q}(s) + k_2(s,r) \overline{\tilde{Q}(s)} \right\} ds = P(r), \quad a < r < b \tag{5}$$

where $\tilde{Q}(s) = Q(t(s))$, $P(r) = N(t(r)) + iT(t(r))$ and regular kernels in (5) assume the form

$$k_1(s,r) = e^{i\theta(s)} e^{-i\theta(r)} \operatorname{Re}\left(\frac{e^{i\theta(r)}}{t(s)-t(r)} \right) - \frac{1}{s-r}$$

$$k_2(s,r) = \frac{1}{2} \frac{e^{-i\theta(s)}}{\overline{t(s)}-\overline{t(r)}} \left(1 - \frac{t(s)-t(r)}{\overline{t(s)}-\overline{t(r)}} e^{-2i\theta(r)} \right) \tag{6}$$

Condition (4) assume the form

$$\int_a^b q(t(s)) e^{i\theta(t(s))} ds = 0 \tag{7}$$

By linear substitution

$$s = \frac{b-a}{2}\xi + \frac{b+a}{2}, \quad |\xi| < 1; \quad r = \frac{b-a}{2}\eta + \frac{b+a}{2}, \quad |\eta| < 1 \tag{8}$$

the problem consisting of SIE (5) and condition (7) assume the following form

$$\frac{1}{\pi}\int_{-1}^{1}\left\{\left[\frac{1}{\xi-\eta} + K_1(\xi,\eta)\right]q(\xi) + K_2(\xi,\eta)\overline{q(\xi)}\right\}\frac{d\xi}{\sqrt{1-\xi^2}} = p(\eta), \quad |\eta| < 1 \tag{9}$$

$$\int_{-1}^{1}\frac{q(\xi)e^{i\theta(\xi)}}{\sqrt{1-\xi^2}}d\xi = 0 \tag{10}$$

Here $q(\xi)$ is a new sought function bounded on [-1,1], it is defined as follows

$$q(\xi) = \widetilde{Q}\left(\frac{b-a}{2}\xi + \frac{b+a}{2}\right)\sqrt{1-\xi^2} \tag{11}$$

Regular kernels are

$$\begin{aligned} K_1(\xi,\eta) &= \frac{b-a}{2}k_1\left(\frac{b-a}{2}\xi + \frac{b+a}{2}, \frac{b-a}{2}\eta + \frac{b+a}{2}\right) \\ K_2(\xi,\eta) &= \frac{b-a}{2}k_2\left(\frac{b-a}{2}\xi + \frac{b+a}{2}, \frac{b-a}{2}\eta + \frac{b+a}{2}\right) \end{aligned} \tag{12}$$

As soon as a solution for q(ξ) is found he stress intensity factors can be calculated as follows

$$K_I^{\pm} - iK_I^{\pm} = q(\pm 1)\sqrt{\pi L} \tag{13}$$

Superscripts “±” corresponds to the right (left) crack tip, i.e. when *s=a*, *s=b* respectively.

3 Numerical implementation

3.1 Discretization of SIE quadratures for singular and regular integrals

The Gauss-Chebyshev quadratures for regular integrals have the form [3,5]

$$\frac{1}{\pi}\int_{-1}^{1}\frac{F_1(\xi)}{\sqrt{1-\xi^2}}d\xi = \frac{1}{n}\sum_{k=1}^{n}F_1(\xi_k), \quad \xi_k = \cos\frac{2k-1}{2n}\pi, \quad k = 1\ldots n \tag{14}$$

$$\frac{1}{\pi}\int_{-1}^{1}\sqrt{1-\eta^2}F_2(\eta)\,d\eta = \frac{1}{n}\sum_{j=1}^{n-1}\left(1-\eta_j^2\right)F_2(\eta_j), \quad \eta_j = \cos\frac{j\pi}{n}, \quad j = 1\ldots n-1 \tag{15}$$

Here ξ_k and η_j are zeroes of the Chebyshev polynomials of the first and second kind correspondingly

$$T_n(\eta) = \cos(n\arccos\eta) \tag{16}$$

$$U_{n-1}(\eta) = \cos(n\arccos\eta)\Big/\sqrt{1-\eta^2} \tag{17}$$

Formulas (14) and (15) are accurate for polynomials of (2*n*-1) and (2*n*-3) degrees correspondingly.

www.witpress.com, ISSN 1743-355X (on-line)

Savruk [3] refers to the following quadrature for the singular integral

$$\frac{1}{\pi}\int_{-1}^{1}\frac{1}{\sqrt{1-\xi^2}}\frac{q(\xi)}{\xi-\eta}d\xi=\frac{1}{n}\sum_{k=1}^{n}\frac{q(\xi_k)}{\xi_k-\eta}+\frac{q(\eta)U_{n-1}(\eta)}{T_n(\eta)} \tag{18}$$

It can be shown that the following quadrature is also valid

$$\frac{1}{\pi}\int_{-1}^{1}\sqrt{1-\eta^2}\frac{q(\eta)}{\eta-\xi}d\eta=\frac{1}{n}\sum_{j=1}^{n-1}\left(1-\eta_j^2\right)\frac{q(\eta_j)}{\eta_j-\xi}-\frac{q(\xi)T_n(\xi)}{U_{n-1}(\xi)} \tag{19}$$

Quadratures (18) and (19) are accurate for polynomials of $2n$ and $(2n\text{-}2)$ degrees respectively.

Application of (14) and (18) to SIE (9) leads to the following functional equation

$$\frac{1}{n}\sum_{k=1}^{n}\left\{\left[\frac{1}{\xi_k-\eta}+K_1(\xi_k,\eta)\right]q_k+K_2(\xi_k,\eta)\overline{q}_k\right\}+\frac{q(\eta)U_{n-1}(\eta)}{T_n(\eta)}=p(\eta),\quad |\eta|<1 \tag{20}$$

The condition of single valuedness assumes the form

$$\sum_{k=1}^{n}q_k e^{i\theta(\xi_k)}=0 \tag{21}$$

Second terms in (18) and (19) formulas vanish if η_j and ξ_k are used as the collocation points correspondingly. In this case (20) assume its usual form for the numerical implementation and it is accurate for polynomials of $(2n\text{-}1)$ degree.

3.2 Quadrature with reduced accuracy

Usual way in calculation of SIFs assumes extrapolation of the obtained solution up to the ends of the interval. This is based on the interpolation formulas, for instance, on the following one [3]

$$q(\eta)=\frac{T_n(\eta)}{n}\sum_{k=1}^{n}\frac{(-1)^k\sqrt{1-\xi_k^2}}{\xi_k-\eta}q_k,\quad |\eta|<1 \tag{22}$$

The use of formulas like this always leads to the loss of accuracy because (22) is accurate for polynomials of $(n\text{-}1)$ degree while the solution obtained for q_k is accurate for polynomials of $(2n\text{-}1)$ degree.

Substitution of (22) into (18) leads to the following quadrature for the singular integral

$$\frac{1}{\pi}\int_{-1}^{1}\frac{1}{\sqrt{1-\xi^2}}\frac{q(\xi)}{\xi-\eta}d\xi=\frac{1}{n}\sum_{k=1}^{n}\frac{1+(-1)^k U_{n-1}(\eta)\sqrt{1-\xi_k^2}}{\xi_k-\eta}q_k,\quad |\eta|<1 \tag{23}$$

Since interpolation is accurate for polynomials of $(n\text{-}1)$ degree, quadrature (23) is accurate for polynomials of the same degree if η is an arbitrary point on [-1,1] (including the ends); at the points η_j (23) is still accurate for polynomials of $2n$ degrees because the additional term in the numerator vanishes. For regular kernels and condition of single valuedness, one can use (14) that is accurate for polynomials of (2n-1) degree. Therefore, the SIE is replaced by the following functional equation on the interval $|\eta|<1$

$$\frac{1}{\pi}\sum_{k=1}^{n}\left\{\left[\frac{1+(-1)^k U_{k-1}(\eta)\sqrt{1-\xi_k^2}}{\xi_k-\eta}+K_1(\xi_k,\eta)\right]q_k+K_2(\xi_k,\eta)\bar{q}_k\right\}=p(\eta) \quad (24)$$

which is accurate for polynomials of

- (n-1) degree for arbitrary η and
- (2n-1) degree if the roots of $U_{n-1}(\eta)$ are used as collocation points (in this case (24) coincides with (20))

Let η_j (j=1…J) be collocation points chosen on [-1,1] such that $J \geq n$-1 then the following system of (J+1) linear algebraic (complex) equations, SLAE, is obtained

$$\sum_{k=1}^{n}\left(D_{kj}q_k+E_{kj}\bar{q}_k\right)=p_j, \quad \sum_{k=1}^{n}e^{i\theta(\xi_k)}q_k=0, \quad j=1\ldots J \quad (25)$$

Here the following notations have been introduced

$$D_{kj}=\frac{1+(-1)^k U_{k-1}(\eta_j)\sqrt{1-\xi_k^2}}{\pi(\xi_k-\eta_j)}+\frac{1}{\pi}K_1(\xi_k,\eta_j), \quad E_{kj}=\frac{1}{\pi}K_2(\xi_k,\eta_j), p_j=p(\eta_j)$$

Complex system (25) is further reduced to a real form by introducing the following array of 2n real unknowns X_k

$$q_k=X_k+iX_{n+k}, \quad k=1\ldots n \quad (26)$$

These unknowns can be placed in an array **X** and then (37) can be presented as

$$\mathbf{AX}=\mathbf{B} \quad (27)$$

Here **A** is (2J+2)x(2n) real matrix and **B** is (2J+2) real array.

Solution of the overspecified system (27) is obtained by the least squares method

$$\mathbf{X}=\left(\mathbf{A}^T\mathbf{A}\right)^{-1}\mathbf{A}^T\mathbf{B} \quad (28)$$

where superscript "T" stands for the transposed matrix.

It should be mentioned that Kim [6] obtained a somewhat similar system for the case of simplest SIE, i.e. when $K_1(\xi,\eta)=K_2(\xi,\eta)\equiv 0$, However he has not acknowledged the fact that the application of Lagrange interpolation reduces the accuracy of discretization and restricted his consideration by the case J=3n-1 in which η_j are zeroes of $U_{3n-1}(\eta)$.

Despite the reduction in accuracy, SLAE (27) may provide better results for some cases of loading then the conventional approach (the latter follows directly from (24) by using (n-1) roots of $U_{n-1}(\eta)$ as collocation points η_j) as, for instance, if continuous load oscillates sufficiently stronger then polynomials of (2n-1) degree or it is a piecewise function. In such cases the reduction of accuracy in conventional approach (due to incomplete account of the right hand side) can affect the results more significantly than the decrease of accuracy caused by the application of quadrature (24) to the left hand side of the SIE. An example supporting this statement is presented in the next section.

4 Examples

4.1 Straight crack under oscillating load

Stress intensity factors, SIFs, for a straight crack on the interval (-*L*,*L*) subjected to normal load, $p=p(t)$ can be calculated exactly as follows

$$K_{I,ideal}^{\pm} = -\frac{1}{\sqrt{\pi L}} \int_{-L}^{L} \sqrt{\frac{L \pm t}{L \mp t}}\, p(t)\,dt \qquad (29)$$

An example of random tractions $p(t)$ is examined further on, see Fig. 1. Values of normal tractions have randomly been generated at 21 equidistant points (including the ends) as follows

P=1,8,9,10,5,10,8,4,2,7,2,10,8,2,8,8,3,5,4,7,8.

After that the continuous tractions are found by either piecewise linear or cubic spline interpolation, which allows to calculate the ideal SIFs by (29) as follows

$K_{I\ ideal}^{+}$=6.289 and $K_{I\ ideal}^{-}$=6.090 for cubic spline interpolation and

$K_{I\ ideal}^{+}$=6.262 and $K_{I\ ideal}^{-}$=5.961 for piecewise linear interpolation.

The following errors have been found for this particular example (n=10)
for cubic spline interpolation: 20.9% and 28.2% in the usual approach and 0.1% and 0.5% for the modified quadrature with J=100;
for piecewise linear interpolation: 20.4% and 28.2% in the usual approach and 0.2% and 0.1% for the modified quadrature with J=100.

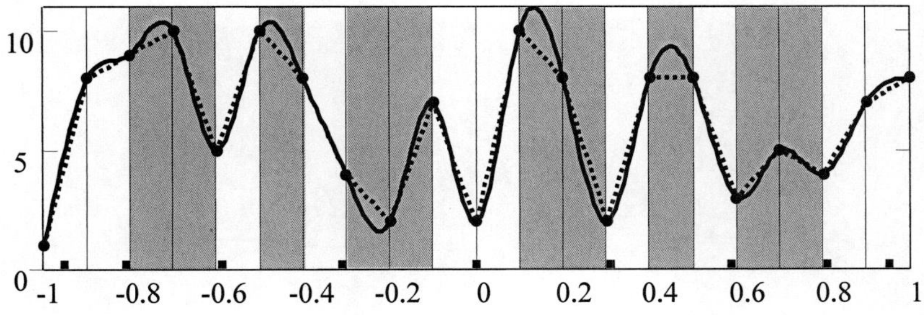

Figure 1: The load: circles are random data; solid line is cubic spline interpolation; dot line is linear interpolation; squares are collocation points, η; shaded areas represent the intervals not covered by the collocation points in the usual method.

These results demonstrate that the modified quadrature provides much better accuracy for the calculation of SIFs although it has less degree of accuracy than the usual scheme. This is explained by the positions of the collocation points in the usual approach. They are depicted by squares in Fig. 1 from which one can see that they do not cover a large part of the entire interval. For illustration, the whole interval is partitioned into 20 subintervals as shown in the Fig. 1 and then one can observe some intervals (shaded in the figure) that do not contain any collocation points. Since neither linear nor spline interpolations are polynomials of the 20th degree (in which case one would have obtained accurate result for n=10 used in the example), the application of the usual approach leads to large errors because it is not able to address all the oscillations. In the meantime, the modified quadrature reflects all particularities of the load shown in the figure because 5 collocation points are placed on every subinterval if J=100.

4.2 Straight crack under step-like loads

The scheme of loading is presented in Fig. 2, where the uniform load is of unit intensity, p=1; it covers both symmetrical and asymmetrical loads. Ideal stress intensity factors are found from (29) as follows

$$\frac{K^{\pm}_{I,ideal}}{p\sqrt{\pi L}} = \arcsin\frac{R_0+\rho}{L} + \arcsin\frac{R_0-\rho}{L} \mp \sqrt{1-\left(\frac{R_0+\rho}{L}\right)^2} \pm \sqrt{1-\left(\frac{R_0-\rho}{L}\right)^2} \quad (30)$$

where ρ=0 for symmetrical loading.

During calculations the number of collocations, J, varies such that at least one collocation point is placed between the neighbouring roots of $U_{n-1}(\eta)=0$. Then J is found through the minimum distance between the neighbouring roots

$$J = \beta\left[\min_{m=1\ldots n-2}\left(\cos\frac{\pi m}{n} - \cos\frac{\pi(m+1)}{n}\right)\right]^{-1} \quad (31)$$

Here parameter β determines the minimum number of collocation points on the smallest interval between m and m+1 zeroes of $U_{n-1}(\eta)$.

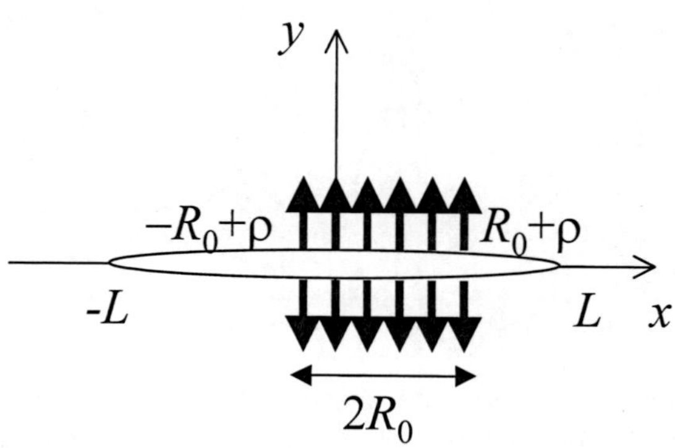

Figure 2: Cracks under step-like loads (symmetric (a) and (b) and asymmetric (c)).

It has been found that in order to obtain SIFs for crack loaded by a step-like load within 1% accuracy it is necessary to increase the number of either nodes, n, or/and collocation points, J. However, in the conventional scheme the accuracy varies depending on n even for large $n>100$. When n is fixed then the increase of J refines SIFs. Some results of calculations are shown below.

Example 1. $R_0=0.7$ $\rho=0$ (load at the ends), $n=12$, error 32.7% for the usual approach. For the modified scheme:

$J=15$ ($\beta=1$) gives qualitatively incorrect result (negative SIF)
$J=45$ ($\beta=3$), 6.15% error
$J=149$ ($\beta=10$), error is 0.3%, no further improvement
$J=445$ ($\beta=30$), 1.5% error (loss of accuracy)

Example 2. $R_0=0.7$ $\rho=0$ (load in the middle), $n=12$, error 32.5% for the usual approach. For the modified scheme:

$J=15$ ($\beta=1$), 146.6% error
$J=45$ ($\beta=3$), 6.3% error
$J=149$ ($\beta=10$) error is 0.3%, no further improvement
$J=445$ ($\beta=30$), 1.5% error (loss of accuracy)

Example 3. Load on (0.3,0.4). The right SIF is zero for $n=57$ (the left SIF has error of 3.1%) in the usual scheme while errors of 0.4% have been found for the modified scheme.

These examples indicate that the usual approach is not reliable and should be rejected for the case of step-like load. The modified quadrature provides reasonable results. The value of $\beta=10$ can be recommended for all calculations, however the increase of approximation is necessary with the decrease of the area where the load is applied. The use of n multiple of 4 can also be recommended for symmetrical cases (agrees with [7]), it gives minimum errors as compared with other close numbers.

4.3 Approximation of curved cracks by polygonal lines

For the case of circular crack depicted in Fig. 3 the exact expressions for SIFs are known, see, for instance, [3].

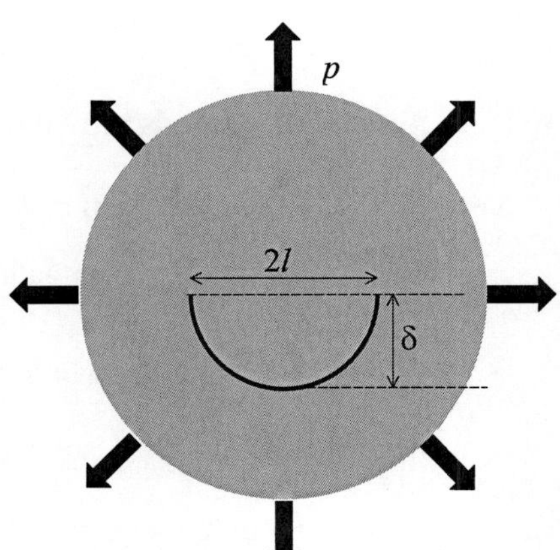

Figure 3: Circular crack in the plate subjected to biaxial uniform tension.

Curvilinear cracks of complex geometry can be approximated by piecewise continuous contours of simple geometry like linear segments and circular arks or by using spline interpolations. We use linear approximation by polygonal lines that may introduce essential errors. However the results obtained demonstrate that the use of quadrature (24) does not induce additional loss of accuracy.

It has been assumed that the crack consists of $2M$ linear segments of the length Δs each. The results presented for the case when $\delta=l$ in which the ideal SIFs are $K_I=K_{II}=0.4714\ l^{1/2}$. In all calculations n=8,12,16,24 and Δs=0.1 and 0.05.

For Δs=0.1 (M=16) maximum of errors in calculations of SIFs was greater than 3% regardless of n and J. Thus, this approximation is not sufficient.

For Δs=0.05 (M=32) maximum of errors in calculations of SIFs was greater than 3% for n=8 regardless of J. For n=12,16 and 24 the maximum errors were 1% for the usual approach and for the modified scheme they were more than 3% for $J=2M$; and 1-3% for $J=4M$, 10M.

5 Conclusions

The main advantage of the proposed numerical scheme is a possibility to calculate the right hand side of the SLAE at the collocation points whose positions are fixed. The number of collocation points is selected considerably greater that the number of nodes, which leads to an overspecified system of SLAE. On one hand this results in reduction of accuracy as the original method is accurate for polynomials of ($2n$-1) degree while the modified method for polynomials of (n-1) degree. On the other hand such accuracy is still better than in the traditional BEM methods, thus, the approach gives an effective way to analyse the crack growth by using the same datasets of the loads acting on the crack surfaces. This is important for modelling of crack propagation in the field of residual stresses or random stress fluctuations.

References

[1] Kalandiya, A.I. 1975. Mathematical methods of two-dimensional elasticity, Mir Publishers, Moscow.
[2] Erdogan, F. and Gupta, G.D. 1972. On the numerical solution of singular integral equations, Quart. Appl. Math. 29, 525-534.
[3] Savruk, M.P. 1981, 2D problems of elasticity for bodies with cracks, Naukova Dumka, Kiev.
[4] Chen, Y.Z. 1995. A survey of new integral equations in plane elasticity crack problem, Eng. Fract. Mech. 51, 97-134.
[5] Abramowitz, M. and Stegun, I.A. 1970. Handbook of mathematical functions. Dover, New York.
[6] Kim, S. 1999. Numerical solutions of Cauchy singular integral equations using generalized inverses, Computers Math. Applic. 38,183-195
[7] Kim, S. 1998. Solving singular integral equations using Gaussian quadrature and overdetermined system, Computers Math. Applic. 35, 63-71.

A tangential differential operator applied to stress and traction boundary integral equations for plate bending including the shear deformation effect

L. Palermo Jr.
Faculty of Civil Engineering, Architecture and Urban Design, State University of Campinas, Brazil

Abstract

Stress boundary integral equations (BIEs) are required in elastic or inelastic analyses of plate bending problems to obtain distributed shear, bending and twisting moments. Traction BIE, which is important to perform fracture analyses, is directly related to stress BIE. The collocation point position and the strategy to treat improper integrals are essential features studied in BIE for tractions or stresses at boundary points. The tangential differential operator (TDO) is used in stress and traction BIEs to reduce the strong singularities in the fundamental solution kernels and remaining singularities can be treated with the Cauchy principal value sense or the first order regularization. This study presents the application of the TDO for stress and traction BIEs used in plate bending models considering the shear deformation effect. The results in bending problems are obtained with traction BIE using TDO, instead of displacement BIE, and are compared to those in the literature where the problem was solved with traction BIE containing the strong singularity or with displacement BIE.

1 Introduction

Distributed shear, bending and twisting moments required in plate bending analyses are computed with stress BIEs. The differentiation in the fundamental solution kernels of displacement BIE, to obtain BIEs for stresses, increases the order of singularities. Strong singularities appear in fundamental solution kernels of stress BIE when values at boundary points are required as well as in those of

WIT Transactions on Modelling and Simulation, Vol 52,
www.witpress.com, ISSN 1743-355X (on-line)
doi:10.2495/BE110031

traction BIE because these equations are related according to the Cauchy formula. The collocation point position and the strategy to treat improper integrals are the essential features studied in numerical implementations for traction or stress BIE in plate bending or in elasticity [1, 2]. The use of the tangential differential operator (TDO) in conjunction with the integration by parts is a way to reduce the order of singularities in stress or traction BIE when Kelvin type fundamental solutions are used. Kupradze [3] first presented an application using the tangential differential operator (TDO) and Sladek and Sladek [4] employed the TDO in a curved crack solution. Regularized boundary element formulations employing TDO for potential and elasticity problems, including fracture mechanics formulations, were presented by Bonnet in [5]. The dual boundary element formulation for two dimensional problems of linear fracture mechanics using TDO in traction BIE was studied in [6]. The strategy presented in [6], which allowed the application of TDO in problems using non-conformal interpolations, was extended in [7] for the traction BIE in three dimensional elasticity.

The purpose of this study is introducing the TDO in stress and traction BIEs for plate bending models including the shear deformation effect. Kelvin type fundamental solution is the main requirement to apply the TDO in conjunction with the integration by parts to reduce singularities in BIE. The motivation to apply the TDO in plate bending models lies on developed studies for [6]. The efficiency of stress BIE for two dimensional elasticity problems was improved when the strong singularity was reduced [8] and the main result was the application for fracture problems in [6]. It is necessary to note the boundary element meshes and positions for collocation points used in [6] were first tested for traction BIEs containing strong singularities in [9] and the results were not changed with reference to those available in the literature. The application of the TDO for plate bending models is similar but not equal to the algebraic manipulation presented in [6] or [7]. The plate bending model represents the equilibrium of out of plane loads by in-plane stresses non-uniformly distributed on the thickness [10].

The boundary element formulation for plates using the TDO in BIE for stresses is extended to traction BIE. TDO was applied in the fundamental solution kernel with the strong singularity whereas other kernels remain unchanged. The results in bending problems are obtained with the traction BIE using TDO instead of displacement BIE and they are compared to those in the literature where the problem was solved with traction BIE containing the strong singularity or with displacement BIE. Fracture problems using the dual boundary element method (DBEM) were not considered because results in DBEM can be changed, too, according to the collocation point position of the displacement BIE, as shown in [11]. Thus, fracture problems will be analyzed in other study.

2 Application of the tangential differential operator

The equilibrium equations for an infinitesimal plate element under a transverse distributed loading $q(x_i)$ are next written with Latin indices considering values {1, 2 and 3} and Greek indices considering values {1, 2}:

$$M_{\alpha\beta,\beta} - Q_\alpha = 0 \tag{1}$$

$$Q_{\alpha,\alpha} + q = 0 \tag{2}$$

The plate has uniform thickness h and the constitutive relations are next written using a unified notation for the Reissner [12] and the Mindlin [13] model:

$$M_{\alpha\beta} = D\frac{1-\nu}{2}\left(\psi_{\alpha,\beta} + \psi_{\beta,\alpha} + \frac{2\nu}{1-\nu}\psi_{\gamma,\gamma}\delta_{\alpha\beta}\right) + \delta_{\alpha\beta}qRE \tag{3}$$

$$Q_\alpha = D\frac{1-\nu}{2}\lambda^2\left(\psi_\alpha + w_{,\alpha}\right) \tag{4}$$

with

$$\lambda^2 = 12\frac{\kappa^2}{h^2} \tag{5}$$

$$RE = \frac{\nu}{\lambda^2(1-\nu)} \tag{6}$$

D is the flexural rigidity, ν is the Poisson ratio, w is the deflection, ψ_α is the plate rotation in the direction α, $\delta_{\alpha\beta}$ is the Kronecker delta. The product qRE in equation (3) corresponds to the linearly weighted average effect of the normal stress component in the thickness direction and should be is considered in the Reissner model but not in the Mindlin model, when it should be considered null in equation (3). Shear parameter κ^2 is equal to 5/6 and $\pi^2/12$ for the Reissner and the Mindlin model, respectively.

A unified displacement BIE for the Reissner and the Mindlin model can be written in terms of generalized displacements and tractions presented in [14]:

$$C_{ij}u_j + \oint_\Gamma \left(T_j^i u_j - U_j^i t_j\right) d\Gamma = \iint_\Omega q\left(U_3^i - U_{\alpha,\alpha}^i RE\right) d\Omega \tag{7}$$

where u_α is ψ_α, u_3 is w, t_α is the product $M_{\alpha\beta}.n_\beta$, t_3 is the product $Q_\alpha.n_\alpha$. U_j^i represents the rotation (j=1,2) or the deflection (j=3) due to a unit couple (i=1,2) or a unit point force (i=3). C_{ij} is an element of the matrix C related to the collocation point position that makes a diagonal matrix with elements equal to 1 for internal collocation points or equal to 0.5 for collocation points on a smooth boundary. The integrand of the domain integral in equation (7) contains the RE factor, which should be cut off for analyses using the Mindlin model.

The application of the TDO will be next presented using Mindlin's model to simplify the presentation because the difference with reference to Reissner's model in the boundary element method corresponds to the introduction of RE parameter which is in an additional domain integral, as shown in equation (7) for the displacement BIE. The BIE for the deflection gradient at an internal point can be written using the differentiation in terms of field variables:

$$u_{3,\gamma} = \oint_\Gamma \left(\frac{\partial}{\partial x_\gamma}\left(T_\beta^3\right)u_\beta + \frac{\partial}{\partial x_\gamma}\left(T_3^3\right)u_3 - \frac{\partial}{\partial x_\gamma}\left(U_\beta^3\right)t_\beta - \frac{\partial}{\partial x_\gamma}\left(U_3^3\right)t_3\right)d\Gamma + \ldots \tag{8}$$

$$\ldots\ldots - \iint_\Omega \frac{\partial}{\partial x_\gamma}\left(U_3^3\right)q\,d\Omega$$

The differentiation is performed over the following fundamental solution kernels in equation (8):

$$T_3^3 = -\frac{1}{2\pi}\frac{1}{r}\frac{\partial r}{\partial n}; \quad U_3^3 = \frac{1}{8\pi D} r^2 (\ln(z) - 1) - \frac{1}{\pi D \lambda^2 (1-\nu)} \ln(z)$$

$$T_\beta^3 = -\frac{1}{4\pi}\left[(1+\nu) n_\beta \ln(z) + (1-\nu)\left(\frac{\partial r}{\partial n} r_\beta - \frac{1}{2} n_\beta\right)\right]; \quad U_\beta^3 = -\frac{1}{4\pi D} r\left(\ln(z) - \frac{1}{2}\right) r_\beta$$

The strong singularity is obtained from the differentiation over T_3^3 whereas 1/r singularities or logarithmic singularities result from differentiation over other terms. The TDO should be applied over the kernel related to T_3^3.

$$\oint_\Gamma \frac{\partial}{\partial x_\gamma}(T_3^3) u_3 d\Gamma = \oint_\Gamma \frac{\partial}{\partial x_\gamma}(Q_\beta^3 n_\beta) u_3 d\Gamma = \oint_\Gamma \left[D_{\beta\gamma}(Q_\beta^3) + n_\gamma \frac{\partial}{\partial x_\beta}(Q_\beta^3)\right] u_3 d\Gamma \quad (9)$$

$D_{bm}(\ )$ is the tangential differential operator, which has the following definition:

$$D_{bm}[f(y)] = n_b(y) f_{,m}(y) - n_m(y) f_{,b}(y)$$

The fundamental solution for a unit point force (i=3) is related to the equilibrium equation (2) in absence of the transverse distributed load q. $Q_{\beta,\beta}^3$ is equal to zero outside the source point which is the second term between brackets of equation (9). The final result from equation (9) is next written after the integration by parts to reduce the order of the singularity in the kernel, which changes the application of the TDO to u_3, and allowing a discontinuity in the boundary line [6]:

$$\oint_\Gamma \frac{\partial}{\partial x_\gamma}(T_3^3) u_3 d\Gamma = \oint_\Gamma Q_\beta^3 D_{\gamma\beta}(u_3) d\Gamma + \left[e_{3\beta\gamma} Q_\beta^3 u_3\right]_0^\Gamma \quad (10)$$

$e_{i\beta\gamma}$ is the permutation symbol.

The term between brackets in equation (10) is cut off when the boundary line is continuous. The final expression for the deflection gradient is given by:

$$u_{3,\gamma} = \oint_\Gamma \left(\frac{\partial}{\partial x_\gamma}(T_\beta^3) u_\beta + Q_\beta^3 D_{\gamma\beta}(u_3) - \frac{\partial}{\partial x_\gamma}(U_\beta^3) t_\beta - \frac{\partial}{\partial x_\gamma}(U_3^3) t_3 \right) d\Gamma + \ldots \quad (11)$$

$$\ldots\ldots - \iint_\Omega \frac{\partial}{\partial x_\gamma}(U_3^3) q d\Omega + \left[e_{3\beta\gamma} Q_\beta^3 u_3\right]_0^\Gamma$$

The algebraic manipulation to introduce the TDO in the BIE for the rotation gradient at an internal point is similar but not equal to that presented for the deflection gradient because the BIE for plates relates the distributed shear and couples with the deflection and rotations in same equation. The forces and displacements are not the same type as in BIEs for two or three-dimensional elasticity and additional care is necessary. The BIE for the rotation gradient is next written using the differentiation in terms of field variables:

$$u_{\rho,\gamma} = \oint_\Gamma \left(\frac{\partial}{\partial x_\gamma}(T_\beta^\rho) u_\beta + \frac{\partial}{\partial x_\gamma}(T_3^\rho) u_3 - \frac{\partial}{\partial x_\gamma}(U_\beta^\rho) t_\beta - \frac{\partial}{\partial x_\gamma}(U_3^\rho) t_3 \right) d\Gamma + \ldots \quad (12)$$

$$\ldots\ldots - \iint_\Omega \frac{\partial}{\partial x_\gamma}(U_3^\rho) q d\Omega$$

The fundamental solution kernels of equation (12), where the differentiation is performed, are next written as was done for equation (8):

$$T_{\beta}^{\rho} = -\frac{1}{\pi}\left[\left(\frac{\lambda K_1(z)}{2} + \frac{(1-\nu)}{4r} + \frac{A(z)}{r}\right)\left[(\delta_{\rho\beta} - 2r_{\beta}r_{\rho})\frac{\partial r}{\partial n} + n_{\rho}r_{\beta}\right] + \ldots\right.$$

$$\left. \ldots + \frac{(1-\nu)}{4r}n_{\beta}r_{\rho} - \frac{A(z)}{r}\left(2r_{\beta}r_{\rho}\frac{\partial r}{\partial n} - n_{\beta}r\right)\right];$$

$$U_{\beta}^{\rho} = \frac{1}{(1-\nu)}\frac{1}{\pi D}\left(B(z)\delta_{\beta\rho} - A(z)r_{\beta}r_{\rho}\right) - \frac{1}{4\pi D}\left[\left(\ln(z) - \frac{1}{2}\right)\delta_{\beta\rho} + r_{\beta}r_{\rho}\right];$$

$$T_3^{\rho} = \frac{1}{2\pi}\lambda^2\left(B(z)n_{\rho} - A(z)\frac{\partial r}{\partial n}r_{\rho}\right); \quad U_3^{\rho} = \frac{1}{4\pi D}r\left(\ln(z) - \frac{1}{2}\right)r_{\rho}.$$

with

$$z = \lambda r; \quad B(z) = K_0(z) + \frac{1}{z}\left[K_1(z) - \frac{1}{z}\right]; \quad A(z) = 2B(z) - K_0(z).$$

An expansion for small arguments should be considered for terms containing modified Bessel functions with real arguments (K_0, K_1) in the analysis of the singularity. The strong singularity is obtained from the differentiation on T_{β}^{ρ} whereas 1/r singularities or logarithmic singularities types result from differentiation over other terms. The TDO is applied on the kernel related to T_{β}^{ρ} of equation (12):

$$\oint_{\Gamma}\frac{\partial}{\partial x_{\gamma}}\left(T_{\beta}^{\rho}\right)u_{\beta}d\Gamma = \oint_{\Gamma}\frac{\partial\left(M_{\alpha\beta}^{\rho}n_{\alpha}\right)}{\partial x_{\gamma}}u_{\beta}d\Gamma = \oint_{\Gamma}\left[D_{\alpha\gamma}\left(M_{\alpha\beta}^{\rho}\right) + n_{\gamma}\frac{\partial\left(M_{\alpha\beta}^{\rho}\right)}{\partial x_{\alpha}}\right]u_{\beta}d\Gamma \quad (13)$$

The fundamental solution for a unit couple (ρ=1, 2) is related to the equilibrium equation (1). $M_{\alpha\beta,\alpha}{}^{\rho}$ is equal to $Q_{\beta}{}^{\rho}$ outside the source point, which is a regular function for the unit couple solution [15]. The final result from equation (13) is next written after the integration by parts to reduce the order of the singularity in the kernel, which changes the application of the TDO to u_{β}, and allowing a discontinuity over the boundary line [6]:

$$\oint_{\Gamma}\frac{\partial}{\partial x_{\gamma}}\left(T_{\beta}^{\rho}\right)u_{\beta}d\Gamma = \oint_{\Gamma}\left[M_{\alpha\beta}^{\rho}D_{\gamma\alpha}\left(u_{\beta}\right) + n_{\gamma}Q_{\beta}^{\rho}u_{\beta}\right]d\Gamma + \left[e_{3\alpha\gamma}M_{\alpha\beta}^{\rho}u_{\beta}\right]_0^{\Gamma} \quad (14)$$

The term between brackets in equation (14) is cut off when the boundary line is continuous. The final expression for the rotation gradient is given by:

$$u_{\rho,\gamma} = \oint_{\Gamma}\left(M_{\alpha\beta}^{\rho}D_{\gamma\alpha}\left(u_{\beta}\right) + n_{\gamma}Q_{\beta}^{\rho}u_{\beta} + \frac{\partial}{\partial x_{\gamma}}\left(T_3^{\rho}\right)u_3 - \frac{\partial}{\partial x_{\gamma}}\left(U_{\beta}^{\rho}\right)t_{\beta} + \ldots\right.$$

$$\left.\ldots - \frac{\partial}{\partial x_{\gamma}}\left(U_3^{\rho}\right)t_3\right)d\Gamma - \iint_{\Omega}\frac{\partial}{\partial x_{\gamma}}\left(U_3^{\rho}\right)q\,d\Omega + \left[e_{3\alpha\gamma}M_{\alpha\beta}^{\rho}u_{\beta}\right]_0^{\Gamma} \quad (15)$$

The BIE for the distributed shear at internal points can be obtained using the constitutive equation (4) together with equations (7) and (11).

$$Q_\gamma = D\frac{(1-\nu)}{2}\lambda^2\left\{\oint_\Gamma\left[\left(T^3_{\beta,\gamma} - T^\gamma_\beta\right)u_\beta + Q^3_\beta D_{\gamma\beta}(u_3) - T^\gamma_3 u_3\right]d\Gamma + \left[e_{3\beta\gamma}Q^3_\beta u_3\right]_0^\Gamma\right\} + \dots \tag{16}$$
$$\dots\dots + \oint_\Gamma\left(Q^\beta_\gamma t_\beta - Q^3_\gamma t_3\right)d\Gamma - \iint_\Omega qQ^3_\gamma d\Omega$$

The algebraic manipulation to get Q_γ^β and Q_γ^3 in equation (16) considered the symmetrical property of U_β^α and that U_β^3 is – U_3^β. Furthermore, the field decomposition presented in [15, 16] can be used to check the symmetry relations notwithstanding the algebraic expressions used for fundamental solutions.

The BIE for bending and twisting moments at internal points is obtained with the constitutive equation (3) together with equation (15). The BIE is next written using the symmetrical property of U_β^α and that U_β^3 is – U_3^β:

$$M_{\alpha\beta} = C_{\alpha\beta\rho\gamma}\left\{\oint_\Gamma\left(M^\rho_{\alpha\beta}D_{\gamma\alpha}(u_\beta) + n_\gamma Q^\rho_\beta u_\beta + T^\rho_{3,\gamma}u_3\right)d\Gamma + \left[e_{3\alpha\gamma}M^\rho_{\alpha\beta}u_\beta\right]_0^\Gamma\right\} + \dots$$
$$\dots\dots + \oint_\Gamma\left(M^3_{\alpha\beta}t_3 - M^\gamma_{\alpha\beta}t_\gamma\right)d\Gamma + \iint_\Omega qM^3_{\alpha\beta}d\Omega \tag{17}$$

with

$$C_{\alpha\beta\rho\gamma} = D\frac{(1-\nu)}{2}\left(\delta_{\alpha\rho}\delta_{\beta\gamma} + \delta_{\alpha\gamma}\delta_{\beta\rho} + \frac{2\nu}{1-\nu}\delta_{\alpha\beta}\delta_{\rho\gamma}\right)$$

The integrals of equations (16) and (17) are regular for internal points and exhibit singularities or singularity type of order 1/r when the field point approaches the collocation point. The BIEs for distributed shear and moments at a boundary point is defined as the limiting form of the corresponding BIE at an internal point when it is led to a point on the boundary. Equations (16) and (17) are next written for the collocation point on a smooth boundary:

$$\frac{1}{2}Q_\gamma = D\frac{(1-\nu)}{2}\lambda^2\left\{\oint_\Gamma\left[\left(T^3_{\beta,\gamma} - T^\gamma_\beta\right)u_\beta + Q^3_\beta D_{\gamma\beta}(u_3) - T^\gamma_3 u_3\right]d\Gamma + \dots \right. \tag{18}$$
$$\left.\dots\dots + \left[e_{3\beta\gamma}Q^3_\beta u_3\right]_0^\Gamma\right\} + \oint_\Gamma\left(Q^\beta_\gamma t_\beta - Q^3_\gamma t_3\right)d\Gamma - \iint_\Omega qQ^3_\gamma d\Omega$$

$$\frac{1}{2}M_{\alpha\beta} = C_{\alpha\beta\rho\gamma}\left\{\oint_\Gamma\left(M^\rho_{\alpha\beta}D_{\gamma\alpha}(u_\beta) + n_\gamma Q^\rho_\beta u_\beta + T^\rho_{3,\gamma}u_3\right)d\Gamma + \left[e_{3\alpha\gamma}M^\rho_{\alpha\beta}u_\beta\right]_0^\Gamma\right\} + \dots$$
$$\dots\dots + \oint_\Gamma\left(M^3_{\alpha\beta}t_3 - M^\gamma_{\alpha\beta}t_\gamma\right)d\Gamma + \iint_\Omega qM^3_{\alpha\beta}d\Omega \tag{19}$$

It is important to note the continuity requirement for the derivative of the displacement function at the collocation point. The BIEs for tractions are obtained from equation (18) and (19) when the distributed shear tensor and the moment tensor at the collocation point on the boundary point x' are multiplied by direction cosines of the outward normal at this point (n'), i.e. the Cauchy formula:

$$\frac{1}{2}t_3(x')=D\frac{(1-\nu)}{2}\lambda^2 n'_\gamma(x')\left\{\oint_\Gamma\left[\left(T^3_{\beta,\gamma}-T^\gamma_\beta\right)u_\beta+Q^3_\beta D_{\gamma\beta}(u_3)-T^\gamma_3 u_3\right]d\Gamma+\ldots\right. \tag{20}$$

$$\left.\ldots..+\left[e_{3\beta\gamma}Q^3_\beta u_3\right]_0^\Gamma\right\}+n'_\gamma(x')\oint_\Gamma\left(Q^\beta_\gamma t_\beta-Q^3_\gamma t_3\right)d\Gamma-n'_\gamma(x')\iint_\Omega qQ^3_\gamma d\Omega$$

$$\frac{1}{2}t_\beta(x')=C_{\alpha\beta\rho\gamma}n'_\alpha(x')\left\{\oint_\Gamma\left(M^\rho_{\alpha\beta}D_{\gamma\alpha}(u_\beta)+n_\gamma Q^\rho_\beta u_\beta+T^\rho_{3,\gamma}u_3\right)d\Gamma+\ldots\right.$$

$$\left.\ldots..+\left[e_{3\alpha\gamma}M^\rho_{\alpha\beta}u_\beta\right]_0^\Gamma\right\}+n'_\alpha(x')\left[\oint_\Gamma\left(M^3_{\alpha\beta}t_3-M^\gamma_{\alpha\beta}t_\gamma\right)d\Gamma+\iint_\Omega qM^3_{\alpha\beta}d\Omega\right] \tag{21}$$

The collocation points must be positioned to satisfy the continuity requirements of BIEs for tractions. The continuity of the displacement function at x' is the necessary condition for the generalized displacement BIE and it is satisfied when the collocation point is placed at the ends of the boundary element or inside the element. The continuities of the displacement and rotations derivatives at x' are required for traction BIEs and they can be satisfied when the collocation point is placed inside the boundary element.

3 Numerical implementation

Linear mapping functions were used to represent displacements and efforts in the boundary elements. The same mapping function was used for conformal and non-conformal interpolations with nodal parameters positioned at the ends of the elements. The collocation points were shifted to the interior of the element at a distance of a sixth part of its length starting from the end. The collocation points were always positioned in the boundary line with the position (ξ'), in the range (-1, 1): i) ξ'= -0.67 for continuous elements; ii) ξ'= -0.67 and ξ'= +0.67 for discontinuous elements. Analytical expressions were used to evaluate singular integrals with the Cauchy principal value sense whereas the Gauss-Legendre scheme was used for regular integrals. An expansion for small arguments was considered for terms containing modified Bessel functions with real arguments (K_0, K_1) [17]. The diagonal terms were directly obtained using the mapping function and the collocation point position on the element. The numerical implementation for the tangential differential operator considered the differentiation of the mapping function according to the following relation [18]:

$$D_{bm}\{f[y(\xi)]\}=n_b(y)f_{,m}(y)-n_m(y)f_{,b}(y)=e_{3\gamma\rho}\frac{1}{|J|}\frac{df}{d\xi}$$

J is the Jacobian of the transformation and ξ is the intrinsic coordinate to perform the integration on the element.

4 Example – the torsion of a cube

The problem was analyzed with traction BIEs using the TDO, equations (20) and (21), replacing the displacement BIEs (equation (7)) for the solution with boundary element method. The torsion of cube is used to show the precision of the formulation and justify the adopted position for the collocation point. A mesh with 128 linear discontinuous elements (256 nodes) along the sides was used. The results were compared with those presented in [1] where the traction BIE containing the strong singularity was used in a mesh containing 128 quadratic discontinuous elements (384 nodes) and with those presented by Weeën [14], that employed symmetry conditions in the solution with the displacement BIE using 8 quadratic continuous elements (16 nodes). The cube has the side length equals to 2a, two opposite faces are free of stress and other faces are under torsion according to Saint-Venant hypotheses (free warping). The generalized displacements were used to introduce torsion:

$$u_3 = \varphi a x_2$$

$$u_2 = \varphi a$$

φ is the prescribed rotation angle

Table 1: Relative values of plate rotations in the normal direction $\left(\psi_n(x_2) \big/ \psi_n(a) \right)$.

(x_2/a)	[12]	[1]	[14]	ξ=0.5	ξ =0.67	ξ =0.75
1.00	1.000	1.000	1.001	1.000	1.000	1.000
0.75	-0.055	-0.051	-0.053	-0.096	-0.058	-0.034
0.50	-0.387	-0.382	-0.386	-0.428	-0.389	-0.366
0.25	-0.292	-0.290	-0.290	-0.317	-0.294	-0.280
0.00	0.000	0.000	0.000	0.000	0.000	0.000

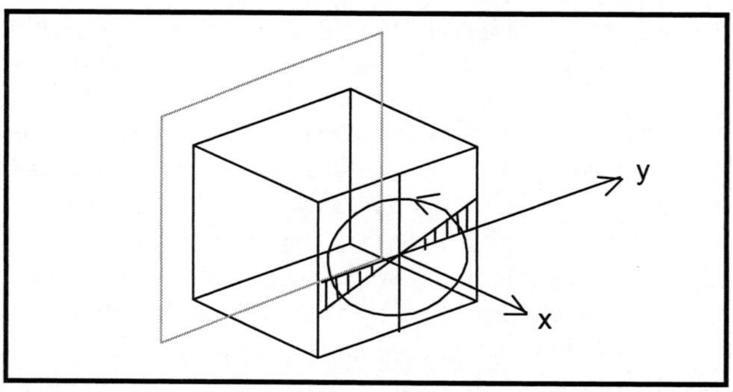

Figure 1: Torsion of a cube.

The greatest differences in results to values obtained by Reissner [12] appeared for plate rotations in the normal direction because these values are indirectly related with prescribed displacements. These rotations were used to qualify the collocation point position. The values for distributed shear and twisting moments are directly related to the constrained displacements, the differences to values obtained by Reissner [12] were low and they cannot be used to qualify the collocation point position or the boundary element solution.

Table 2: Relative values of distributed shear $\left(Q_n(x_2) / Q_n(a)\right)$.

(x_2/a)	[12]	[1]	[14]	ξ=0.5	ξ=0.67	ξ=0.75
1.00	1.000	1.000	0.999	1.000	1.000	1.000
0.75	0.638	0.638	0.638	0.640	0.640	0.636
0.50	0.377	0.377	0.377	0.377	0.378	0.375
0.25	0.174	0.174	0.174	0.175	0.175	0.174
0.00	0.000	1.000	0.000	0.000	0.000	0.000

Table 3: Relative values of twisting moments $\left(M_t(0) / M_t(x_2)\right)$.

(x_2/a)	[12]	[1]	[14]	ξ=0.5	ξ=0.67	ξ=0.75
1.00	0.000	0.000	0.035	0.000	0.000	0.000
0.75	0.485	0.485	0.469	0.487	0.486	0.484
0.50	0.785	0.785	0.809	0.786	0.786	0.785
0.25	0.948	0.948	0.938	0.949	0.948	0.948
0.00	1.000	1.000	1.018	1.000	1.000	1.000

5 Conclusions

The use of the tangential differential operator in traction BIE for plate bending models considering the shear deformation effect allowed the order reduction of the singularities without reducing the accuracy of the boundary element method. The use of fundamental solutions having singularity at the source point is the main request to apply the TDO with the integration by parts. The results have agreed with those obtained in [12, 1 and 14]. Thus, the same efficiency shown for two dimensional problems of the linear fracture mechanics in [6] and for three dimensional elasticity problems in [7] was also shown for plate bending models including the shear deformation effect. Furthermore, the use of derivatives of the adopted shape function for displacement without using other interpolation for TDO was an interesting alternative without reducing the expected precision. The next step will be the study of fracture problems in plate bending using the DBEM with the TDO.

Acknowledgement

The author is grateful to CNPq for support for the development of research on plates.

References

[1] Rashed YF, Aliabadi MH, Brebbia CA. Hypersingular boundary element formulation for Reissner plates. Int J Solids Struct, 35(18), 2229–49, 1998.

[2] Guiggiani, M. Hypersingular formulation for boundary stresses evaluation. Eng. Anal. Bound. Elem., 14, 169-179, 1994.
[3] Kupradze, V.D., Three-dimensional problems of the mathematical theory of elasticity and thermoelasticity. North Holland, 1979.
[4] Sladek, J.; Sladek, V., Three-dimensional curved crack in an elastic body, Int. J. Solids Struct., 19, 425-436, 1983.
[5] Bonnet, M. Boundary Integral Equation Methods for Solids and Fluids, John Wiley & Sons Ltd, 1999.
[6] Palermo, Jr., L., Almeida, L.P.C.P.F. and Gonçalves, P.C., The Use of the Tangential Differential Operator in the Dual Boundary Element Equation, Structural Durability & Health Monitoring, vol.2, no.2, pp.123-130, Tech Science Press, 2006.
[7] Palermo, Jr., L., Almeida, L.P.C.P.F., On the Use of the Tangential Differential Operator in the Traction Boundary Integral Equation of the Dual Boundary Element Method for Three Dimensional Problems, ICCES, vol.7, no.2, pp.83-87, 2008.
[8] Gonçalves, P.C., Uma Alternativa para a Solução de Problemas de Elasticidade pelo Método dos Elementos de Contorno, MS of Eng. Dissertation (in Portuguese), University of Campinas, 2006.
[9] Almeida, L.P.C.P.F.; Palermo Jr., L. On the Implementation of the Two Dimensional Dual Boundary Element Method for Crack Problems, 5th Int. Conference on Boundary Elements Techniques, Lisboa, Portugal, 2004.
[10] Love A.E.H.. A treatise on the mathematical theory of elasticity, 4th ed. New York, Dover Publications; 1944.
[11] Palermo Jr., L., Gonçalves, P.C., Figueiredo, L.G., A simple implementation of the dual boundary element method using the tangential differential operator for plane problems, BEM 32, Editor C.A. Brebbia, 2010.
[12] Reissner, E., The Effect of Transverse Shear Deformation on the Bending of Elastic Plates, Journal of Applied Mechanics, 1945.
[13] Mindlin, R.D., Influence of rotatory inertia and shear on flexural motions of isotropic elastic plates, Journal of Applied Mechanics, 1951.
[14] Weeën, F., Application of the boundary integral equation method to Reissner's plate model, International Journal for Numerical Methods in Engineering, vol. 18, 1-10, 1982 .
[15] Palermo Jr., L., A strategy to perform the Reissner-Mindlin's theory, BEM 24, Editors C.A. Brebbia, A. Tadeu, V. Popov, 2002.
[16] Palermo Jr., L., Plate Bending analysis using the classical or the Reissner-Mindlin models, Engineering Analysis with Boundary Elements, 2003.
[17] Abramowitz M, Stegun I. Handbook of mathematical functions, New York: Dover Publications; 1972.
[18] Kane, J. H., Boundary Element Analysis in engineering mechanics. Prentice Hall, 1994.

The BEM for buckling analysis of viscoelastic plates modelled with fractional derivatives

J. T. Katsikadelis & N. G. Babouskos
School of Civil Engineering,
National Technical University of Athens, Greece

Abstract

In this paper the buckling of viscoelastic plates is studied. The constitutive equations of the viscoelastic material are expressed in differential form using fractional derivatives. The proposed analysis is illustrated with the fractional Kelvin-Voigt and fractional Standard solid models. Plates of arbitrary shape with any type of boundary conditions under interior and edge conservative membrane loads are considered. *The principle of the analog equation* is applied to convert the original equation into a plate equation (biharmonic) under a fictitious load. Subsequent application of the BEM enables the spatial discretization resulting thus an initial value problem for the values of the fictitious load, which is a system of linear Fractional Differential Equations (FDEs) with respect to time. Using a property of the Mittag-Leffler function a dynamic criterion is established and the eigenvalue problem for the evolution equations is converted into an eigenvalue problem of linear algebra, which permits the evaluation of the buckling loads of the viscoelastic plate. Several plate problems are studied and interesting conclusions on the effect of viscoelasticity on bucking of thin plates are drawn.
Keywords: thin plates, viscoelastic, fractional derivative models, buckling, boundary element method, analog equation method.

1 Introduction

The buckling of viscoelastic structures modelled with integral or integer order differential constitutive equations has been investigated by several authors [1–3]. There are papers dealing with the stability of viscoelastic beams [4–7] and viscoelastic plates under conservative loads [8–10] using such models. In the last

WIT Transactions on Modelling and Simulation, Vol 52, © 2011 WIT Press
www.witpress.com, ISSN 1743-355X (on-line)
doi:10.2495/BE110041

years many researchers have shown that the response of real viscoelastic structures is described accurately using fractional derivatives with much less parameters than the classical models. However their use leads to partial fractional differential equations, whose mathematical treatment is very difficult. Therefore the literature is limited on viscoelastic structures modelled with fractional derivatives [11–13]. Recently, Katsikadelis developed a numerical method to solve ordinary fractional differential equations [14] that was extended to partial ones. This paved the way to analyse viscoelastic structures described with fractional derivative models. Using this technique Katsikadelis and Babouskos [15] studied recently the postbuckling response of viscoelastic plates modelled with fractional derivatives.

In this paper the linear buckling of viscoelastic plates modelled with generalized fractional differential constitutive equations is investigated. The plate has an arbitrary shape and is subjected to any type of boundary conditions under interior and edge conservative membrane loads. The governing equation and the boundary conditions are first written in terms of stress resultants. Then application of the differential operator of the viscoelastic constitutive equations [1, 7] yields an evolution partial differential equation in terms of the transverse displacement. The proposed analysis is illustrated with two fractional viscoelastic models, namely the fractional Kelvin-Voigt model and the fractional Standard solid model. The *principle of the analog equation* [16] is applied to convert the original equation into a plate equation (biharmonic) under a time dependent fictitious load, unknown in the first instance. Then application of the BEM yields the semi-discrete evolution equations, which is a system of linear ordinary FDEs. Assuming the Mittag-Leffler function [17] as the solution of homogeneous system, we obtain an eigenvalue problem of linear algebra, which permits the evaluation of the buckling loads. Several plate problems are examined and interesting conclusions on the effect of viscoelasticity on bucking of thin plates are drawn.

2 Governing equations

We consider a thin elastic plate of uniform thickness h occupying the two dimensional multiply connected domain Ω of the xy plane with boundary $\Gamma \cup_{i=0}^{K} \Gamma_i$ (Fig. 1). The non intersecting curves Γ_i $(i = 0,1,2,...,K)$ may be piece-wise smooth. There are N_c corner points along the boundary. The boundary may be simply supported, clamped, free or elastically supported with transverse stiffness $k_T(\mathbf{x})$ and rotational stiffness $k_R(\mathbf{x})$, $\mathbf{x}:(x,y) \in \Gamma$. The plate is subjected to inplane loads p_x, p_y inside Ω as well as edge inplane forces N_n^*, N_t^*.

Applying the principle of virtual work, we obtain

$$\begin{aligned} &M_{x,xx} + 2M_{xy,xy} + M_{y,yy} + N_x w_{,xx} \\ &\quad + 2N_{xy} w_{,xy} + N_y w_{,yy} - p_x w_{,x} - p_y w_{,y} = 0 \end{aligned} \quad \text{in } \Omega \qquad \text{(1a)}$$

$$V_n + N_n^* w_{,n} + N_t^* w_{,t} + k_T w = 0 \text{ or } \quad w = 0 \quad \text{on } \Gamma \tag{2a}$$

$$M_n - k_R w_{,n} = 0 \text{ or } \quad w_{,n} = 0 \quad \text{on } \Gamma \tag{2b}$$

$$k_T^{(k)} w^{(k)} - [\![M_{nt}]\!]_k = 0 \text{ or } w^{(k)} = 0 \quad \text{at corner point k} \tag{2c}$$

where w is the transverse deflection of the plate and M_x, M_y, M_{xy} are the bending and twisting moments of the plate; V_n and M_n are the equivalent shear force and the normal bending moment along the boundary and $[\![M_{nt}]\!]_k = M_{nt}^+ - M_{nt}^-$ is the discontinuity jump of the twisting moment at the corner point k. N_x, N_y, N_{xy} are the stress resultants due to the membrane loading obtained from the solution of plane stress problem and are assumed to be time independent, that is they are obtained from the elastic solution. In this investigation, the plane stress problem is solved using the conventional BEM [18].

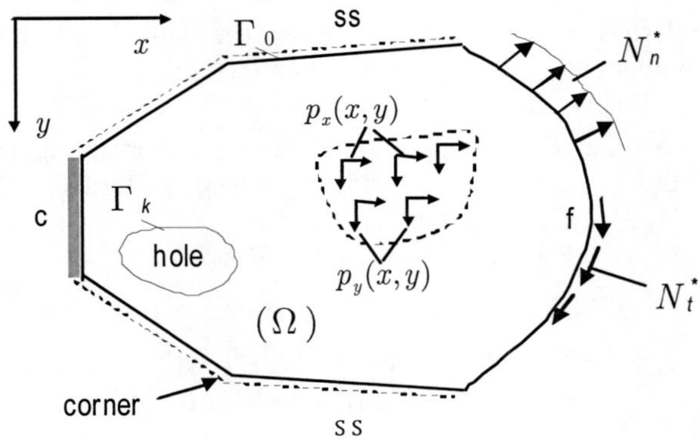

Figure 1: Plate geometry and supports (c=clamped, ss=simply supported, f=free).

For the elastic material the stress resultants are written as

$$M_x = -D(w_{,xx} + \nu w_{,yy}), \; M_y = -D(\nu w_{,xx} + w_{,yy}) \tag{3a, b}$$

$$M_{xy} = -D(1-\nu) w_{,xy} \tag{3c}$$

$$V_n = Vw, \quad M_n = Mw, \quad M_{nt} = Tw \tag{4a, b, c}$$

where $D = Eh^3/(1-\nu^2)$ is the flexural rigidity of the plate. The operators V, M and T are given as [19]

$$V = -D\left[\frac{\partial}{\partial n}\nabla^2 + (1-\nu)\frac{\partial}{\partial s}\left(\frac{\partial^2}{\partial s \partial n} - \kappa\frac{\partial}{\partial s}\right)\right] \tag{5a}$$

$$M = -D\left[\nabla^2 - (1-\nu)\left(\frac{\partial^2}{\partial s^2} + \kappa\frac{\partial}{\partial n}\right)\right], \; T = D(1-\nu)\left(\frac{\partial^2}{\partial s \partial n} - \kappa\frac{\partial}{\partial s}\right) \tag{5b, c}$$

www.witpress.com, ISSN 1743-355X (on-line)

The constitutive equations of the viscoelastic material are written in differential form [1, 20] and can be generalized by replacing the integer order derivatives with fractional derivatives [11, 12]. Thus for the plate bending problem the viscoelastic constitutive equations can be written as

$$\mathbf{P}M_x = -D\mathbf{Q}\left(w_{,xx} + \nu w_{,yy}\right) \tag{6a}$$

$$\mathbf{P}M_y = -D\mathbf{Q}\left(\nu w_{,xx} + w_{,yy}\right) \tag{6b}$$

$$\mathbf{P}M_{xy} = -D(1-\nu)\mathbf{Q}w_{,xy} \tag{6c}$$

where $\mathbf{P},\mathbf{Q}$ are fractional differential operators with respect to time defined as

$$\mathbf{P} = \sum_{k=0}^{l} p_k D_c^{\alpha_k}\ ,\ \mathbf{Q} = \sum_{k=0}^{m} q_k D_c^{\alpha_k} \tag{7a, b}$$

in which $D_c^{a_k}$ is the Caputo fractional derivative of order α_k, with $\alpha_0 = 0$, and p_k, q_k are the viscoelastic parameters, which should satisfy certain conditions resulting from second law of thermodynamics [11]. Note that for $l = m = 0$ it is $\mathbf{P} = \mathbf{Q} = 1$ and eqns (6) give the constitutive equations (3) of the elastic material. Similarly the equivalent shear force, the normal bending moment and the twisting moment along the boundary can be written as

$$\mathbf{P}V_n = \mathbf{Q}Vw\ ,\ \ \mathbf{P}M_n = \mathbf{Q}Mw\ ,\ \ \mathbf{P}M_{nt} = \mathbf{Q}Tw \tag{8a, b, c}$$

Applying operator $\mathbf{P}$ on eqns (1) and (2) we obtain the equations and the boundary conditions of the plate bending problem in terms of the transverse deflection w [1, 7]. Thus we obtain

$$\begin{aligned}&\mathbf{P}M_{x,xx} + 2\mathbf{P}M_{xy,xy} + \mathbf{P}M_{y,yy} + N_x\mathbf{P}w_{,xx}\\ &\quad + 2N_{xy}\mathbf{P}w_{,xy} + N_y\mathbf{P}w_{,yy} - p_x\mathbf{P}w_{,x} - p_y\mathbf{P}w_{,y} = 0\end{aligned} \quad \text{in } \Omega \tag{9}$$

$$\mathbf{P}V_n + N_n^*\mathbf{P}w_{,n} + N_t^*\mathbf{P}w_{,t} + k_T\mathbf{P}w = 0 \text{ or } w = 0 \quad \text{on } \Gamma \tag{10a}$$

$$\mathbf{P}M_n + k_R\mathbf{P}w_{,n} = 0 \text{ or } w_{,n} = 0 \quad \text{on } \Gamma \tag{10b}$$

$$k_T^{(k)}\mathbf{P}w^{(k)} - [\![\mathbf{P}M_{nt}]\!]_k = 0 \text{ or } w^{(k)} = 0 \quad \text{at corner point k} \tag{10c}$$

which by virtue of eqn (6) and (8) become

$$\begin{aligned}&\mathbf{Q}D\nabla^4 w - N_x\mathbf{P}w_{,xx} - 2N_{xy}\mathbf{P}w_{,xy}\\ &\quad - N_y\mathbf{P}w_{,yy} + p_x\mathbf{P}w_{,x} + p_y\mathbf{P}w_{,y} = 0\end{aligned} \quad \text{in } \Omega \tag{11}$$

$$\mathbf{Q}Vw + N_n^*\mathbf{P}w_{,n} + N_t^*\mathbf{P}w_{,t} + k_T\mathbf{P}w = 0 \text{ or } w = 0 \quad \text{on } \Gamma \tag{12a}$$

$$\mathbf{Q}Mw + k_R\mathbf{P}w_{,n} = 0 \text{ or } w_{,n} = 0 \quad \text{on } \Gamma \tag{12b}$$

$$k_T^{(k)}\mathbf{P}w^{(k)} - [\![\mathbf{Q}Tw]\!]_k = 0 \text{ or } w^{(k)} = 0 \quad \text{at corner point k} \tag{12c}$$

The above equations are fractional differential equations with respect to time. The number of the initial conditions that are needed for the solution of eqn (11)-(12) depends on the order of the viscoelastic constitutive equation. Thus in the general case the initial conditions can be written as

$$\frac{d^k w(\mathbf{x},0)}{dt^k} = g_k(\mathbf{x}),\ (k = 0,1,2..,n) \tag{13}$$

where $g_k(\mathbf{x})$ is a known function and $n \in \mathbb{N}$ with $n-1 < \alpha_k < n$.

For the fractional Kelvin-Voigt model (Fig 2a) it is $l=0$, $m=1$ and $p_0=1$, $q_0=1$ with $q_1=\eta$ being the viscous parameter. Thus eqn (11) becomes

$$D\nabla^4 w+\eta D D_c^\alpha \nabla^4 w-N_x w_{,xx} \\ -2N_{xy}w_{,xy}-N_y w_{,yy}+p_x w_{,x}+p_y w_{,y}=0 \quad (14)$$

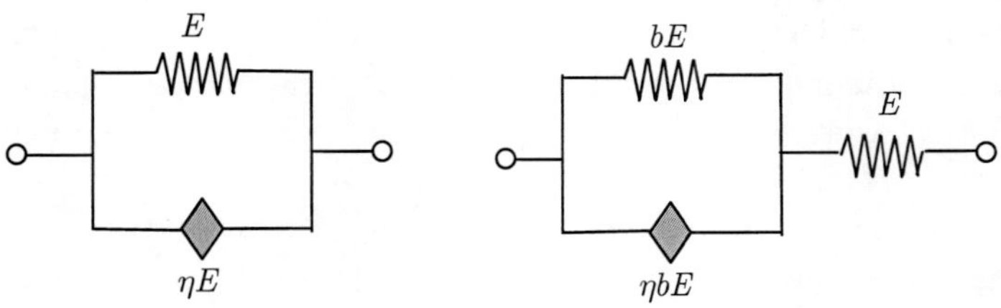

Figure 2: (a) The fractional Kelvin-Voigt model and (b) the fractional Standard solid model in uniaxial representation.

For the fractional Standard solid model (Fig. 2b) it is $l=m=1$ and $p_0=1$, $p_1=\eta b/(1+b)$, $q_0=b/(1+b)$, $q_1=\eta b/(1+b)$ [21] where b,η are parameters of the viscoelastic material. Thus eqn (11) becomes

$$q_0 D\nabla^4 w+q_1 D D_c^\alpha \nabla^4 w-N_x\left(w_{,xx}+p_1 D_c^\alpha w_{,xx}\right) \\ -2N_{xy}\left(w_{,xy}+p_1 D_c^\alpha w_{,xy}\right)-N_y\left(w_{,yy}+p_1 D_c^\alpha w_{,yy}\right) \\ +p_x\left(w_{,x}+p_1 D_c^\alpha w_{,x}\right)+p_y\left(w_{,y}+p_1 D_c^\alpha w_{,y}\right)=0 \quad (15)$$

Equations (14) and (15) are fractional differential equations of order $0<\alpha\leq 1$ with respect to time, therefore only one initial condition, namely the initial displacement, is required. Since we study the stability of the plate, we suppose that the plate is initially undeformed. Thus the initial condition is

$$w(\mathbf{x},0)=0 \quad (16)$$

In this investigation the time dependent terms in the boundary conditions of the viscoelastic plate are neglected. It can be shown that when the plate is initially undeformed and for $k_T=k_R=0$ the time dependent terms do not appear in the boundary conditions. The influence of these terms will be the subject of a forthcoming paper. Thus, the boundary conditions for both viscoelastic models are given by eqn (12) with $\mathbf{P}=\mathbf{Q}=1$.

3 The solution procedure

The initial boundary value problem (11)-(12) is solved using the AEM. The analog equation for the problem at hand is

www.witpress.com, ISSN 1743-355X (on-line)

$$\nabla^4 w = b(\mathbf{x}, t), \quad \mathbf{x} = \{x, y\} \in \Omega \tag{17}$$

where $b(\mathbf{x}, t)$ represents a fictitious load, unknown in the first instance. Eqn (17) under the boundary conditions (12) is solved using the BEM as presented in [15]. This yields

$$w_{,pqr}(\mathbf{x}, t) = \sum_{k=1}^{M} b_k(t) W_{k,pqr}(\mathbf{x}), \; p, q, r = 0, x, y \;\; \mathbf{x} \in \Omega \tag{18}$$

where $W_{k,pqr}$ are known functions and $b_k(t)$ the values of the fictitious load at the M collocation points inside Ω. Note that the above notation implies $w_{,000} = w$, $w_{,0y0} = w_{,y}$, etc. Applying eqn (18) for the M internal nodal points we obtain

$$\mathbf{w}_{,pqr} = \mathbf{W}_{,pqr}\,\mathbf{b} \tag{19}$$

Further, collocating the governing equations at the M internal nodal points and using eqn (19) to express the involved derivatives in terms of $\mathbf{b}$ we obtain:
(i) For the elastic plate

$$(\mathbf{K} - \mu \mathbf{F})\mathbf{b} = 0 \;\text{ in } \Omega \tag{20}$$

where μ is the buckling parameter and $\mathbf{K}, \mathbf{F}$ are $M \times M$ known matrices defined as

$$\mathbf{K} = D\mathbf{I} \tag{21a}$$

$$\mathbf{F} = \mathbf{N_x W}_{,\mathbf{xx}} + 2\mathbf{N_{xy}} W_{,xy} + \mathbf{N_y W}_{,yy} - \mathbf{p_x} W_{,x} - \mathbf{p_y} W_{,\mathbf{y}} \tag{21b}$$

with $\mathbf{I}$ being the identity matrix and $\mathbf{N_x}, \mathbf{N_{xy}}, \mathbf{N_y}, \mathbf{p_x}, \mathbf{p_y}$ are $M \times M$ diagonal matrices including the values of the membrane forces and membrane loads at the M internal nodal points. It should be noted that in this case the fictitious load $\mathbf{b}$ does not depend on time. Eqn (20) constitutes an eigenvalue problem of linear algebra, which can be solved for the critical values of the load parameter μ.

(ii) For the fractional Kelvin-Voigt plate

$$\eta \mathbf{K} D_c^\alpha \mathbf{b} + \mathbf{K} - \mu \mathbf{F} = 0 \tag{22}$$

(iii) For the fractional Standard solid plate

$$(q_1 \mathbf{K} - p_1 \mu \mathbf{F}) D_c^\alpha \mathbf{b} + q_0 \mathbf{K} - \mu \mathbf{F} = 0 \tag{23}$$

For $\mathbf{b}(0) = 0$, eqns (22) and (23) constitute eigenvalue problems for the buckling parameter μ. The stability of the fractional differential equations (22) and (23) can be studied by introducing a *dynamic criterion* analogous to that for elastic structures. Thus, observing that for Mittag-Leffler function $E_{\alpha,1}(\lambda t^\alpha)$ holds [17]

$$D_c^\alpha \left[E_{\alpha,1}(\lambda t^\alpha)\right] = \lambda E_{\alpha,1}(\lambda t^\alpha) \tag{24}$$

we can assume a solution of eqns (22) and (23) of the form

$$\mathbf{b}(t) = \boldsymbol{\beta} E_{\alpha,1}(\lambda t^\alpha) \tag{25}$$

where $\boldsymbol{\beta}$ is a constant vector, and $\lambda \in \mathbb{C}$ is a parameter. Note that for $\alpha = 1$ it is $E_{1,1} = e^{\lambda t}$. Therefore, the parameter λ plays here the role of the eigenfrequency of free vibrations. Substituting eqn (25) into eqn (22) and taking into account eqn (24) we obtain

$$(\eta\lambda\mathbf{K} + \mathbf{K} - \mu\mathbf{F})\boldsymbol{\beta} = 0 \tag{26}$$

Similarly eqn (23) becomes

$$[\lambda(q_1\mathbf{K} - p_1\mu\mathbf{F}) + q_0\mathbf{K} - \mu\mathbf{F}]\boldsymbol{\beta} = 0 \tag{27}$$

Equations (26) and (27) constitute generalized eigenvalue problems of linear algebra. For a given value of the load parameter μ the solution of the eigenvalue problems (26) and (27) give the eigenvalue λ which in general is a complex number. The stability of the plate is determined by writing the eigenvalue λ in polar form $\lambda = \rho e^{i\phi}$ and taking into account the asymptotic response of the Mittag-Leffler function as $t \to \infty$ [17]:

(a) $E_{\alpha,1}(t^\alpha \rho e^{i\phi}) \to 0$ when $|\phi| > \alpha\pi / 2$

(b) $E_{\alpha,1}(t^\alpha \rho e^{i\phi})$ remains bounded when $|\phi| = \alpha\pi / 2$

(c) $|E_{\alpha,1}(t^\alpha \rho e^{i\phi})| \to \infty$ when $|\phi| < \alpha\pi / 2$

Hence, by increasing the load parameter μ the plate becomes unstable if $|\phi| < \alpha\pi / 2$ (Fig. 3). The threshold of instability is $|\phi| = \alpha\pi / 2$. If the eigenvalue λ is a real number the plate becomes unstable when $\lambda > 0$.

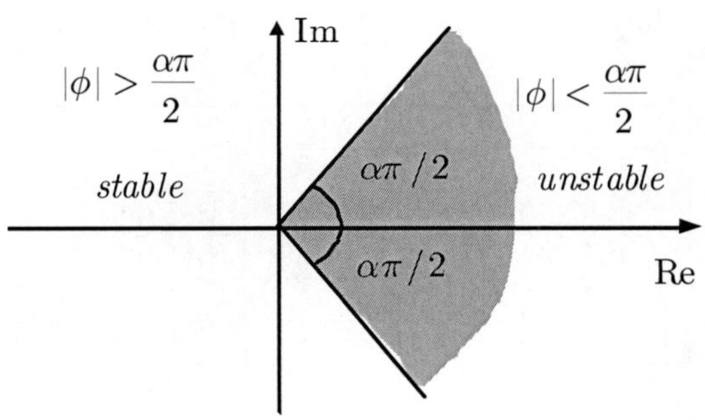

Figure 3: Stability region.

4 Examples

4.1 Example 1

The simply supported square plate of uniform thickness $h = 0.1\text{m}$ of Fig. 4 is subjected to a uniform membrane load along the edges with $x = 0$ and $x = 4$. Due to inplane boundary conditions the produced membrane forces are $N_x = -P$ and $N_{xy} = N_y = 0$. First the elastic plate is studied. The elastic parameters are $E = 21 \times 10^6\ \text{kN/m}^2$ and $\nu = 0.25$. The results were obtained with $N = 200$ boundary elements.

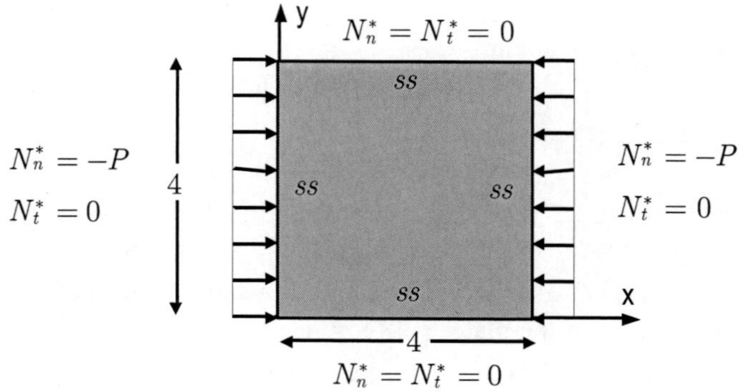

Figure 4: Boundary conditions of plate in example 1.

Table 1 presents the first four buckling loads of the elastic plate for various values of the internal nodal points and the results are compared with the exact ones [22]. Next the response of the Kelvin-Voigt plate is studied. The results were obtained with $M = 236$ internal nodal points resulting from 312 linear triangular domain elements. Fig. 5 presents the first four eigenvalues λ as the load P increases for $\eta = 1$. In this case the eigenvalues are real numbers and

Table 1: Buckling loads of the elastic plate for various values of the internal nodal points.

M	Buckling loads			
	1	2	3	4
145	4682.40	7483.87	13789.24	19620.16
236	4651.44	7371.86	13402.42	19144.91
329	4636.32	7312.76	13202.77	18901.36
401	4629.79	7288.81	13118.50	18800.78
Exact [22]	4605.82	7196.59	12793.93	18423.26

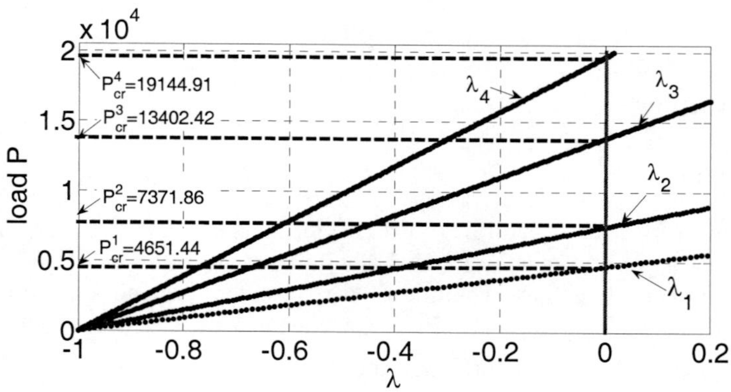

Figure 5: Load-eigenvalue curves for Kelvin-Voigt plate ($\eta = 1$).

plate becomes unstable when $\lambda > 0$. Fig. 6 presents the load -1st eigenvalue curves for various values of the viscous parameter η. The critical loads of the Kelvin-Voigt plate are the same with that of the elastic plate. Similar results were obtained in [1, 7] who studied the stability of the viscoelastic beams. Finally, the stability results do not depend on the order of the fractional derivative α. However, the postbuckling response of the viscoelastic plate is affected by the fractional derivative α as it is shown in [15].

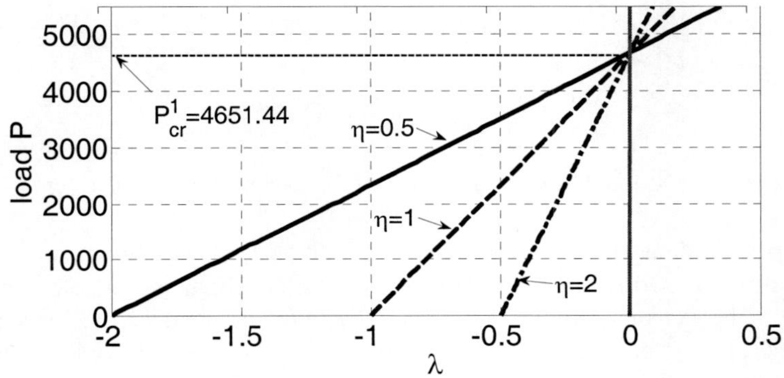

Figure 6: Load-1st eigenvalue curves for various values of η for Kelvin-Voigt plate.

4.2 Example 2

The stability of the fractional Standard solid plate of fig. 4 is studied. Fig. 7 presents the 1st eigenvalue as the load P increases for $b = 2$ and for various values of η. The eigenvalues are real numbers and instability occurs when $\lambda > 0$. The buckling load does not depend on the parameter η or on the order of the fractional derivative α. Fig. 8 presents the load-eigenvalue curves for $\eta = 1$ and various values of the parameter b. It is observed that the buckling load increases as the parameter b increases.

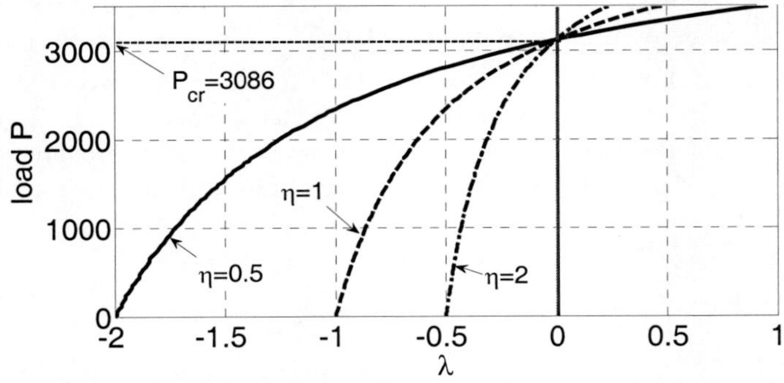

Figure 7: Load-1st eigenvalue curves for various η ($b = 2$).

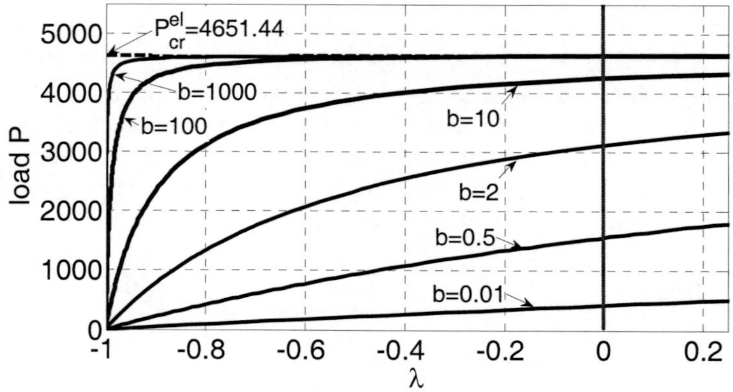

Figure 8: Load-1st eigenvalue curves for various values of b $(\eta = 1)$.

For $b = 0$ the buckling load is $P_{cr} = 0$. In this case the Standard solid model is transformed to the Maxwell viscoelastic model which is suitable for fluid materials [1]. As the parameter $b \to \infty$ the buckling load of the viscoelastic plate approaches the buckling load of the elastic plate (see Example 1). In this case the behaviour of the Standard solid model approaches the behaviour of the elastic material. The postbuckling response of the Standard solid model is more complicated and can be studied by integration of the evolution equations [1, 7].

5 Conclusions

The linear buckling problem of viscoelastic plates is studied. The plate has an arbitrary shape and is subjected to any type of boundary conditions under interior and edge conservative membrane loads. The governing equations are partial fractional differential equations. The AEM is applied to convert the initial boundary value problem of the viscoelastic plate to an initial value problem. Based on a property of the Mittag-Leffler function a dynamic criterion is developed, which converts the initial value problem into an algebraic eigenvalue problem. The latter permits the evaluation of the buckling loads.

It is observed that the stability of the plate does not depend on the order of the fractional derivative α or the viscous parameter η of the viscoelastic model, contrary to the postbuckling response which is affected by the viscoelastic model [7, 15]. For the Standard solid model the parameter b changes significantly the buckling load of the plate. An advantage of the presented procedure is that it avoids the direct time integration of the evolution equations to establish the buckling loads

References

[1] Flügge, W., *Viscoelasticity, Second Revised Edition*, Springer-Verlag, Berlin/Heidelberg, 1975.

[2] Bazant, Z. & Cedolin, L., Stability of Structures, Elastic, Inelastic, Fracture, and Damage Theories, Oxford University Press, 1991.

[3] Minahen, T. M. & Knauss, W.G., Creep buckling of viscoelastic structures. *Int. J. Solids Structures*, **30(8)**, pp. 1075-1092, 1993.

[4] Chang, W.P., Creep buckling of nonlinear viscoelastic columns. *Acta Mech.*, **60**, pp. 199-215, 1986.

[5] Vinogradov, A.M., Buckling of viscoelastic beam columns. *AIAA Journal*, **25**, pp. 479-483, 1987.

[6] Drawshi, M. & Cederbaum, G., Stability of multiloaded viscoelastic nonlinear beams. *Computers & Structures*, **46(2)**, pp. 215-218, 1993.

[7] Shirahatti, U.S & Sinha, S.C., On the stability of perfect viscoelastic columns. *Journal of Sound and Vibration*, **174**, pp. 57-68, 1994.

[8] Hewitt, J.S. & Mazumdart, J., Buckling of viscoelastic plate. *AIAA Journal*, **15**, pp. 451-452, 1977.

[9] Drozdov, A.D. & Zhukhovitskii, D.M., Stability of circular plates from ageing viscoelastic material. *Journal of Applied Mathematics and Mechanics*, **50(2)**, pp. 208-212, 1986.

[10] Peng, F., Fu, Y. & Liu, Y., On the durable critic load in creep buckling of viscoelastic laminated plates and circular cylindrical shells. *Sci China Ser G-Phys Mech Astron*, **51(7)**, pp. 873-882, 2008.

[11] Atanackovic, T.M., A modified Zener model of a viscoelastic body. *Continuum Mech. Thermodyn*, **14**, pp. 137-148, 2002.

[12] Schmidt, A. & Gaul, L., Finite element formulation of viscoelastic constitutive equations using fractional time derivatives. *Nonlinear Dynamics*, **29(1-4)**, pp. 37-55, 2002.

[13] Galucio, A.C., Deu, J.–F. & Ohayon, R., Finite element formulation of viscoelastic sandwich beams using fractional derivative operators. *Computational Mechanics*, **33**, pp. 282-291, 2004.

[14] Katsikadelis, J.T., Numerical solution of multi-term fractional differential equations. *ZAMM Zeitschrift für Angewandte Mathematik und Mechanik*, **89(7)**, pp. 593 – 608, 2009.

[15] Katsikadelis, J.T. & Babouskos, N.G., Postbuckling analysis of viscoelastic plates with fractional derivative models. *Engineering Analysis with Boundary Elements*, **34**, pp. 1038-1048, 2010.

[16] Katsikadelis, J.T., The analog equation method. A powerful BEM–based solution technique for solving linear and nonlinear engineering problems. *Boundary Element Method XVI*, ed. C.A. Brebbia, CMP, Southampton, pp. 167-182, 1994.

[17] Diethelm, K., The Analysis of Fractional Differential Equations, An Application-Oriented Exposition Using Differential Operators of Caputo Type, Springer-Verlag Berlin Heidelberg, 2010.

[18] Katsikadelis, J.T., *Boundary Elements: Theory and Applications*, Elsevier Science: London, 2002

[19] Katsikadelis, J.T. & Armenakas, A.E., A new boundary equation solution to the plate problem. *ASME Journal Applied Mechanics*, **56**, 364-374, 1989.

www.witpress.com, ISSN 1743-355X (on-line)

[20] Christensen, R.M., *Theory of Viscoelasticity, Second Edition*, Academic Press, New York, 1982.
[21] Mesquita, A.D. & Coda, H.B., Boundary integral equation method for general viscoelastic analysis. *International Journal of Solids and Structures*, **39**, pp. 2643-2664, 2002.
[22] Bulson, P.S., *The stability of flat plates*, Chatto & Windus, London, 1970.

Section 2
Fluid flow applications

BEM simulation of transient fluid flow phenomena

J. Ravnik & L. Škerget
University of Maribor, Faculty of Mechanical Engineering, Slovenia

Abstract

In this paper a Boundary Element Method based numerical algorithm is presented for the simulation of three-dimensional unsteady fluid flow and heat transfer. Four different time discretization techniques are considered and compared on a model unsteady heat diffusion problem. Analytical solution of the problem is used to designate the three point second-order finite different approximation of the accumulation term of the transport equations as the most accurate. This choice is incorporated into the flow solver and the developed algorithm is used to simulate Rayleigh–Bénard convection. Oscillatory and chaotic behaviour of the flow field and heat transfer are observed. Temperature slices and velocity vectors are presented. Heat flux is presented in terms of the Nusselt number.
Keywords: Boundary Element Method, Rayleigh–Bénard convection, velocity-vorticity formulation, time discretization.

1 Introduction

The Boundary Element Method (BEM) is a numerical technique that has been successfully applied for the solution of a wide variety of engineering problems and natural phenomena. In this paper we are focused on unsteady natural convection phenomena. A transient simulation is required, which must be able to capture the oscillatory and chaotic nature of the flow problem.

The paper presents a BEM based laminar viscous flow solver and focuses on the time discretization. Three implicit finite difference approximations and an explicit approximation of the accumulation term of the transport equations are compared against an analytical solution of a model problem. The three point second order finite difference approximation is then incorporated into the flow solver and used to obtain an unsteady solution of the Rayleigh–Bénard convection problem.

WIT Transactions on Modelling and Simulation, Vol 52,
www.witpress.com, ISSN 1743-355X (on-line)
doi:10.2495/BE110051

2 Governing equations

We consider laminar viscous fluid flow coupled with heat transfer in three-dimensional setting. The fluid properties are assumed constant. Buoyancy is modelled within the Boussinesq approximation, where density variation with temperature is considered only in the momentum source term. The non-dimensional velocity-vorticity formulation of Navier–Stokes equations for simulation of laminar viscous fluid flow coupled with heat transfer consists of the kinematics equation, the vorticity transport equation and the energy equation:

$$\nabla^2\vec{v} + \vec{\nabla} \times \vec{\omega} = 0, \tag{1}$$

$$\frac{\partial\vec{\omega}}{\partial t} + (\vec{v} \cdot \vec{\nabla})\vec{\omega} = (\vec{\omega} \cdot \vec{\nabla})\vec{v} + Pr\nabla^2\vec{\omega} - PrRa\vec{\nabla} \times T\vec{g}, \tag{2}$$

$$\frac{\partial T}{\partial t} + (\vec{v} \cdot \vec{\nabla})T = \nabla^2 T. \tag{3}$$

Here, the velocity field is denoted by $\vec{v}$, the vorticity by $\vec{\omega}$ and temperature by T. The flow and heat transfer of a fluid is defined by specifying the Rayleigh Ra and Prandtl Pr number values. They are defined as

$$Pr = \frac{\nu\rho c_p}{\lambda}, \tag{4}$$

where ν is the kinematic viscosity of the fluid, ρ is the density, c_p thermal capacity and λ heat conductivity. The Rayleigh number is defined by

$$Ra = Pr\frac{g\beta\Delta T\mathcal{L}^2}{\nu^2}, \tag{5}$$

where β is thermal expansion coefficient, $g = 9.81m/s^2$, ΔT is the characteristic temperature difference and $\mathcal{L}$ is the characteristic length scale of the system.

3 Time discretization

The main aim of this work was to study transient phenomena. We chose Rayleigh–Bénard convection for the test case. When fluid is heated from below, it very soon becomes unsteady and exhibits a wide range of time dependent phenomena. In order to estimate the accuracy of time discretization, we compared three implicit approximations of the time derivative and an explicit Runge–Kutta scheme.

In order to be able to compare the accuracy of different schemes, we chose the following heat transfer example. Consider a thermally isolated thin thread of length π. At the beginning of the simulation the thread has a certain temperature profile. As time passes the thread cools by emitting heat flux through both ends. Since the thread is isolated, there is no heat flux through the casing, and the problem can be treated as one-dimensional. The material properties of the thread are such, that the following equation is valid

$$\frac{\partial T}{\partial t} = \frac{\partial^2 T}{\partial x^2}. \tag{6}$$

The equation includes only accumulation and diffusion terms and is a simplification of (3), where the advection term is also present. At $t = 0$ a triangular temperature profile is applied to the thread.

$$T(x,0) = \begin{cases} 2\frac{x}{\pi}, & x \in (0, \pi/2); \\ 2\frac{\pi - x}{\pi}, & x \in (\pi/2, \pi). \end{cases} \tag{7}$$

A Dirichlet boundary condition is applied on both sides, i.e. $T(x = 0) = 0$ and $T(x = \pi) = 0$. Figure 1 shows the sketch of the problem along with boundary and initial conditions. The analytical solution of the problem gives temperature dependence of x and time. It is (Weisstein [1]):

$$T(x,t) = \frac{2}{\pi}\sum_{n=0}^{\infty}\left\{2\int_0^{\pi/2}\sin(nx)\frac{x}{\pi}dx + 2\int_{\pi/2}^{\pi}\sin(nx)x\frac{\pi - x}{\pi}dx\right\}\cdot$$
$$\cdot \sin(nx)e^{-n^2 t}. \tag{8}$$

The time dependent diffusion equation (6) is discretizised using subdomain BEM technique. Denoting the subdomain with Ω and its boundary with Γ, the following integral equation is obtained

$$c(\vec{\xi})T(\vec{\xi}) + \int_\Gamma T\vec{\nabla}u^* \cdot \vec{n}d\Gamma = \int_\Gamma u^* q_T d\Gamma + \int_\Omega f(T, T', T'', \Delta t)u^* d\Omega, \tag{9}$$

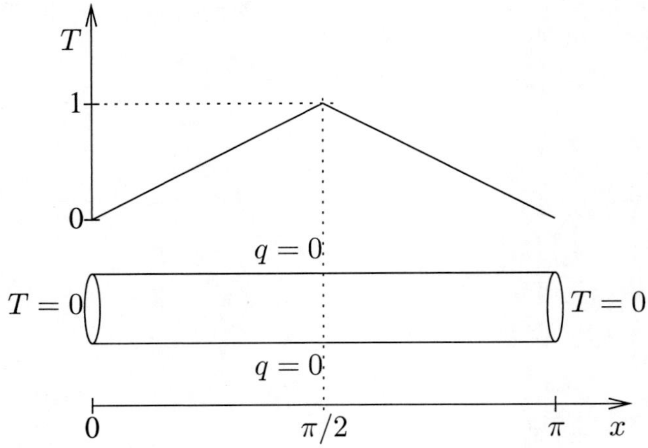

Figure 1: Boundary and initial conditions for time depended diffusion test case.

where $\vec{\xi}$ is the source or collocation point, $\vec{n}$ is a vector normal to the boundary, pointing out of the domain and $u^\star$ is the fundamental solution for the diffusion operator: $u^\star = 1/4\pi|\vec{\xi} - \vec{r}|$. q_T is the heat flux defined by $q_T = \vec{\nabla} T \cdot \vec{n}$. $c(\vec{\xi})$ is the geometric factor defined as $c(\vec{\xi}) = \alpha/4\pi$, where α is the inner angle with origin in $\vec{\xi}$.

Function $f(T, T', T'', \Delta t)$ denotes the discretization of the time derivative;

$$\frac{\partial T}{\partial t} = f(T, T', T'', \Delta t). \tag{10}$$

The discrete version of the partial time derivative may depend on the time step Δt, temperature in the next time step T, temperature in the current time step T' and temperature in the previous time step T''.

Firstly, we consider first order *backward Euler* approximation of the time derivative defined by

$$\frac{\partial T}{\partial t} \approx \frac{T - T'}{\Delta t}, \tag{11}$$

secondly the second order *trapezoid* scheme

$$\frac{\partial T}{\partial t} \approx \frac{1}{2}\frac{T - T'}{\Delta t} + \frac{1}{2}\frac{T' - T''}{\Delta t} = \frac{1}{2}\frac{T - T''}{\Delta t}, \tag{12}$$

and thirdly a *three point* second order scheme defined by

$$\frac{\partial T}{\partial t} \approx \frac{3T - 4T' + T''}{2\Delta t}. \tag{13}$$

The three finite difference approximation of the time derivative are all implicit, i.e. they include a temperature value that must be calculated. This contribution is added to the system matrix when the system of linear equations is formed.

Furthermore, we consider an explicit Runge–Kutta type scheme. Since the scheme is explicit, we estimate the partial time derivate in the present time step and use its value to get the function value at the next time step. Fourth order scheme was used, given by the following algorithm

1. $T^{(1)} = T' + \frac{1}{4}\Delta t \left[a\frac{\partial^2 T}{\partial x^2}\right]'$
2. $T^{(2)} = T' + \frac{1}{3}\Delta t \left[a\frac{\partial^2 T}{\partial x^2}\right]^{(1)}$
3. $T^{(3)} = T' + \frac{1}{2}\Delta t \left[a\frac{\partial^2 T}{\partial x^2}\right]^{(2)}$
4. $T = T' + \Delta t \left[a\frac{\partial^2 T}{\partial x^2}\right]^{(3)}$

Time step size for the Runge–Kutta scheme is limited by the Courant–Friedrichs–Lewy stability condition:

$$\Delta t \leq \frac{\mu \Delta x}{\delta}, \tag{14}$$

where $\mu < 1$ is the Courant–Friedrichs–Lewy number, δx the mesh size and δ a characteristic velocity.

Table 1: Comparison of temperature in the middle of the domain; $T(\pi/2, 1)$.

Δt	No. of time steps	Euler	Trapezoid	3 point	Runge–Kutta
0.1	10	0.3127	0.3263	0.2995	
0.01	100	0.2997	0.3012	0.2982	
0.001	1000	0.2984	0.2985	0.2982	
0.0001	10^4				0.2982
Analytical		0.2982	0.2982	0.2982	0.2982

The four time discretization schemes using different time steps were compared against the analytical solution. A time of $t = 1$ was chosen to compare temperature profiles. A mesh of 100 equidistant quadratic elements having 201 nodes in x direction was used to solve the problem. Table 1 provides temperature values in the middle of the domain (at $x = \pi/2$) for different time steps along with the analytical solution. Figures 2 and 3 show the comparison of temperature profiles at $t = 1$.

The results clearly show, that the three point scheme (13) gives the most accurate results. The explicit scheme also yields very accurate results, but unfortunately requires a very large number of time steps, due to the Courant-Friedrichs-Lewy (14) stability condition.

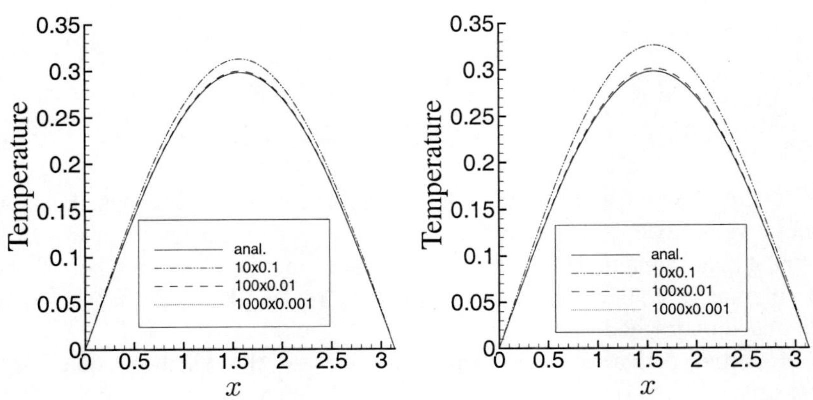

Figure 2: Time dependent solution of diffusion equation. Comparison of analytic and simulated temperature profiles for different time step lengths. Left backward Euler scheme and right trapezoid scheme.

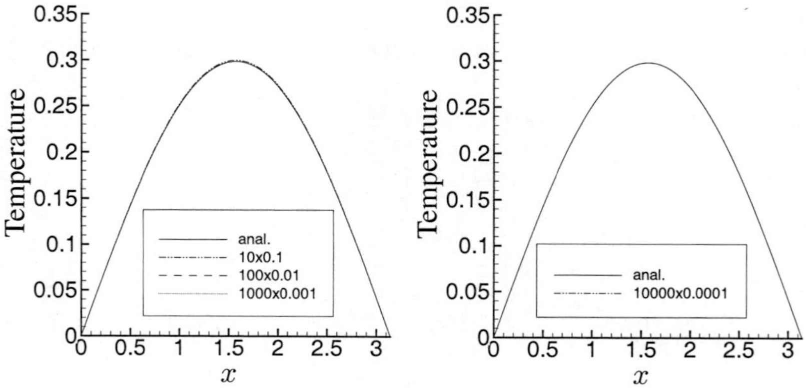

Figure 3: Time dependent solution of diffusion equation. Comparison of analytic and simulated temperature profiles for different time step lengths. Left three point scheme and right Runge Kutta scheme.

4 Numerical method

With the choice of the time discretization established, we proceeded to rewrite the governing equations (1)–(3) in integral form. Three-dimensional solver capable of steady simulating flow and heat transfer by solving velocity-vorticity formulation of Navier–Stokes equations by a combination of single and sub-domain BEM was developed by Ravnik et al. [2, 3]. Integral equations were written in the same manner with an addition of the discretization of the accumulation term.

The kinematics equation is

$$c(\vec{\xi})\vec{n}(\vec{\xi}) \times \vec{v}(\vec{\xi}) + \vec{n}(\vec{\xi}) \times \int_\Gamma \vec{v}\vec{\nabla}u^\star \cdot \vec{n}d\Gamma$$
$$= \vec{n}(\vec{\xi}) \times \int_\Gamma \vec{v} \times (\vec{n} \times \vec{\nabla})u^\star d\Gamma + \vec{n}(\vec{\xi}) \times \int_\Omega (\vec{\omega} \times \vec{\nabla}u^\star)d\Omega. \qquad (15)$$

In order to write a linear system of equations for the unknown boundary vorticity values, we set the source point into every boundary node of the whole computational domain. This yields a full system matrix where number of rows and columns is equal to number of boundary nodes. It is solved using a LU decomposition method.

The partial derivative with respect to time in the kinetics equations is approximated by second order three point finite difference scheme. The final forms of vorticity transport and energy equation are

$$c(\vec{\xi})\omega_j(\vec{\xi}) + \int_\Gamma \omega_j \vec{\nabla}u^* \cdot \vec{n}d\Gamma = \int_\Gamma u^* q_j d\Gamma +$$
$$+\frac{1}{Pr}\left(\int_\Gamma \vec{n} \cdot \{u^*(\vec{v}\omega_j - \vec{\omega}v_j)\}\, d\Gamma - \int_\Omega (\vec{v}\omega_j - \vec{\omega}v_j) \cdot \vec{\nabla}u^* d\Omega\right)$$

$$-Ra\left(\int_\Gamma (u^\star T\vec{g}\times\vec{n})_j d\Gamma + \int_\Omega (T\vec{\nabla}\times u^\star\vec{g})_j d\Omega\right) + \frac{1}{Pr}\frac{1}{2\Delta t}\int_\Omega (3\omega_j - 4\omega_j' + \omega_j'')u^* d\Omega, \quad (16)$$

$$c(\vec{\xi})T(\vec{\xi}) + \int_\Gamma T\vec{\nabla}u^*\cdot\vec{n}d\Gamma = \int_\Gamma u^* q_T d\Gamma + \int_\Gamma \vec{n}\cdot\{u^*(\vec{v}T)\}\,d\Gamma - \int_\Omega (\vec{v}T)\cdot\vec{\nabla}u^* d\Omega + \frac{1}{2\Delta t}\int_\Omega (3T - 4T' + T'')u^* d\Omega, \quad (17)$$

where ω_j is the jth component of vorticity.

In the subdomain BEM method we make a mesh of the entire domain Ω and name each mesh element a subdomain. Equation (16) is written for each of the subdomains. In order to obtain a discrete version of (16) we use shape functions to interpolate field functions and flux across the boundary and inside of the subdomain. In this work we used hexahedral subdomains, which enable continuous quadratic interpolation of field functions. On each boundary element we interpolate the flux using discontinuous linear interpolation scheme. By using discontinuous interpolation we avoid flux definition problems in corners and edges. A function, e.g. temperature, is interpolated over a boundary elements as $T = \sum \varphi_i T_i$, inside each subdomain as $T = \sum \Phi_i T_i$, while flux is interpolated over boundary elements as $q = \sum \phi_i q_i$. The following integrals must be calculated:

$$[H] = \int_\Gamma \varphi_i\vec{\nabla}u^\star\cdot\vec{n}d\Gamma, \quad [G] = \int_\Gamma \phi_i u^\star d\Gamma, \quad [\vec{A}] = \int_\Gamma \varphi_i\vec{n}u^\star d\Gamma, \quad (18)$$

$$[B] = \int_\Omega \Phi_i u^\star d\Omega, \qquad [\vec{D}] = \int_\Omega \Phi_i\vec{\nabla}u^\star d\Omega. \quad (19)$$

The square brackets denote integral matrices. In order to calculate the integrals, a Gaussian quadrature algorithm is used. Calculation of the free coefficient $c(\vec{\xi})$ is preformed indirectly considering rigid body movement problem solution. The calculated $c(\vec{\xi})$ are added to the diagonal terms of the $[H]$ matrix.

The source point is set to all function and flux node in each subdomain. Since neighbouring subdomains share nodes, the resulting systems of linear equations are over-determined. After taking into account the boundary conditions, we solve them using a least squares solver (Paige and Saunders [4]). All integrals depend only on the shape of subdomains and as such may be calculated only once, prior to the start of the nonlinear iterative process.

The algorithm used to solve the set of governing equations (1)–(3) is devised as follows. Either Dirichlet or Neumann type boundary conditions for velocity and temperature must be known. In this paper we use the no-slip boundary condition on all solid walls and prescribe temperature or temperature flux. Boundary conditions for vorticity are unknown and are calculated as a part of the algorithm. The following steps are performed.

Within each time step do the following. Firstly, calculate vorticity values on the boundary by single domain BEM from the kinematics equation (1). Secondly calculate velocity values by sub-domain BEM from the kinematics equation (1). Next calculate temperature values by sub-domain BEM from the energy equation (3). Next calculate vorticity values in the domain by sub-domain BEM from the vorticity transport equation (2). Finally check convergence. If all flow fields converged to 10^{-6} stop, else go to 2. When a time step has converged, advance the flow fields for one time step, i.e. use $T'' = T'$ and $T' = T$ and repeat the procedure for the next time step. Convergence was accelerated using an algorithm for adaptive setting of solver accuracy, developed by Ravnik et al. [5].

5 Rayleigh–Bénard convection

In order to test the capability of the method to simulate unsteady flows, we simulated the Rayleigh–Bénard convection (Solomon and Gollub [6], Shan [7]). The domain was a cubic cavity, where the bottom wall was heated to a constant temperature and the top wall cooled to an also constant but lower temperature. The temperature difference between the walls defines the Rayleigh number for this case. The vertical walls of the cavity are insulated. No-slip velocity boundary conditions are applied on all walls. Boundary conditions are sketched in Figure 4.

Meshes of 16^3 elements having in total 33^3 nodes and mesh of 20^3 elements with 41^3 nodes were used. Simulation were run with air ($Pr = 0.71$) as the working fluid and Rayleigh number values of $Ra = 10^5$ and $Ra = 10^6$. Nondimensional time step of $\Delta t = 0.001$ was used. Preliminary results on the coarse mesh proved significantly different from the results obtained on the fine mesh, thus all simulations were preformed on the fine mesh.

In both cases ($Ra = 10^5$ and $Ra = 10^6$) the flow field was unsteady exhibiting a variety of structures in the domain. Figures 5 and 6 show two times instants, giving

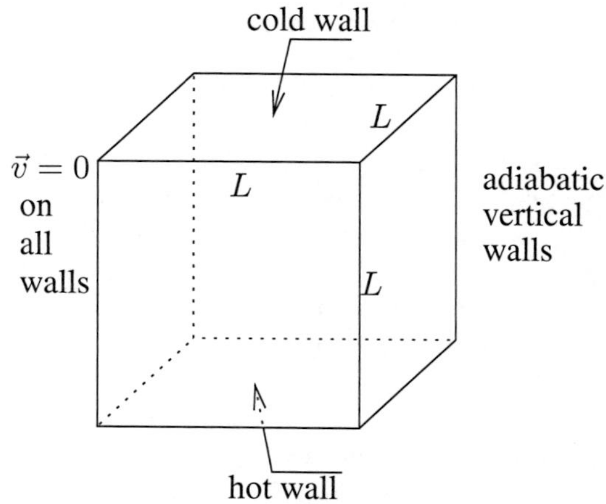

Figure 4: Boundary conditions for the simulation of Rayleigh–Bénard convection.

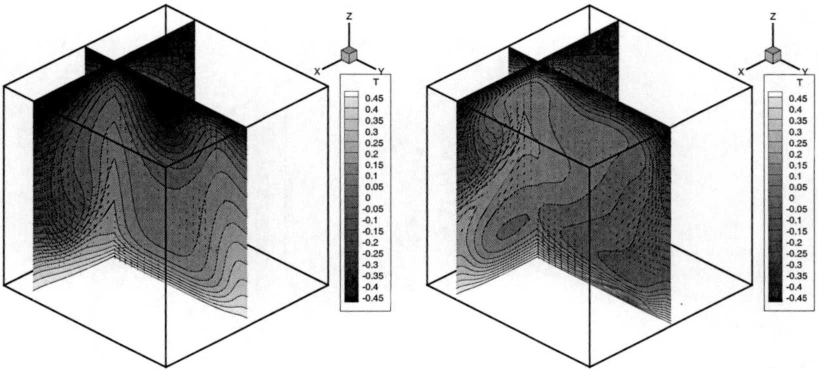

Figure 5: Two slices showing the temperature field and velocity vectors at $Ra = 10^5$. There are 199 time steps between the left and right figure.

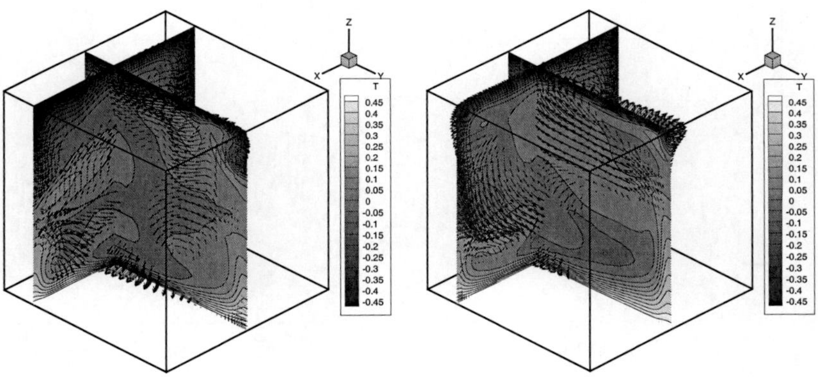

Figure 6: Two slices showing the temperature field and velocity vectors at $Ra = 10^6$. There are 80 time steps between the left and right figure.

an impression of the changing temperature and velocity fields. For $Ra = 10^5$ these changes are slow compared to the rapid and unpredictable flow and temperature pattern at $Ra = 10^6$. After a long time, flow at $Ra = 10^5$ stabilised and reached steady state. The final, steady flow field is shown in Figure 7.

During the simulation heat transfer through the top and bottom walls was measured in terms of Nusselt number value. Usually, the heat flux $\dot{Q}$ is expressed in terms of fluid thermal conductivity, characteristic flow scale and a non-dimensional Nusselt number, i.e. $\dot{Q} = \lambda \mathcal{L} \Delta T \cdot Nu$. The Nusselt number, Nu, is defined as the integral of the temperature flux through a wall:

$$Nu = \int_\Gamma \vec{\nabla} T \cdot \vec{n} d\Gamma, \tag{20}$$

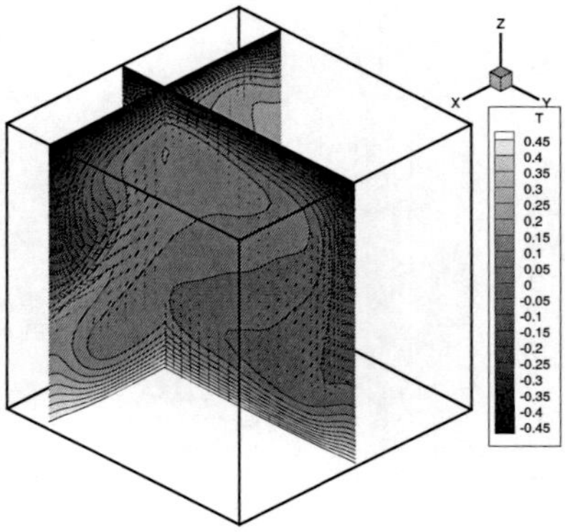

Figure 7: Two slices showing the temperature field and velocity vectors at $Ra = 10^5$ for the final steady flow configuration.

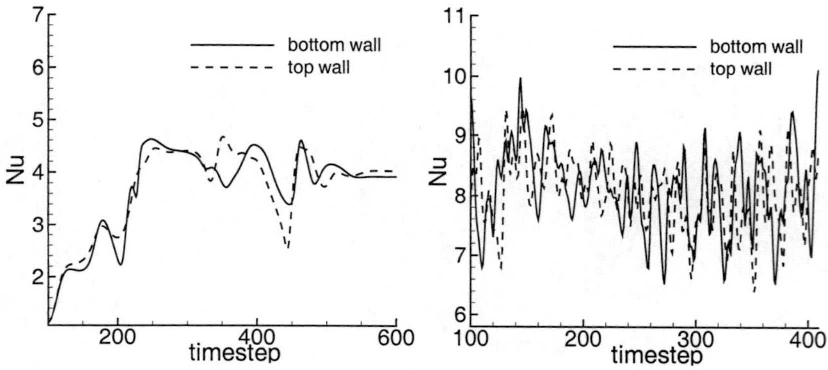

Figure 8: Development of heat flow through the top and bottom walls in the Rayleigh–Bénard convection case; left $Ra = 10^5$, right $Ra = 10^6$.

where Γ is the surface through which we calculate the heat flux and $\vec{n}$ is a unit normal to this surface. We also study local variation of heat flux using the local Nusselt number defined as $Nu_l(x, y, z) = \vec{\nabla} T \cdot \vec{n}$.

The development of heat flux through time is shown in Figure 8. In the $Ra = 10^5$ case the heat flux slowly varies in time and reaches steady state after about 600 time steps. At steady state, heat fluxes through top and bottom walls are equal, reaching $Nu = 3.98$. In the $Ra = 10^6$ case steady state is not reached. The chaotic nature of the graphs indicates that the flow regime is past the oscillatory unsteady phase in a chaotic regime heading towards turbulence at higher Rayleigh number values.

Figure 9 shows heat flux variation through top and bottom walls for the $Ra = 10^5$ case. The heat flux is expressed using the local Nusselt number. Symmetric distribution of fluxes is observed. This is consistent with the fact that the flow

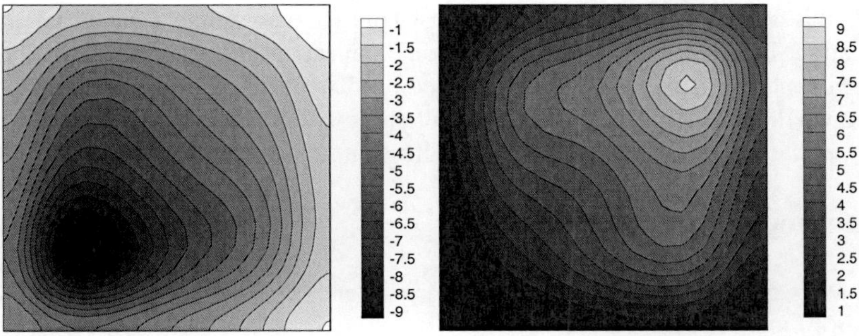

Figure 9: Heat flux (local Nusselt number) through the top (left) and bottom (right) walls at $Ra = 10^5$.

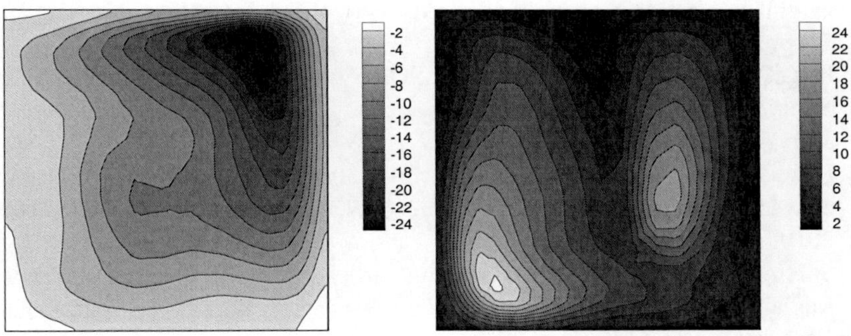

Figure 10: Heat flux (local Nusselt number) through the top (left) and bottom (right) walls at $Ra = 10^6$.

reached steady state and that the same amount of heat entering the domain through the bottom wall also exits through the top wall. Figure 10 presents the $Ra = 10^6$ case at some time during the simulation. Here steady state is not reached and the heat fluxes through the top an bottom wall are distinctly different.

6 Summary and outlook

We presented a boundary element based method for simulation of unsteady laminar viscous flows coupled with heat transfer. The method solves the velocity-vorticity formulation of Navier–Stokes equations using a combination of single domain BEM and subdomain BEM. A study of different approximations of time derivate of the governing equations showed that the second order three point finite difference approximation yields the most accurate results at a given time step size. The algorithm was successfully used for simulation Rayleigh–Bénard convection, showing a unsteady behaviour on the verge of chaos.

In this work a diffusion type fundamental solution was used to write the integral equation. In future, we plan to use the time dependent diffusion-advection type fundamental solution, with which the need for discretization of the accumulation term would be avoided. First successful attempts in this direction were done in 2D by Škerget et al. [8] using the parabolic diffusion fundamental solution.

References

[1] Weisstein, E.W., Heat conduction equation. *MathWorld–A Wolfram Web Resource*, 2005.

[2] Ravnik, J., Škerget, L. & Žunič, Z., Combined single domain and subdomain BEM for 3D laminar viscous flow. *Eng Anal Bound Elem*, **33**, pp. 420–424, 2009.

[3] Ravnik, J., Škerget, L. & Žunič, Z., Velocity-vorticity formulation for 3D natural convection in an inclined enclosure by BEM. *Int J Heat Mass Transfer*, **51**, pp. 4517–4527, 2008.

[4] Paige, C.C. & Saunders, M.A., LSQR: An algorithm for sparse linear equations and sparse least squares. *ACM Transactions on Mathematical Software*, **8**, pp. 43–71, 1982.

[5] Ravnik, J., Škerget, L. & Hriberšek, M., Analysis of three-dimensional natural convection of nanofluids by BEM. *Eng Anal Bound Elem*, **34**, pp. 1018–1030, 2010.

[6] Solomon, T. & Gollub, J., Chaotic particle transport in time-dependent Rayleigh–Benard convection. *Physical Review A*, **38(12)**, pp. 6280–6286, 1988.

[7] Shan, X., Simulation of Rayleigh–Benard convection using a lattice Boltzmann method. *Physical Review E*, **55(3)**, pp. 2780–2788, 1997.

[8] Škerget, L., Hriberšek, M. & Žunič, Z., Natural convection flows in complex cavities by BEM. *Int J Num Meth Heat & Fluid Fl*, **13**, pp. 720–735, 2003.

Devised numerical criteria for calculating the density diffusion in a water reservoir

M. Kanoh[1] & T. Kuroki[2]
[1]*Department of Civil Engineering, Kyushu Sangyo University, Japan*
[2]*Baikoen 1-15-21 Dazaifu, Fukuoka, Japan*

Abstract

In an earlier study to numerically represent the phenomenon described below, we proposed the density diffusion caused by the liquid density ρ, the gravity acceleration g, and the time increment Δt. During field studies, an attempt was made to ameliorate the concentration of oxygen in the lower layer of a few water reservoirs by using a machine that supplies dissolved oxygen (DO), which led to reports of a phenomenon in which the distance reached by the DO-rich water was more than 300 metres despite the very low velocity of the water flow In this report, we investigate the numerical criteria for calculating the density diffusion in a water reservoir by using two-dimensional convective-diffusion equations. Using the signs of the space division h (= Δs), the time increment k (= Δt), the diffusion parameter Γ (= $D^*k/(h)^2$), and the Curant number C_r (= V*k/h), we discuss the order estimate for calculating the density diffusion.

1 Introduction

If the DO concentration equals 100 mg/L (milligrams per litre), the liquid density (ρ) becomes 1.0001 Kg/L (kilograms per litre). To analyse the two-dimensional convective-diffusion equation in the problem described above, we combine the finite difference scheme of the order of four-degree accuracy ($O(h^4)$) for the diffusion term and the UTOPIA scheme of the order of three-degree accuracy ($O(h^3)$) for the convection term. When the above combined method is adopted, the space division h should be less than $0.0001^{1/3}$ (= 0.046) m for the estimation of the liquid density ρ of 1.0001 Kg/L. If the methods of the order of two-degree accuracy are used, the space division h becomes less than $0.0001^{1/2}$ (= 0.01) m. The weighted finite difference method (WFDM) is newly investigated to

WIT Transactions on Modelling and Simulation, Vol 52, © 2011 WIT Press
www.witpress.com, ISSN 1743-355X (on-line)
doi:10.2495/BE110061

introduce the scheme of the order of three-degree accuracy ($O(h^3)$) for both the diffusion and the convection terms to calculate the density diffusion. The WFDM and the FDM may greatly reduce the CPU time for problem analysis (when comparing with the meshless and the BE methods) since the schemes of both methods are expressed explicitly. We tried to upgrade the convergence and the stability or the order of the accuracy of the finite element method (FEM), the meshless method, and the boundary element method (BEM) by introducing the upwind shape function to the FEM, the various basis functions to the meshless method, and the especial fundamental solution to the BEM. It is expected that the fundamental solution of the BEM (Kanoh et al. [2]) may yield accurate and convergent solutions in the analysis of the two-dimensional convective-diffusion equation since the fundamental solution includes the velocity term. The newly developed meshless, BE, and FE methods are used, and the order of the accuracy of each method is investigated numerically. The solutions calculated by these methods are compared with the observed results in our model simulation, and the effect and accuracy of the alternative numerical methods are estimated.

2 Governing equations

Equations (1), (2), and (3) govern the diffusion of the concentration of oxygen in a water reservoir in the vertical (x_1, x_2) plane, as illustrated in Fig. 1,

$$C_{,t} + u_1 \cdot C_{,1} + u'_2 \cdot C_{,2} - D_1 \cdot C_{,11} - D_2 \cdot C_{,22} = 0 \qquad (1)$$

where C is the concentration of dissolved oxygen (DO), $C_{,t}$ is the time derivative of C, u_1 and u'_2 are the velocities of the x_1 and x_2 directions, respectively, and D_1 and D_2 are the diffusion coefficients of the x_1 and x_2 directions, respectively. Here, $C_{,1}$ and $C_{,2}$ describe the derivatives of C differentiated with respect to x_1 and x_2, respectively, and $C_{,11}$ and $C_{,22}$ are the derivatives of C differentiated twice with respect to x_1 and x_2, respectively. The velocity u'_2, which is shown in the above Equation (1), is defined as written in Equation (2),

$${}_{t+\Delta t}u'_2 = {}_t u_2 + (1 - \rho_{DO}/\rho_0)\rho g \Delta t = {}_t u_2 + \omega g \Delta t$$

in the finite difference scheme expression (2)

where ${}_{t+\Delta t}u'_2$ and ${}_t u_2$ are the velocities at time (t+Δt) and time (t) in the vertical direction, respectively. The second term (ωgΔt) of the right-hand side of Equation (2) means that the DO concentration increases the velocity of the vertical direction, and ω describes the density of the liquid that dissolves DO. The velocity increase is caused by the liquid density ρ, the gravity acceleration g, and the time increment Δt. We refer to the velocity increase as the density diffusion, since the velocity increase seems to allow the area of DO diffusion to increase in the vertical direction, as described above, and expect that the velocity increase in the convective diffusion can be used as a device or evidence to explain the phenomenon in which the distance reached by the DO-rich water was more than 300 metres in spite of the very low velocity of the water flow. Here, the density ρ is connected to the DO concentration C, as written in Equation (3), where ρ_0 and ρ_{DO} describe the densities of pure water and dissolved oxygen, respectively.

$$\rho = \rho_0 + C \cdot 10^{-6}(1 - \rho_{DO}/\rho_0) \tag{3}$$

3 Numerical methods for calculating the density diffusion

We applied the meshless method, the FDM, the FEM and the BEM to analyse the density diffusion in the unsteady state in a water area, as shown in Fig. 1.

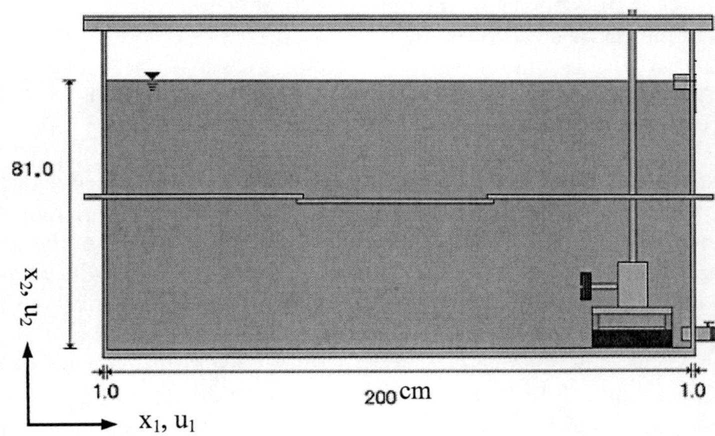

Figure 1: Analytical domain and a DO-supplying machine in a constructed model of a water reservoir.

3.1 Meshless method formulation for concentration analysis

The concentration in the steady state is expressed as Equation (4) with Equation (5) (Sakamoto et al. [1]),

$$C = \gamma_j Xj\ ,\ Xj = (r^2+c^2)^{-1/2} \ \text{ or }\ Xj = \exp(-c\, r^2)\ \text{ or }\ Xj = 1/\log(r^2+c^2)^{1/2} \tag{4}$$

$$\left\{\left(u_1\frac{\partial X_j}{\partial x_1} + u_2\frac{\partial X_j}{\partial x_2}\right) - (D_1\frac{\partial^2 X_j}{\partial {x_1}^2} + D_2\frac{\partial^2 X_j}{\partial {x_2}^2})\right\}\gamma_j = 0 \tag{5}$$

where r^2 equals $(x\text{-}x_j)^2+(y\text{-}y_j)^2$ and c is the constant. The transient convective-diffusion equation is then rewritten as follows,

$$C_{,t} + L(C) = 0 \tag{6}$$

www.witpress.com, ISSN 1743-355X (on-line)

where $C_{,t}$ is the time derivative of C and L(C) has the terms of convection and diffusion in the steady state. Applying the finite difference scheme, Equation (6) yields

$$(C^{t+\Delta t} - C^{t})/\Delta t + \{L^{t+\Delta t}(C) + L^{t}(C)\}/2 = 0 \tag{7}$$

$$C^{t+\Delta t} + L^{t+\Delta t}(C)\cdot \Delta t/2 = C^{t} - L^{t}(C)\cdot \Delta t/2 \tag{8}$$

where $C^{t+\Delta t}$ and C^{t} are the concentrations at time (t+Δt) and time (t), respectively, and $L^{t+\Delta t}$ and L^{t} are the terms of convection and diffusion at time (t+Δt) and time(t), respectively. Finally, using Equations (3), (4), and (8), the meshless method can be used to analyse the DO concentration in the unsteady state using the global expansion function Xj (= $(r^2+c^2)^{-1/2}$ or $\exp(-c\, r^2)$) or $1/\log((r^2+c^2)^{1/2})$ of the mesh-free RBF collocation method (Devo et al. [3]) or the radial basis functions of the Gaussians (Powell [6]).

3.2 Finite difference and weighted finite difference methods for convective-diffusion analysis

With respect to the analysis of the one-dimensional convective-diffusion equation, we combine the finite difference scheme of the order of four-degree accuracy ($O(h^4)$) for the diffusion terms and the UTOPIA scheme of the order of three-degree accuracy ($O(h^3)$) for the convective terms (Leonard [4]). To analyse the DO concentration by using the two-dimensional convective -diffusion equation, we propose that the analysis of the two-dimensional density diffusion be calculated by using the one-dimensional convective-diffusion analysis twice. The weighted finite difference scheme of the order of three-degree accuracy ($O(h^3)$) for both the diffusion and the convective terms is proposed and applied to the calculation of the two-dimensional density diffusion by using also the one-dimensional convective-diffusion analysis twice. The WFDM and FDM can be expressed explicitly.

3.3 Finite element method for convective-diffusion analysis

The upwind shape function of the Ptrov-Galerkin weighting (ZienKiewicz and Taylor [3]) is tested to analyse the two-dimensional density diffusion using the FEM. To reduce the CPU time of the FEM, the technique of the skyline solver and the conjugate gradient method are combined and applied to our upwind FEM.

3.4 Boundary element method for convective-diffusion analysis

Using the ordinary fundamental solution, our BEM does not yield any advantage for calculating the density diffusion over the FE, the FD and the meshless methods. Namely, the convection terms should be treated on the right hand side of our BEM, where the solution cost is significantly larger due to the full and non-symmetry of the coefficient matrices. On the other hand, since the especial fundamental solution of the BEM (Kanoh et al. [2]) is expected to yield accurate and convergent solutions in the analysis of the two-dimensional convective - diffusion equation, we need to mathematically integrate the fundamental solution in the time domain.

4 Model simulation

The analogy between the differences in the water temperature and the DO concentration was proved by referring to the concentration distribution in the reservoir model visualised using a pigment and a VTR (Sakamoto, et al. [1]). Using the analogy, we reproduce, in our model simulation, the density flow and convective diffusion of the DO concentration in the lower layer of a water reservoir at a depth of about 50 metres. In reference to the observed some velocity vectors and the distributions of the DO concentration in the model, we tried to obtain some evidence to explain the phenomena that the distance reached by the DO-rich water was more than 300 metres in a reservoir in spite of the small velocity of the water flowing out.

5 Results and discussion

As described above, we introduced a concept in the simulation model developed in our laboratory and were able to observe some velocity vectors and obtain the distributions of the DO concentration in the model. In reference to the observed results, we tried to produce some evidence to explain the phenomena that the distance reached by the DO-rich water was more than 300 metres in a reservoir in spite of the small velocity of the water flowing out. The numerical results of the meshless method, the BEM, the FEM, and the WFDM are also discussed in this section in order to investigate the numerical criteria for calculating the density diffusion in a water reservoir using two-dimensional convective diffusion equations.

5.1 Observed values in a model around a DO-supplying machine

5.1.1 Concentration distribution of DO in a model

Figure 2 is an illustration of the concentration distribution of the temperature difference, -0.1, caused by a DO-supplying machine in a reservoir model visualised using a pigment (methylene blue) and a VTR. We consider that the analogy between the differences of the water temperature and the DO concentration can also be proved by using the observed concentration distribution, since both profiles of the temperature difference (Fig. 2) and the DO-concentration difference seemed almost identical (the figure to illustrate the concentration distribution of the DO-concentration difference, 23 mg/L, in the reservoir model was omitted).

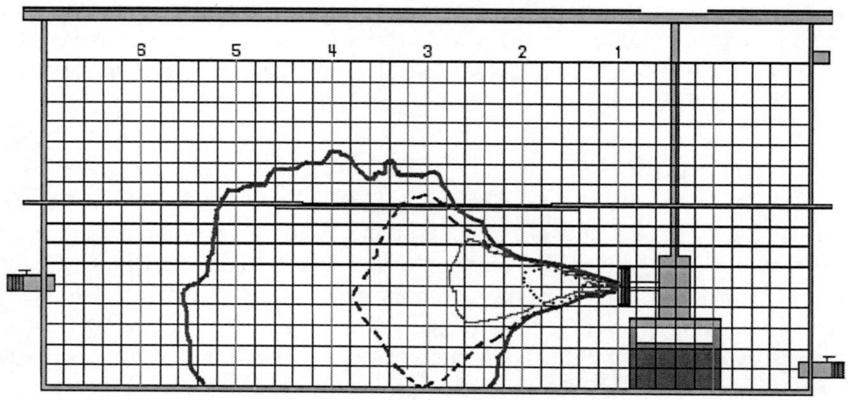

Figure 2: Observed areas of diffusion of pigment (Temperature difference:-0.1).

5.2 Flow analysis in the model of a water reservoir

For computing the density diffusion, the velocity data are important and have significant influence on the calculated results. However, we would like to focus on the numerical criteria for calculating the density diffusion in this paper. The figures to illustrate the calculated velocity vectors were omitted, and discussions of flow analysis are limited to their influence for computing the density diffusion.

5.3 DO-concentration analysis in the model of a water reservoir

5.3.1 Time required by the four numerical methods for the DO analysis

The table shows the time required by the four numerical methods for analysing the DO concentration in the model. When the number of divisions of the analytical domain was 9,075, the WFDM, the FDM, and the meshless method needed almost 18.3, 27.5, and 495 times the time required by the upwind FEM, respectively. For the purposes of saving time, the upwind FEM was the best; the WFDM was second best; the FDM was third best; and the meshless method was the poorest performer. We believe that the reason that the upwind FEM was the best, the WFDM was second best, and the FDM was third best is that the coefficient matrix of the upwind FEM was developed to be suitable for employing both the skyline solver and the conjugate gradient method, and the WFDM and FDM can be easily applied to an explicit scheme.

Table 1: The time required by the four methods for analysing the unsteady convective diffusion of DO for 240 seconds in a model of a water reservoir.

Relative computational time / Numerical method	Number of divisions: 9,075	Time increment: Δt (sec)
FEM	1.00	0.5
WFDM	18.3	0.0005
FDM	27.5	0.0005
Meshless method	495	5.0

5.3.2 FEM calculation of the concentration distribution

Figures 3(a) and 3(b) are illustrations of the concentration distribution calculated using the FEM with the upwind shape function, in which the duration of convective-diffusion of DO is 120 and 240 sec, respectively. Figure 4 shows the concentration distribution calculated using the FEM, in which the number of divisions in the FEM 9,075. Comparing Fig. 3(b) (in which the adopted number of divisions is 4,800 and the space division (h) is 0.013 m) with Fig. 4 (in which the adopted number of divisions is 9,075 and the space division (h) is 0.010 m), it was noted that the both space divisions (h=0.013 and 0.010 m) made the areas of the DO distribution wide in the vertical direction and the effect of the density diffusion conspicuous in the upwind FEM analysis where the concentration of the flowing-out DO was 100 mg/L. Now, we discuss the order estimate for calculating the density diffusion with the upwind FEM, where the DO concentration equals 100 mg/L and the liquid density (ρ) becomes 1.0001 Kg/L. As described above, when the methods of the order of three-degree accuracy are used, the space division h should be less than $0.0001^{1/3}$ (= 0.046) m, and in case the methods of the order of two-degree accuracy are used, the space division h becomes less than $0.0001^{1/2}$ (= 0.010) m for the estimation of the liquid density ρ of 1.0001 Kg/L. We consider that the upwind FEM has the order of 2.14 degree accuracy ($O(h^{2.14})$) in the density diffusion analysis. The reason is that 0.013 to the 2.14 power is less than 0.0001 ($0.013^{2.14}$= 0.000097 < 0.0001). With respect to the influence of flow analysis for computing the density diffusion, the upwind shape function could make it possible to set the optimum value of ν and obtain the stable velocity distribution in the FEM flow analysis.

5.3.3 Meshless calculation of the concentration distribution

Figures 5(a), 5(b), and 5(c) are illustrations of the concentration distribution calculated using the meshless method, in which the term of the velocity increase (ωgΔt) is adopted, the number of divisions in the meshless method is 4,800 or 9,075. Here, we investigated the convergence and the accuracy of three kinds of the global expansion function Xj, $(r^2+c^2)^{-1/2}$ or $\exp(-c\, r^2)$ or $1/\log((r^2+c^2)^{1/2})$, and

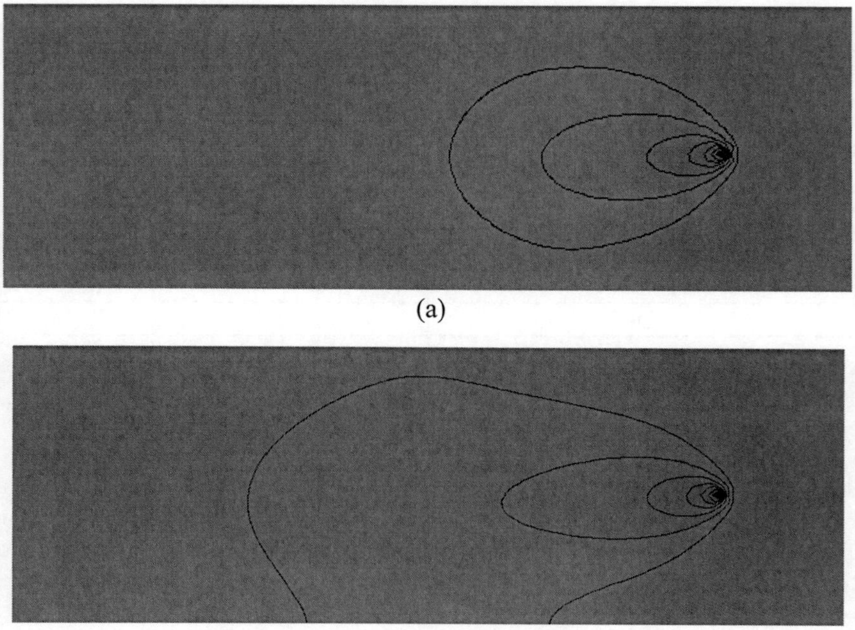

(a)

(b)

Figure 3: (a) DO-concentration distribution calculated using the upwind FEM with 4,800 elements (4,961 points, h=0.013m) [t=120 sec]. (b) [t=240sec].

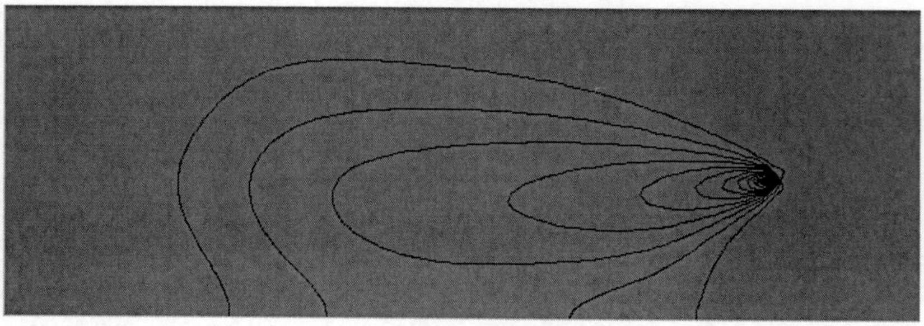

Figure 4: DO-concentration distribution calculated using the upwind FEM with 9,075 elements (9,296 points, h=0.010m) [t=240sec].

adopted the logarithm type of the mesh-free RBF collocation method because of the convenience of determining the appropriate value of c in the function Xj and the convergence of it. Comparing Fig. 5(a) (in which the adopted space division (h) is 0.013 m) with Fig. 5(b) (in which the adopted space division (h) is 0.010 m), it was noted that the former space division (h=0.013) made the areas of the DO distribution slightly wide, but the latter space division (h=0.010 m) made it wider in the vertical direction. Furthermore, the effect of the density diffusion was conspicuous in the meshless method analysis where the concentration of the

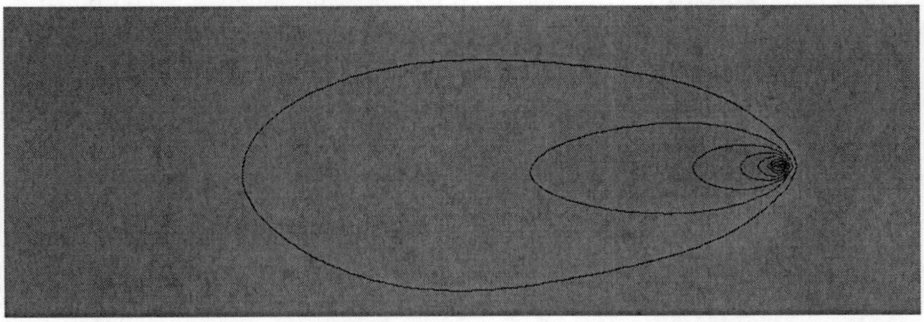

(a)

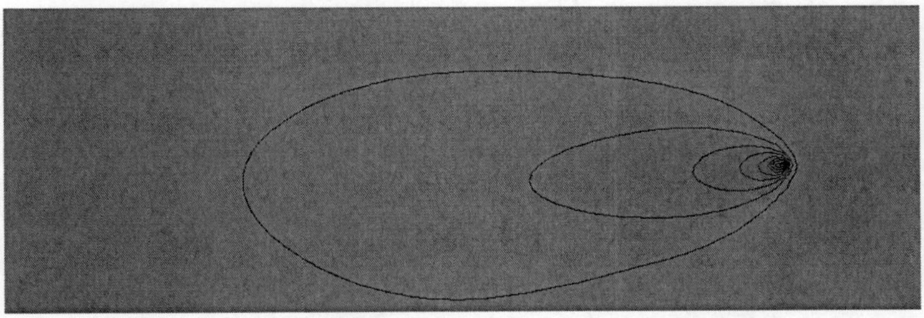

(b)

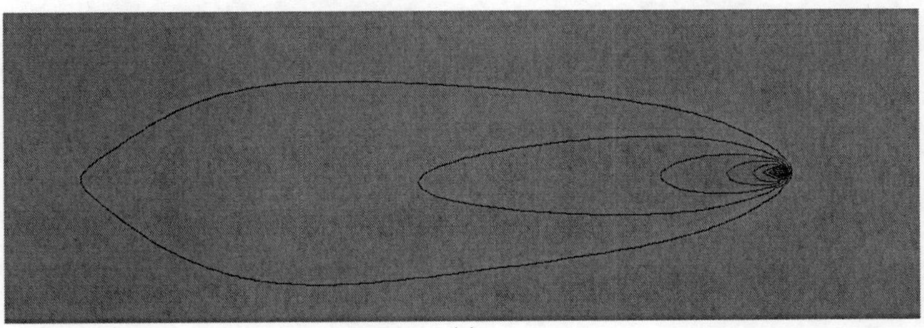

(c)

Figure 5: (a) DO-concentration distribution using the meshless method with 4,961 points (h=0.013m) where the flowing–out DO is 100 mg/L [t=240sec]. (b) DO-concentration distribution using the meshless method with 9,296 points (h=0.010m) where the flowing–out DO is 100 mg/L [t=240sec]. (c) DO-concentration distribution using the meshless method with 4,961 points (h=0.013m) where the flowing–out DO is 120 mg/L [t=240sec].

flowing-out DO was 100 mg/L. Figure 5(c) shows that the larger space division (h=0.013 m) could make it wider in the vertical direction, and the effect of the density diffusion was conspicuous in the meshless method analysis when the concentration of the flowing-out DO was 120 mg/L. We consider that our meshless method has the order of 2.1 degree accuracy ($O(h^{2.1})$) in the density diffusion analysis. The reason for this is that 0.013 to the two point first power is less than 0.00012 ($0.013^{2.10} = 0.000115 < 0.000120$). Namely, the combination of the order of 2.1 degree accuracy ($O(h^{2.1})$) and the larger space division (h=0.013 m) in our meshless method can yield the value of 0.000115 that is less than 0.000120, which is caused by the concentration of the flowing-out DO 120 mg/L.

5.3.4 FDM calculation of the concentration distribution

Figure 6 is an illustration of the concentration distribution calculated using the FDM, in which the finite difference scheme of the order of four-degree accuracy ($O(h^4)$) for the diffusion terms and the UTOPIA scheme of the order of three-degree accuracy ($O(h^3)$) for the convective terms are combined. Here, the number of divisions in the FDM is 4,800. The term ($\omega g \Delta t$) of the density diffusion seemed to make the areas of the diffusion wider in the vertical and flowing-out directions and the speed of the convective diffusion higher than those in the analyses of the FDM when this density diffusion was not applied. We conclude that the FDM satisfied the order of three-degree accuracy ($O(h^3)$).

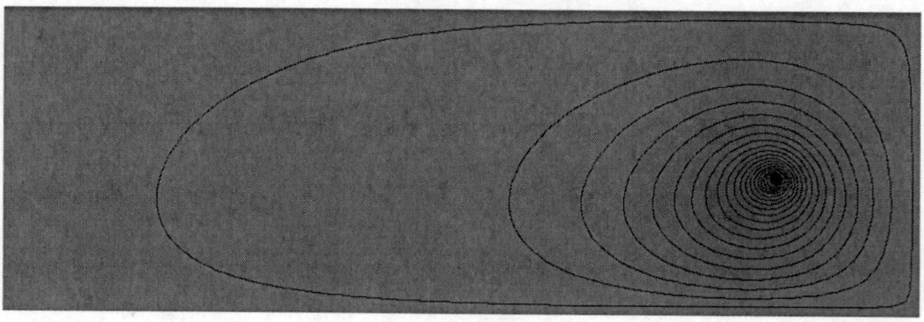

Figure 6: DO-concentration distribution calculated using the FDM with 4,961 points (h=0.013m) where the flowing–out DO is 100 mg/L [t=240sec].

5.3.5 BEM calculation of the concentration distribution

Using the ordinary fundamental solution of our BEM, we could not reproduce the density difference of the high DO concentration (80 or 100 mg/L) of the water reservoirs with the space division (h) of 0.013 m. We concluded that the convergence and accuracy of the upwind FE, the FD, the WFD, the BE, and the meshless methods for this problem were satisfactory (the figures to illustrate the concentration distribution calculated using the WFD and the BE methods have been omitted).

6 Conclusion

In summary, (1) in this study, the meshless method, the BEM, the FEM, the FDM, and the WFDM were developed and applied to the analysis of the density diffusion that was caused by the water density ρ, the gravity acceleration g, and the time increment Δt; (2) introducing the radial basis functions of the Gaussians or the multiquadric to the meshless method, the special fundamental solution to the BEM and the upwind shape function to the FEM, we upgraded the order of the accuracy of these methods; (3) using the technique of the ADI method, the finite difference scheme of the order of four-degree accuracy ($O(h^4)$) for the diffusion terms and the UTOPIA scheme of the order of three-degree accuracy ($O(h^3)$) for the convective terms were successfully combined for the FDM; (4) the weighted finite difference scheme of the order of three-degree accuracy ($O(h^3)$) for both the diffusion terms and the convective terms was defined; (5) the density diffusion could make the areas of the diffusion wider in the vertical and outflow directions and make the speed of the convective diffusion higher than it was in the analyses of these methods when this velocity increase was not applied; (6) the stability and convergence of the five kinds of analysis using these newly developed methods seemed satisfactory; (7) the degrees of accuracy of the four methods (the meshless, the FE, the FD, and the WFD methods) were upgraded sufficiently to calculate the density diffusion in case the space division (h) was 0.0133 or 0.01 m; (8) the technique of the skyline solver and the conjugate gradient method were applied to our upwind FEM, and the CPU time of the FEM was greatly reduced; (9) the analogy between the differences of the water temperature and the DO concentration was used to reproduce the density-difference of the high DO concentration (80 or 100 mg/L) of the water reservoirs in our model simulation; and (10) the developments and ideas described above were investigated, and the numerical criteria for calculating the density diffusion in a water reservoir using the two-dimensional convective -diffusion equation were discussed.

References

[1] Sakamoto, K., Kanoh, M., &., Kuroki T., Numerical criteria for calculating the density diffusion in a water reservoir, *Proc. of the 32nd World Conf. on Boundary Elements and Other Mesh Reduction Methods*, ed. C.A. Brebbia, WIT Press, Southampton, UK, pp.41-52, 2010.

[2] Kanoh, M. Kuroki T., Aramaki, G. and Onishi, K., Boundary element method for salt water wedge diffusion., *Proc. 1st Joint Japan/U.S. Sym. on BEM,* pp.277-284, Pergamon Press, 1988.

[3] O.C. ZienKiewicz and R.L. Taylor, 13.7.2 Direct, steady-state, solution and the Petrov-Galerkin weighting, The Finite Element Method, Fourth Edition, Volume 2, MacGraw-Hill, p539-p541, 1991.

[4] Leonard, B.P., The QUICK algorithm, Computer Methods in Fluids, ed. K. Morgan, C. Taylor, C.A. Brebbia, WIT PRESS, Pentech Press, p.159, 1980.

[5] Divo E., Kassab A., and Zahab El., Parallel domain decomposition meshless modeling of dilute convection-diffusion of species, *Proc. of 27th World Conf. on Boundary Elements and Other Mesh Reduction Methods*, ed. C.A. Brebbia, WIT Press, Florida, USA, pp.79-89, 2005.

[6] Powell, M.J.D, The theory of radial basis function approximation in 1990, *Advances in Numerical Analysis*, Vol. II, ed. W. Light, Oxford Science Publications: Oxford, pp.105-210, 1992.

[7] Kanoh, M., Kuroki T., Fujino K., and Ueda, T., Weighted Finite Difference and Boundary Element Methods Applied to Groundwater Pollution Problems, *Wat. Sci. Tech. Vol. 23, Kyoto,* pp. 517-524, 1991.
[8] Kanoh, M. and Kuroki T., Upwind finite element analysis of the inverse problem for groundwater resources, *Proc. of 2nd International Symposium on Inverse Problems - ISIP '94, Paris,* ed. Bui, H.D., Tanaka M. et al., pp. 29-32, 1994.

Application of the boundary element method to 2D and 3D bubble dynamics

Z. Fu & V. Popov
Wessex Institute of Technology, Environmental and Fluid Mechanics, Southampton, UK

Abstract

The Boundary Element Method (BEM) was applied to the solution of bubble dynamics in an ideal fluid. Two-dimensional (2D) and three-dimensional (3D) models have been developed and the results have been compared to the analytical solution. The results show that only the 3D model can correctly represent the physics of the problem. The influence of the model parameters on the solution has been investigated for a single bubble.
Keywords: BEM, numerical simulation, bubble dynamics.

1 Introduction

A great amount of work has been conducted in the area of bubble dynamics which can be seen in the review papers by Plesset and Prosperetti [1] and Feng and Leal [2]. During the course of bubble oscillation, jet formation is a common phenomenon. As pointed out by Tsiglifis and Pelekasis [3], even when the bubble shape is initially very close to spherical, asymmetric collapse happens at the following stage and it gives rise to jet formation. The bubbles in the acoustic field can oscillate or they can collapse producing high temperatures and possibly jets. The creation of high temperature spots can help chemical reactions and this phenomenon is used in sonochemical reactors.

With advancement of the computational technology, numerical studies have been carried out in order to understand the bubble dynamics. One of the numerical techniques used is the Boundary Element Method (BEM) which possesses unique advantages in respect to other numerical techniques for these types of problems. Indirect BEM has been employed by Wang and Khoo [4] who pointed out that both source and dipole distribution methods encounter

WIT Transactions on Modelling and Simulation, Vol 52,
www.witpress.com, ISSN 1743-355X (on-line)
doi:10.2495/BE110071

difficulties arising from hyper-singular integrals. By contrast, direct BEM solves the unknown (either potential or derivative) on the surface of the bubble in a direct way. One problem by applying the direct BEM is that the solution is only provided for the normal velocity on the bubble surface while the tangential components, which are also required, are not provided. Zhang et al. [5] applied a finite difference scheme to evaluate the tangential velocity (2D) or two tangential velocities (3D), thereby obtaining the material velocity on the bubble surface. This approach cannot guarantee sufficient accuracy for the velocity. When the simulation involves time history, errors accumulate and may result in computational instability.

In this work a BEM code has been developed to simulate bubble dynamics. The problem includes moving boundaries of the pulsating sphere which with time accumulates errors leading to numerical instabilities. A scheme has been applied in order to deal with the numerical instability and the model has been verified by comparison with the Rayleigh-Plesset equation.

2 Description of the problem

Let us consider a single bubble where the surrounding fluid represents the domain for numerical computation. The boundary of the problem is defined with the surface of the bubble, the fluid free surface and the solid structure(s) that may enclose the fluid or alternatively an imaginary boundary at infinity.

Bubble dynamics is essentially a time-dependent process, involving bubble expansion and collapse, bubble translation, jet formation, and possibly bubble separation and merging (for multiple bubbles). However, all of these mechanisms can be summarised as the consequence of the geometric evolution of the bubble surface, which is driven directly by the pressure difference between the bubble interior and exterior. In such sense, the problem is generally focused on the shape variation and moving of the bubble surface, which represents a moving boundary problem. The nature of the problem of bubble dynamics justifies the choice of the BEM numerical approach.

The boundary conditions involve generally the potentials and velocities. In case of the solid boundary, the normal velocities are set to be zero. The conditions on the bubble surface govern the bubble evolution. The subtle change of the boundary exerts highly sensitive outputs of velocities on the boundary; on the other hand, the normal velocities generated directly from BEM and the tangential velocities obtained from the potentials are used to update the geometry of the bubble surface. Hence both the geometry and the velocities on the boundary influence each other. As a result, the precision of boundary conditions determines how stable and accurate the numerical simulation is, as even small errors will be accumulated and will give rise to a bubble surface error as the time progresses.

The pressure governs the bubble surface evolution in time. Usually the pressure at the bubble interior can be described with a gas state equation. The pressure at the bubble exterior may be due to atmospheric pressure, hydraulic pressure, buoyancy forces, as well as the acoustic forcing.

www.witpress.com, ISSN 1743-355X (on-line)

3 Theoretical foundation

3.1 Boundary integral equation

The fluid is assumed to be incompressible, homogeneous, inviscid and irrotational. The idealised fluid flow is governed by Laplace equation, i.e.:

$$\nabla^2\phi = 0\,, \tag{1}$$

where ϕ is the velocity potential of the fluid and the normal velocity on the surface of the bubble can be defined as $-\frac{\partial\phi}{\partial\vec{n}}$. Hence the above Laplace equation has the following representation in the integral form:

$$c(\mathbf{p})\phi(\mathbf{p}) = \int_{\Omega}\left(G(\mathbf{p},\mathbf{q})\frac{\partial\phi(\mathbf{q})}{\partial\vec{n}} - \frac{\partial G(\mathbf{p},\mathbf{q})}{\partial\vec{n}}\phi(\mathbf{q})\right)d\Gamma\,\cdot \tag{2}$$

The coefficient c(**p**) is a function of the solid angle of the boundary at the collocation point **p** (see Figure 1). With **p** located inside the domain, c(**p**) is fixed as 1.0 and for **p** on a smooth boundary c(**p**) becomes 1/2. However, the boundary is more realistically of irregular shape, which requires repeated laborious computation for c(**p**). Alternatively, c is calculated in an indirect way by physically considering a constant potential, which is explained in more detail in the book by Brebbia and Dominguez [6]. G stands for the fundamental solution:

$$G(\mathbf{p},\mathbf{q}) = \begin{cases} -\dfrac{1}{2\pi} In(|\,\mathbf{p}-\mathbf{q}\,|) & \text{for 2D} \\ \dfrac{1}{4\pi\,|\,\mathbf{p}-\mathbf{q}\,|} & \text{for 3D} \end{cases} \tag{3}$$

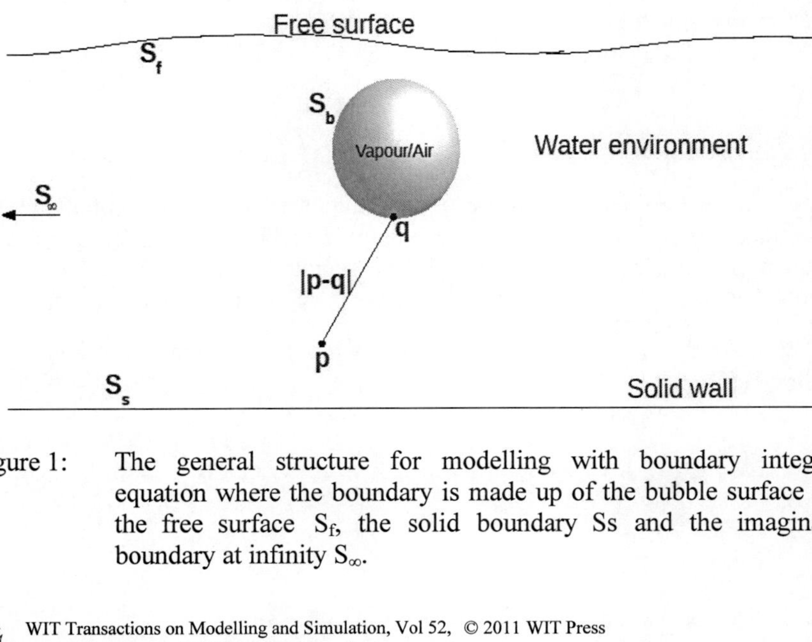

Figure 1: The general structure for modelling with boundary integral equation where the boundary is made up of the bubble surface S_b, the free surface S_f, the solid boundary Ss and the imaginary boundary at infinity S_∞.

3.2 Bernoulli equation

The boundary integral equation is capable of solving the bubble problems in a stationary state. When it comes to bubble oscillation with time, the modified Bernoulli equation should be introduced to describe the dynamic boundary conditions at the interface. By neglecting the effects of viscosity, buoyancy force and surface tension, the Bernoulli equation is expressed as follows:

$$\frac{D\tilde{\phi}}{D\tilde{t}} = -\frac{1}{2}|\nabla\tilde{\phi}|^2 + \frac{1}{\tilde{\rho}_l}\Delta\tilde{p}, \tag{4}$$

where $\Delta\tilde{p}$ is the pressure difference on the boundary. Note that the symbol '~' represents the physical quantity of a parameter; without '~' it indicates a dimensionless quantity. For any point $\mathbf{q}\in S$ (see Figure 1), the particle velocity is:

$$\tilde{\mathbf{u}} = \frac{D\tilde{\mathbf{q}}}{D\tilde{t}} = -\nabla\tilde{\phi}. \tag{5}$$

Meanwhile, the pressure difference at the bubble interface follows:

$$\Delta\tilde{p} = \tilde{p}_b - \tilde{p}_a - \tilde{p}_\infty, \tag{6}$$

where $\tilde{p}_b$, $\tilde{p}_a$ and $\tilde{p}_\infty$ are the pressure at the bubble interior, the standard atmospheric pressure and the pressure at infinity.

We consider a bubble that embraces air (assumed to be ideal gas) and vapour. The bubble interior is assumed uniform in terms of pressure, temperature and composition. In addition, the idealised gas has a polytrophic behaviour. Then the internal pressure follows the gas state equation:

$$\tilde{p}_b = \tilde{p}_v + \tilde{p}_0\left(\frac{\tilde{V}_0}{\tilde{V}}\right)^\gamma, \tag{7}$$

where $\tilde{p}_v$ is the vapour pressure assumed constant, $\tilde{p}_0$ is the initial pressure contributed by air with the corresponding initial bubble volume $\tilde{V}_0$, and γ is the ratio of specific heat (see Figure 2). The standard atmosphere pressure $\tilde{p}_a$ is applied as the pressure scale. In addition, we prescribe the length and time scales as $\tilde{V}_0^{\frac{1}{3}}$ and $\tilde{V}_0^{\frac{1}{3}}\sqrt{\tilde{\rho}_l/\tilde{p}_a}$, respectively, then Bernoulli equation is obtained in the following form:

$$\frac{D\phi}{Dt} = -\frac{1}{2}|\nabla\phi|^2 + p_0\left(\frac{1}{V}\right)^\gamma - 1 - p_\infty. \tag{8}$$

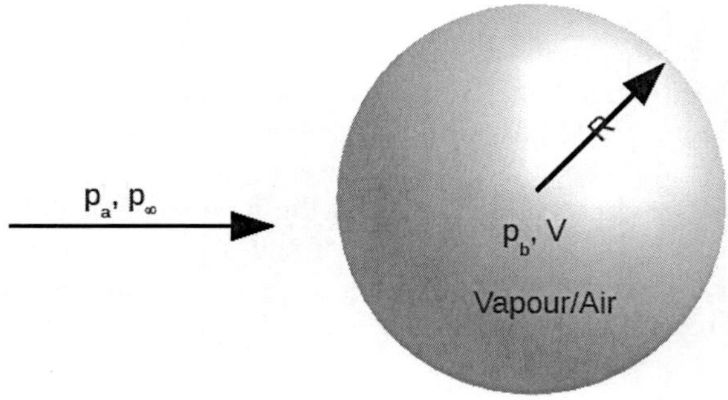

Figure 2: A spherical bubble composed of vapour and air with the pressure, temperature and composition evenly distributed.

4 Numerical scheme

4.1 Numerical procedure

The initial state of the bubble(s) should be prescribed to carry out the numerical simulation. In case of a single spherical bubble, Blake and Gibson [7] proposed a mathematical formula to obtain a reasonable initial potential on the boundary. This formula can be derived from the Rayleigh-Plesset equation:

$$\tilde{\phi}_0 = \tilde{R}_0 \left\{ \frac{2}{3} \left(\frac{\tilde{p}_\infty - \tilde{p}_b}{\tilde{\rho}_l} \right) \left[\left(\frac{\tilde{R}_m}{\tilde{R}_0} \right)^3 - 1 \right] \right\}^{\frac{1}{2}}, \tag{9}$$

where $\tilde{R}_m$ is the maximum radius that the bubble can reach under the constant pressure difference $\Delta\tilde{p} = \tilde{p}_b - \tilde{p}_\infty$. The above equation sets both the initial radius and the maximum radius as known and it is beneficial as a predictable maximum bubble size can be used as an important parameter for reference during the bubble oscillation. Nonetheless, the formula is only restricted to the vapour bubble with constant internal pressure which is driven by a constant pressure forcing from outside. More generally, it is more reasonable to prescribe the initial conditions of the bubble(s) as stationary, i.e. $\mathbf{u}=-\nabla\phi=0$, or written in another way, the normal potential derivatives $\frac{\partial \phi}{\partial \vec{n}}$ on the boundary are known. Hence the corresponding potentials can be obtained by discretising the boundary and constructing the matrices so as to solve the integral equations with Neumann boundary conditions. Thereafter, the Runge-Kutta method is employed to update the potentials on the boundary. Once new potentials on the boundary are determined, the normal potential derivatives can be calculated. Both the potentials and their derivatives are used to obtain the material velocities on the bubble surface and the geometry of the bubble surface is ready for update. On the same principle to update both the potentials and the bubble geometry, repetition is carried out likewise and the bubble oscillation with a clear time history is depicted. However, special care should be taken during implementation of the numerical techniques in order to guarantee the accuracy of results and the stability of the model.

The developed programme allows us to select different orders and types of elements. Herein, the quadratic quadrilateral element is put forward for illustration. When the boundary is discretised, the information is created on the number of nodes and elements, together with the arrangement of nodes in each individual element. The employed order of nodes in an element can be seen in Figure 3, which determines not only the order of interpolation coefficients Φ_i, but also the direction of the unit normal. The interpolation coefficients are given as:

$$\begin{aligned}
\Phi_1 &= \frac{1}{4}\eta_1(\eta_1-1)\eta_2(\eta_2-1), \quad \Phi_2 = \frac{1}{2}(1-\eta_1^2)\eta_2(\eta_2-1) \\
\Phi_3 &= \frac{1}{4}\eta_1(1+\eta_1)\eta_2(\eta_2-1), \quad \Phi_4 = \frac{1}{2}\eta_1(1+\eta_1)(1-\eta_2^2) \\
\Phi_5 &= \frac{1}{4}\eta_1(1+\eta_1)\eta_2(1+\eta_2), \quad \Phi_6 = \frac{1}{2}(1-\eta_1^2)\eta_2(1+\eta_2) \\
\Phi_7 &= \frac{1}{4}\eta_1(\eta_1-1)\eta_2(1+\eta_2), \quad \Phi_8 = \frac{1}{2}\eta_1(\eta_1-1)(1-\eta_2^2) \\
\Phi_9 &= (1-\eta_1^2)(1-\eta_2^2)
\end{aligned} \quad , \qquad (10)$$

where both η_1 and η_2 are in the range of (-1, 1). Meanwhile, the right-hand rule is applied for the direction of the unit normal, as shown in Figure 3. The unit normal vector with respect to the surface is given as:

$$\frac{\partial \mathbf{q}(\eta_1,\eta_2)}{\partial \vec{n}} = \left(\sum_{i=1}^{9}\frac{\partial \Phi_i}{\partial \eta_1}\cdot \mathbf{q}_i\right) \times \left(\sum_{i=1}^{9}\frac{\partial \Phi_i}{\partial \eta_2}\cdot \mathbf{q}_i\right). \qquad (11)$$

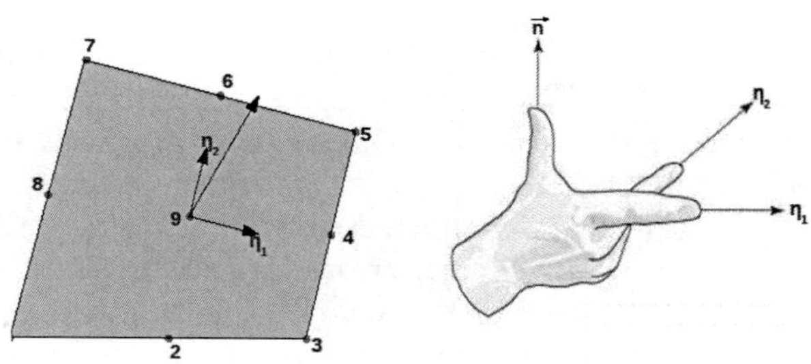

Figure 3: The position of nodes on an element and the normal vector determined with the right-hand rule.

After dicretisation of the boundary, the system matrix coefficients can be evaluated and the matrix can be assembled. With the application of quadratic quadrilateral elements, the potential and the derivatives ought to be interpolated from (10). Correspondingly, (2) has the following discretised form:

$$c(\mathbf{p})\phi(\mathbf{p}) = \sum_{i=1}^{m}\left\{\int_{s_i}\left\{G(\mathbf{p},\mathbf{q})\sum_{j=1}^{9}\left[\Phi_j\frac{\partial\phi_j}{\partial\vec{n}}\right]\right\}d\Gamma\right\} - \sum_{i=1}^{m}\left\{\int_{s_i}\left\{\frac{G(\mathbf{p},\mathbf{q})}{\partial\vec{n}}\sum_{j=1}^{9}\left[\Phi_j\phi_j\right]\right\}d\Gamma\right\}. \tag{12}$$

where m stands for the number of elements on the boundary. Integration on each element S_i is operated by isoparametric Gaussian quadrature. When the collocation point is allocated on the integrated element as a singularity point, an extra measure should be taken. Chahine and Perdue [8] evaluated the strongly singular integrals analytically. In this work, the third-degree coordinate transformation proposed by Telles [9] has been employed.

Once the unknown field variables are obtained from (12), the evaluation for the next time step can begin. The time scheme is constructed by the Bernoulli equation with the form of ordinary differential equations. To secure high accuracy, a fifth-order Runge-Kutta formula is employed. The step size is adaptive rather than constant, which depends on the embedded criterion (see Cash and Karp [10]). The scheme is given in (14):

$$\begin{gathered} k_1 = hf(x_n, y_n) \\ k_2 = hf(x_n + a_2 h, y_n + b_{21}k_1) \\ \cdots \\ k_6 = hf(x_n + a_6 h, y_n + b_{61}k_1 + \cdots + b_{65}k_5) \\ y_{n+1} = y_n + c_1k_1 + c_2k_2 + c_3k_3 + c_4k_4 + c_5k_5 + c_6k_6 + O(h^6) \end{gathered}, \tag{13}$$

where the coefficients a, b and c are constant and known, for more details see Press et al. [11]. However, (8) tells us that $\frac{D\phi}{Dt}$ cannot be updated between the current and next time steps, as both the velocity and pressures have yet to be determined. Hence, one more physical criterion is established as follows:

$$\Delta t_{\max} = \frac{d\phi}{\max\left\{\left|-\frac{1}{2}|\nabla\phi|^2 + p_0\left(\frac{1}{V}\right)^{\gamma} - 1 - p_{\infty}\right|\right\}}, \tag{14}$$

where the denominator on the right hand side of the equation above is the maximum value of $\frac{D\phi}{Dt}$ along the whole boundary. With the predefined maximum potential jump $d\phi$, the Δt is available as the maximum step size to the next time step. Generally $d\phi$ is set at the level of $O(10^{-2})$, as shown previously by Taib [12] who used 0.08 and by Wang et al. [13] who used 0.03.

4.2 The smoothing scheme

The implementation of surface smoothing is one of the key points to maintain the stability during the simulation. We introduce a spherical single bubble. When it is subjected to symmetric pressure in a infinitely open field, the spherical shape is preserved. Nonetheless, the inevitable tiny errors for each time step accumulate and the bubble gradually loses its spherical shape. In case of 2D, the curve becomes saw-teeth shaped, whist convex and concave appear alternately on the surface of the 3D bubble (see Figure 4).

The implemented smoothing scheme is based on the technique of least squares. The basic idea is to predefine the smoothing function in advance with a certain number of unknown coefficients. Then a greater number of nodes are picked up around the targeted nodes to be smoothed. Once the least squares problem is solved, the new coordinate is determined by the smoothing function.

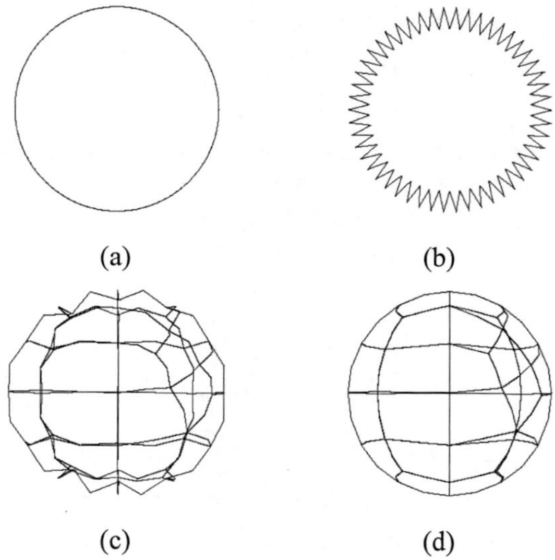

Figure 4: The meshes for a single bubble in two dimension and three dimension: (a) 2D before smoothing; (b) 2D after smoothing; (c) 3D before smoothing; (d) 3D after smoothing.

For 2D, the function is in the form of a polynomial representation referred to a local coordinate system:

$$y_l' = a_0 + a_1 x_l + a_2 x_l^2 , \tag{15}$$

while the function for 3D is represented by a biquadratic equation as:

$$z_l' = a_0 + a_1 x_l + a_2 y_l + a_3 x_l^2 + a_4 x_l y_l + a_5 y_l^2 , \tag{16}$$

where the symbol '′' indicates values after smoothing. In the case of a 2D problem, three unknown coefficients are included in the function; hence we pick up four neighbouring nodes in addition to the targeted node itself. By contrast, the way to select nodes in the 3D case is a bit more complex. More than six nodes ought to be selected in accordance with the unknowns in the smoothing function. We employed the approach proposed by Zhang et al. [5]. Firstly, we find l nodes which are immediately neighbouring to the targeted nodes. Then with reference to each of the l nodes, the neighbouring nodes in the second level are found, adding up m nodes. Finally, the third level of neighbouring nodes are found in the same way based on the m nodes at the second level, in this case we denote them with n. Therefore, we have in total l + m + n + 1 nodes to smooth the boundary involving targeted nodes.

To guarantee the serviceability of smoothing, the principle should be strictly followed that the new normal vector at the targeted point should be the same or highly similar to the old one before smoothing. Zhang et al. [5] suggested an iterative method to compute the new normal vector, based on which the local frame is determined.

Moreover, the potential is also required for smoothing to avoid a 'jump' from one node to its neighbouring nodes for which the same scheme applies. Figure 4 displays comparison of the bubble before and after smoothing.

5 Verification and analyses

5.1 Verification of the model

The developed model is capable of simulating bubble oscillation in both two and three dimensions. Before carrying out the investigation on the bubble dynamics, the model should be verified. Herein the Rayleigh-Plesset equation is introduced as an analytical solution for the bubble oscillation which is compared to the one obtained by using the model.

Further we refer to the Rayleigh-Plesset equation as the 'Rayleigh-Plesset bubble'. The Rayleigh-Plesset bubble is assumed to be perfectly spherical with a radius R_0 at the initial state t =0. The Rayleigh-Plesset equation is represented by:

$$R\ddot{R} + \frac{3}{2}\dot{R}^2 = p_b - 1 - p_\infty \cdot \tag{17}$$

Substituting p_b with the dimensionless form of (7), the equation above is modified as:

$$R\ddot{R} + \frac{3}{2}\dot{R}^2 = p_v + p_0\left(\frac{R_0}{R}\right)^{3\gamma} - 1 - p_\infty \cdot \tag{18}$$

To compare the results, input values are used as shown in Table 1. The radius variation with time is shown in Figure 5. The results show that the bubble oscillation based on the 3D model is in agreement with that of the Rayleigh-Plesset bubble. However, the bubble from the 2D model oscillates with apparently different frequencies and phases, despite of the similar amplitude. Therefore, the 3D model proves its applicability for modelling the bubble dynamics, whilst the 2D model apparently cannot be used, as it represents oscillation of an infinite cylinder rather than a 3D sphere.

Table 1: Input values for verification of bubble oscillation.

R_0	0.1	γ	1.4
p_v	1.0	p_0	2.0
p_∞	2.0		

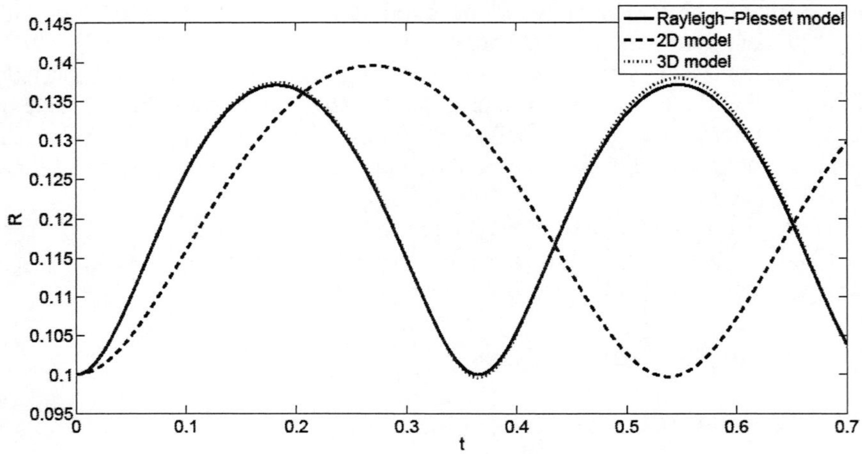

Figure 5: The radius variation of the oscillating bubble against time where three curves correspond to results from the Rayleigh-Plesset equation, 2D model and 3D model.

5.2 Oscillation of a single bubble

This example investigates the oscillation of a single bubble located in an infinite domain. The following factors are investigated: the initial bubble size, the air pressure at the bubble interior and the initial bubble-fluid interface velocity (i.e. $\nabla\phi$).

Figure 6 depicts bubble oscillations of different initial radii. The amplitude tends to be bigger when the initial radius of the bubble increases. Also, the bigger initial bubble possesses a lower frequency. Meanwhile the oscillating frequency decreases when the pressure amplitude increases (Figure 7).

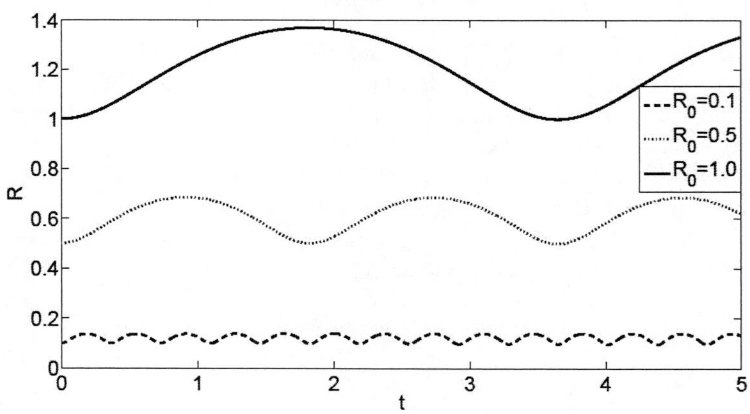

Figure 6: The oscillation of single bubbles with different initial radii (p_0=2.0 and $\nabla\phi$=0).

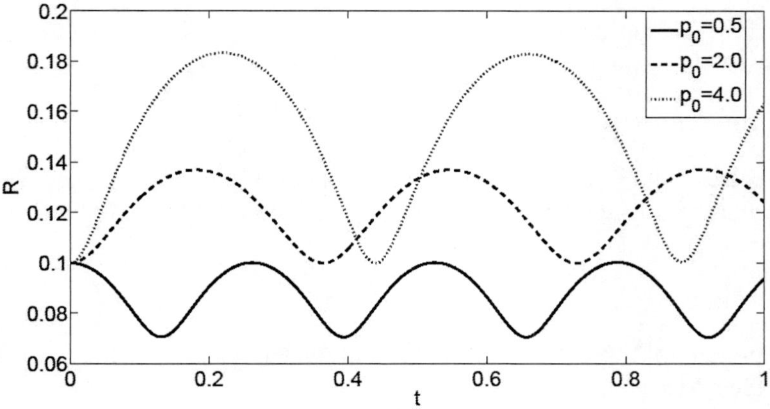

Figure 7: The oscillation of single bubbles with different initial air pressures (R_0=0.1 and $\nabla\phi$ =0).

The influence of the initial bubble-fluid interface velocity on the bubble dynamics may be comparatively more complicated. Even the same amplitude of the velocity may not guarantee an identical process. As shown in Figure 8, plot (a) is for initial normal velocity forcing the bubble to expand, and the higher normal velocity produces higher oscillation amplitude. When the interface velocity direction is set towards the bubble interior, the bubble initially compresses corresponding to the velocity direction and then it expands. Both plots show that the initial velocity of the interface enlarges the amplitude of the oscillations.

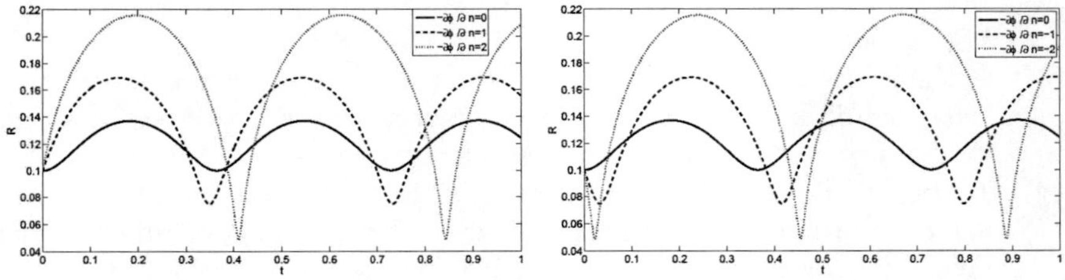

(a) (b)

Figure 8: The oscillation of a single bubble with different initial interface velocities (p_0=2.0 and R_0=0.1). (a) Velocity towards exterior; (b) velocity towards interior.

6 Conclusions

A numerical model has been developed for the problem of bubble dynamics. The model has been developed for the 2D and 3D cases. The model has been verified by using the Reyleigh-Plesset equation for oscillation of single bubble. The comparison of the results of the 2D and 3D models with the Reyleigh-Plesset equation shows that the 2D model cannot be used to accurately predict the behaviour of the bubbles, while the 3D model is in good agreement with the analytical solution. Once verified the model has been used to investigate the influence of several parameters on the single bubble dynamics.

Acknowledgement

The present study was supported by the SONO project, contract number: 228730, as part of the Seventh Framework Programme (FP7-NMP-2008-Large-2).

References

[1] Plesset, M.S. & Prosperetti, A., Bubble dynamics and cavitation. Ann. Rev. Fluid Mech., 9, pp. 145–185, 1977.

[2] Feng, Z.C. & Leal, L.G., Nonlinear bubble dynamics. Annu. Rev. Fluid Mech., 29, pp. 201–243, 1997.

[3] Tsiglifis, K. & Pelekasis, N.A., Numerical simulations of the aspherical collapse of laser and acoustically generated bubbles. Ultrasonics Sonochemistry, 14, pp. 456–469, 2007.

[4] Wang, C. & Khoo, B.C., An indirect boundary element method for three-dimensional explosion bubbles. Journal of Computational Physics, 194, pp. 451–480, 2004.

[5] Zhang, Y.L., Yeo, K.S., Khoo, B.C. & Wang, C., 3d jest impact and toroidal bubbles. Journal of Computational Physics, 166, pp. 336–360, 2001.

[6] Brebbia, C.A. & Dominguez, J., Boundary elements an introductory course. Computational Mechanics Publications, Southampton, UK, 1989.

[7] Blake, J.R. & Gibson, D.C., Growth and collapse of a vapour cavity near free surface. J. Fluid Mech., 111, pp. 124–140, 1981.

[8] Chahine, G.L. & Perdue, T.O., Simulation of the three-dimensional behaviour of an unsteady large bubble near a structure. In Proc. Third Int. Colloquium on bubbles and drops, Monterey, California, 1988.

[9] Telles, J.C.F., A self-adaptive co-ordinate transformation for efficient numerical evaluation of general boundary element integrals. International journal for numerical methods in engineering, 24, pp. 959–973, 1987.

[10] Cash, J.R. & Karp, A.H., A variable order Runge-Kutta method for initial value problems with rapidly varying right-hand sides. ACM Transactions on Mathematical Software, 16, pp. 201–222, 1990.

[11] Press, W.H., Teukolsky, S.A., Vetterling, W.T. & Flannery, B.P. Numerical recipes in Fortran 77. Cambridge University Press, 2nd edition edition, 2001.

[12] Taib, B.B., Boundary integral method applied to cavitation boundary dynamics. PhD thesis, The University of Wollongong, Wollongong, New South Wales, USA, 1985.

[13] Wang, Q.X., Yeo, K.S., Khoo, B.C. & Lam, K.Y., Nonlinear interaction between gas bubble and free surface. Comput. Fluids, 25, pp. 607–628, 1996.

Calculation of propeller's load noise using LES and BEM numerical acoustics coupling methods

Q. Yang[1], Y. Wang[1] & M. Zhang[2]
[1]*College of Naval Architecture and Marine Power, Naval University of Engineering, Wuhan, China*
[2]*College of Electronic Engineering, Naval University of Engineering, Wuhan, China*

Abstract

Coupling large eddy simulation (LES) and boundary element numerical acoustics methods in the frequency domain to predict the underwater non-cavitation far-field load noise in the ship wake flow, which presents a new method to measure the propeller noise level in the ship engineering. The Dynamic Smagorinsky-Lilly model (DSM) is applied to subgrid-scale (SGS) stress tensor in the process of unsteady field LES simulation. Then the sound pressure field is determined solely by the pressure distribution and normal velocity distribution on the closed surface including blades and hub surface after node numbers conservative transfer of the acoustic nodal sources to the acoustic grid. The predicted results show that the steady thrust and torque coefficients of the propeller differ by less than 2% on design point to the experiment data, and the pressure coefficients fit very well. The unsteady calculation can present precisely the blade passing frequency (BPF), its harmonics and wake axis passing frequency (APF) information. Under low frequency, the concentrated area contributing mainly to the noise lies in the blade tip and trailing edge near the root section, whereas for high frequency the source intensity transfers to the blade surface with a centralized pressure load. The spherical surface sound pressure distribution and wake point frequency spectra line of the propeller just get quantitative agreement with the reference calculated results because of the slight difference of ship wake distribution, and the calculated total sound pressure level below 200Hz is 126.4dB.
Keywords: propeller, load noise, boundary element method, integral method, frequency domain.

WIT Transactions on Modelling and Simulation, Vol 52,
www.witpress.com, ISSN 1743-355X (on-line)
doi:10.2495/BE110081

1 Introduction

The principle mechanisms of underwater radiated noise of the ship propeller can be divided into four parts, thickness noise, which dues to the displacement of the water by the blade profile, steady load noise, which points to the pressure difference between the suction and pressure sides of the blade when they are rotating, unsteady load noise, which caused by the periodic fluctuation and the random vibration of the blades induced by spatially non-uniform wake flow behind the vessel, and the broadband spectrum, which associates with the interaction noise between the turbulent inflow and the leading edge of the blades, the trailing edge vortex shedding noise and cavitation radiated noise caused by the bubbles collapse process [1-3]. The radiated noise of non-cavitating propeller in a uniform flow condition mainly consists of monopole thickness noise and dipole Gutin sound [3], and as for the ship propeller, the Gutin sound often contributes little to the overall noise in the far field and can be negligible [1]. Whereas in a ship wake condition, the dominating source comes from unsteady blade loading acting as a dipole source, or named as load noise directly [4, 5].

In the phases of the propeller design process nowadays, it general uses the empirical wake fraction to get the optimum circulation distribution, then aims for the highest possible level hydrodynamic performances, efficiency, for instance. However, the effects of the real wake flow and its effects on the radiated noise haven't been involved into the design loop [6, 7]. So the investigation into the acoustic performances for a propeller working in a spatially non-uniform wake flow will be productive for the hull-behind low noise propeller design focusing on its acoustics target directly.

For flows with a relatively low Mach number, propeller operation, for instance, feedback effects of the sound to the source fluctuated flow is negligible in general, and the turbulent quadrupole noise source can be neglected too comparing with the rotating load noise [8]. It is thus possible to predict the far-field radiated noise of the propeller by the weak coupling method, or named as hybrid CFD/CA method, of using computational fluid dynamics (CFD) to predict the source flow field and computational acoustics (CA) to analyze the resulting acoustical field. As a simplification, the CFD calculation to obtain the time-domain fluctuating pressure on the blades can be included in the hypothesis of incompressibility due to the high Reynolds number and low Mach number for the flow, while the retarded time effects due to the finite sound velocity must be accounted for in the propagation [9, 10].

To get an accurate prediction of the local unsteady noise sources, available CFD techniques includes the unsteady Reynolds-Averaged Navier-Stokes (URANS), Scale-Adaptive Simulation (SAS), Detached Eddy Simulation (DES), and Large Eddy Simulation (LES) for the engineering in general, and their ability seems to promise gradually [11]. The most practical applications for the radiated noise prediction in the marine engineering are surface integrals method in the time domain based on Ffowcs Williams-Hawkings equation (FWH), including both of the porous FWH equation and Kirchhoff FWH equation [2, 4, 5, 9, 10]. The time domain method must work by time series, and at least including the points shifting in a whole circle. A frequency domain method that aims to predict the harmonic acoustic intensity directly replaces the rotating source distribution on a propeller blade with a stationary distribution over the propeller disc, which has the advantage of eliminating the rotational component of the source motion from the calculations. So the moving source distribution as a function of time can be replaced by its Fourier series as a function of azimuth and the acoustic modes is introduced into solving that is easy to be analyzed [10]. As for the unbounded exterior acoustic calculation, the direct boundary element method (DBEM) in frequency domain can be a good selection for the applications [12, 13].

Seol et al. [4, 5] used the separate computations of the flow field being analyzed with non-viscous potential-based panel method and the noise propagation being predicted using time-domain FWH equation (Farassat formulation 1A) to investigate the no-skewed DTRC4119 propeller's non-cavitation and DTMB4381 propeller's blade sheet cavitation overall noise and the thickness and load noise components contribution and their noise directivity characteristics. The results showed that thickness noise was much smaller than the load noise component under the non-cavitation condition. Considering the determination of the source flow field for noise prediction and the limit of panel method, the attention in this paper is to enhance the precision of the viscous flow field prediction as high as possible, so the most promising LES simulation is adopted to predict the same 4119 propeller flow in a wake flow, and then the source fluctuations are fed to the BEM numerical acoustical computation as input data, which aims to demonstrate the capability of the new approach for propeller noise judging.

The LES and BEM numerical acoustics theory will be presented in section 2. The unsteady flow field calculation and the acoustic prediction will be completed in section 3. Section 4 will summarize the results that have been obtained in this study.

2 LES and BEM acoustics methodology

Fig. 1 shows the general solving steps and typical available hybrid methods for the flow noise and propeller noise prediction. Herewith, Ω_F denotes the flow area calculational domain and confined by Γ_F where the acoustic sources are interpolated from the fluid simulation to the acoustic computation. Subscripts S, F, and A denote the source, flow and acoustics respectively. Fig. 2 presents the common schematic for the flow-induced noise coupled computation in time and frequency domain, and the sequentially coupled approach in frequency domain is adopted in this paper.

LES is currently widely used for solving the turbulent fluids required for acoustic analogy computations. In the LES method the large scales are directly resolved and the effect of the small scales on the large scales are modeled by the sub-grid scale (SGS) model, and the turbulent models used in the RANS simulation are not needed again. At the same time, it requires a relatively high computational cost, and the mesh space and time scale's requirement is also the

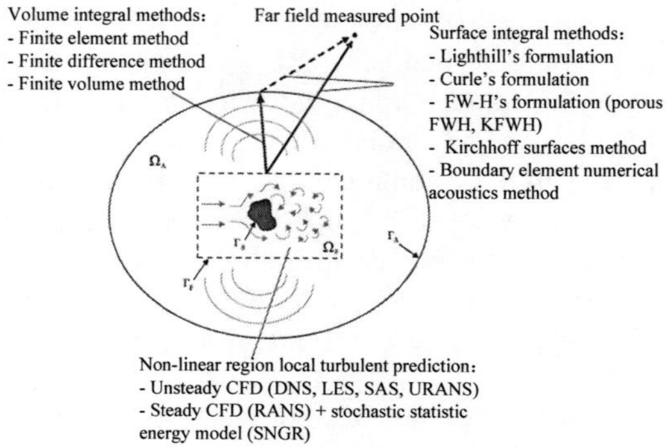

Figure 1: The general solving steps and typical available hybrid methods for the flow noise prediction.

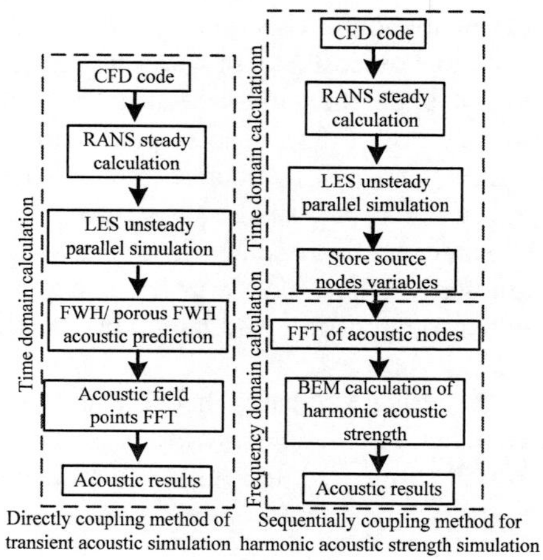

Figure 2: The solving loop for flow-acoustics hybrid prediction.

highest. It is obvious that the spatial filtering function of the incompressible flow equations to define the large eddy and the subgrid modelling problem mainly affect the predict errors [14, 15]. Reference to the LES simulation construction for propeller crashback in Vysohlid Ph.D dissertation (2007), the dynamic Smagorinsky model with modification by Lilly (1992) (DSM-Lilly) is used as the SGS in this paper, and it reads

$$\tau_{ij} = \frac{1}{3}\delta_{ij}\tau_{kk} - 2\mu_t \overline{S}_{ij} \tag{1}$$

where δ_{ij} is the Kronecker symbol, τ_{kk} is isotropic SGS stress, μ_t is the SGS turbulent viscosity and $\overline{S}_{ij}$ the large-scale strain rate tensor.

$$\mu_t = (C\Delta)^2\left|\overline{S}\right|, \quad \overline{S}_{ij} = \frac{1}{2}\left(\frac{\partial \overline{u}_i}{\partial x_j} + \frac{\partial \overline{u}_j}{\partial x_i}\right), \quad \left|\overline{S}\right| = \sqrt{2\overline{S}_{ij}\overline{S}_{ij}}$$

where C is the Smagorinsky coefficient, which adjusts automatically to the flow type corresponding to the size of the test-filtered width. The local grid size Δ is one third of the element volume.

After the fluctuated pressure and normal velocity distribution on the blades in the time series has been obtained by the CFD calculation, the weighted interpolation transfer of the CFD nodes to the acoustic nodes should be done to map the input data of the BEM calculation. The variables transformation will get a little effect on the source strength. So the mapping is commonly completed between the same grids if the computer allowed, called one to one transformation. If the BEM grids are too much, counter-refinement of the acoustic nodes to some extent is also ok, named many to one, and the weighted interpolation is needed for this case of course [12,13]. For the 4119 propeller is no-skewed and is relatively easy to make the mesh topology to control the mesh density, so the one to one model is chosen in this paper.

After accounting for the effects of blades rotating, the wave equation changes to

$$\frac{\partial^2 \rho'}{\partial t^2} - c_0^2 \frac{\partial^2 \rho'}{\partial x_i^2} = \frac{\partial^2 T_{ij}}{\partial x_i \partial x_j} + \frac{\partial f_s}{\partial x_i} + \frac{\partial m_s}{\partial t} \tag{2}$$

where ρ' relates to the acoustic component of the density, c_0 is the sound speed, and T_{ij} is the Lighthill stress tensor that reads

$$T_{ij} = \rho v_i v_j + (p' - c_0^2 \rho')\delta_{ij} - \sigma_{ij}$$

where $\rho v_i v_j$ is the instantaneous Reynolds stress, σ_{ij} is the viscous stress term that is generally negligible because of its extremely inefficient octupole nature as a noise source. The second term is also often thought to be small in absence of strong temperature inhomogeneities, the propeller flow for example. So, as for the low Mach number flow, the equation (1) can be simplified into

$$T_{ij} \approx \rho v_i v_j \tag{3}$$

In the CFD solving process, the variable T_{ij} and pressure and three velocity components are all used as residuals monitor, so that it can be sure that the convergence is completed. But the T_{ij} term refers to the quadrupoles is neglected in the acoustic calculation [4, 5, 9, 10]. And so does the third $\frac{\partial m_s}{\partial t}$ term too that is equivalent to monopole thickness noise. The remained $\frac{\partial f_s}{\partial x_i}$ term in equation (2) is just the main non-cavitation load noise and will be analyzed in this paper.

When using the BEM numerical acoustic method in the frequency domain for the propeller, the variable f_s points to the fluctuated pressure on the blades and hub surface. Specifically, the exterior propeller radiated noise problem (see Fig. 3) can be solved by the DBEM equation [13],

$$C(\vec{r}) \cdot p(\vec{r}) = \int_{\Omega_a} \left(p(\vec{r}_a) \cdot \frac{\partial G(\vec{r}, \vec{r}_a)}{\partial v} + j\rho\omega G(\vec{r}, \vec{r}_a) \cdot v_v(\vec{r}_a) \right) d\Omega(\vec{r}_a) \tag{4}$$

where $C(\mathbf{r})=\begin{cases}1,\mathbf{r}\in\mathbf{V},\\ 1/2,\mathbf{r}\in\Omega_a\\ 0,\vec{r}\notin\mathbf{V}.\end{cases}$ is the sound pressure coefficient, V is the volume of the closed domain for surface integral, $p(\vec{r})$ is the sound pressure at position $\vec{r}$, $p(\vec{r}_a)$ is the pressure distribution on the closed surface, $G(\vec{r},\vec{r}_a)=\dfrac{e^{-jk|\vec{r}-\vec{r}_a|}}{4\pi|\vec{r}-\vec{r}_a|}$ is the Green's kernel function, ν denotes for the normal direction of the closed surface Ω_a with a positive orientation into the unbounded domain (ν=-$\vec{n}$), and $v_\nu(\vec{r}_a)$ is the normal velocity component on position $\vec{r}_a$, see Fig. 4. Hence, the sound pressure $p(\vec{r})$ for any position $\vec{r}$ outside the closed surface Ω_a in the unbounded domain can by solely determined by the $p(\vec{r}_a)$ and $v_n(\vec{r}_a)$ on the Ω_a surface. The Ω_a surface for the propeller is shown in Fig. 3.

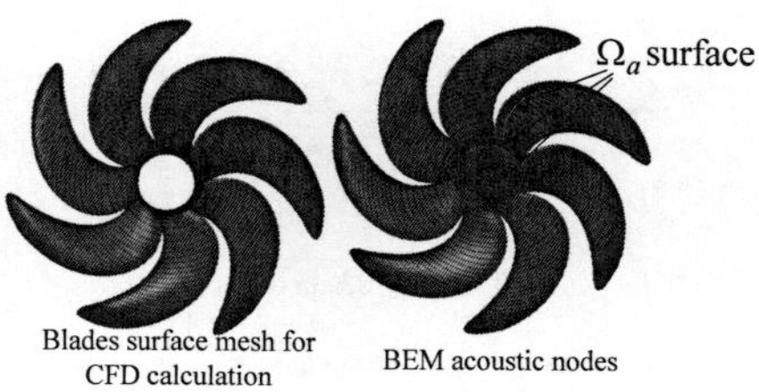

Figure 3: The CFD mesh and acoustics nodes.

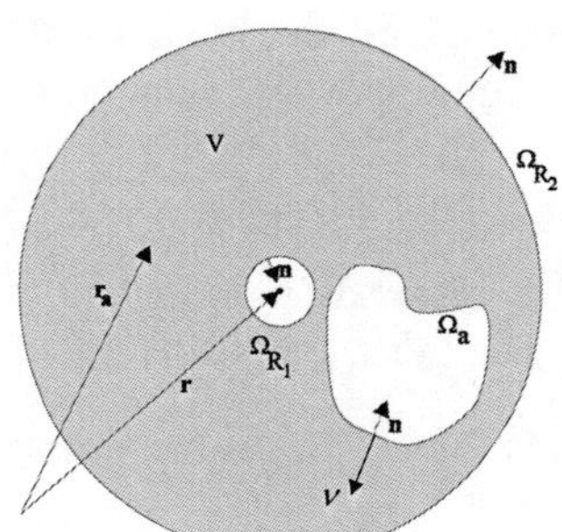

Figure 4: DBEM integral formulation variables definition for unbounded domain [13].

3 Propeller noise numerical simulation

This section will present the non-cavitation load noise of the DTRC4119 using the coupling method interpreted above, and do the qualitative and quantitative compare with the references [4] and [5], so as to validate indirectly the ability of the coupling method using in propeller non-cavitation noise prediction.

3.1 LES simulation of the unsteady flow field

To satisfy the requirement of the smallest mesh size for LES simulation, and quicken the solution, special care should be paid to the mesh creation for the propeller to get the mesh quality as high as possible, high determinant, for instance, especially for the blade leading and trailing edge and their vicinity and the tip area geometry. Fig. 5 shows the propeller CAD geometry and hex mesh topology structure adopted for LES simulation at last and the local surface mesh refinement details, and the blade leading edge, trailing edge, and tip area are associated and refined by independent blocks. The O-grid topology around the blade surface wall is used to capture the boundary layer flow structure through to the blades. The mesh determinant, which can be a comprehensive mesh quality index, also is shown in Fig. 5. A determinant value of 1 will indicate a perfectly regular hex/tet mesh element, 0 will indicate an element degenerate in one or more edges, and negative values will indicate inverted elements, which are not allowed in the commonly used commercial CFD solvers. To the best of author's experience [18–20], when the smallest mesh determinant value is bigger than 0.2, the solving convergence will be good and fast. If the reasonable mesh topology, good turbulence model and effective wall function (the first node adjacent to the wall should be fit to the limit of y+ value) are added in addition, it will get a promising result for the RANS simulation.

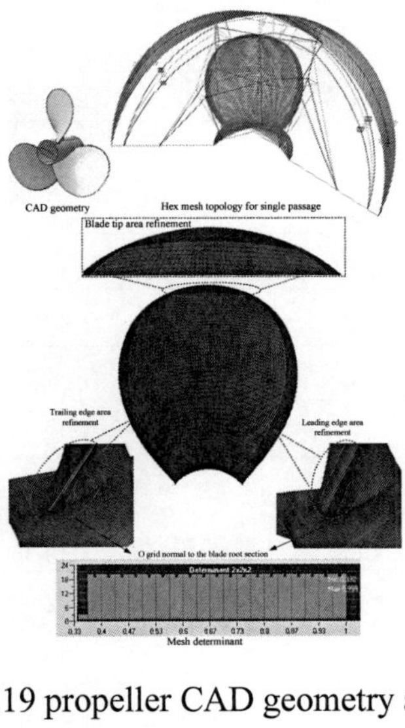

Figure 5: DTRC4119 propeller CAD geometry and hex mesh topology.

Mesh sensitive analysis is an essential step in the CFD calculation process. Fig. 6 shows different mesh topologies used in validation of the hydrodynamics of DTRC4119 propeller. The calculated open water thrust and torque coefficients of the mesh series are shown in Table 1, and the pressure coefficient distribution at 0.7*R* section is shown in Fig. 7 for the G3 and G4 mesh. According to Table 1 and Fig. 7, it is obvious that the G4 mesh topology and grid density is optimum and being the prototype for the LES simulation shown in Fig. 5.

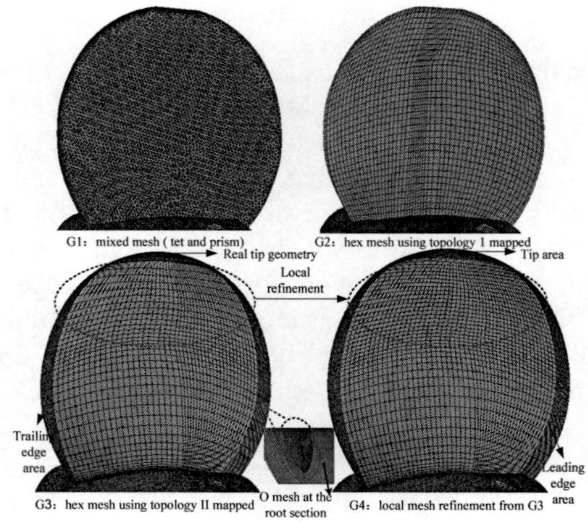

Figure 6: Different mesh topologies construction for propeller 4119.

Table 1: Mesh sensitive analysis for RANS simulation of DTMB4119 propeller.

J	Mesh	K_t_Cal	K_t_Exp	Error(%)	$10K_q$_Cal	$10K_q$_Exp	Error(%)
0.5	G1	0.2775	0.285	-2.617	0.4583	0.477	-3.923
	G2	0.2858	0.285	0.289	0.4615	0.477	-3.245
	G4	0.2858	0.285	0.281	0.4615	0.477	-3.249
0.7	G1	0.1998	0.2	-0.119	0.3524	0.36	-2.105
	G2	0.2035	0.2	1.755	0.3522	0.36	-2.170
	G4	0.2035	0.2	1.750	0.3521	0.36	-2.194
0.833	G1	0.1484	0.146	1.637	0.2827	0.28	0.964
	G2	0.1510	0.146	3.415	0.2797	0.28	-0.106
	G4	0.1479	0.146	1.301	0.2824	0.28	0.857
0.9	G1	0.1213	0.12	1.077	0.2456	0.239	2.746
	G2	0.1246	0.12	3.857	0.2429	0.239	1.615
	G4	0.1184	0.12	-1.333	0.2401	0.239	0.460
1.1	G1	0.0329	0.034	-3.191	0.1114	0.106	5.121
	G2	0.0420	0.037	13.581	0.118	0.106	11.420
	G4	0.0318	0.033	-3.636	0.108	0.106	1.859

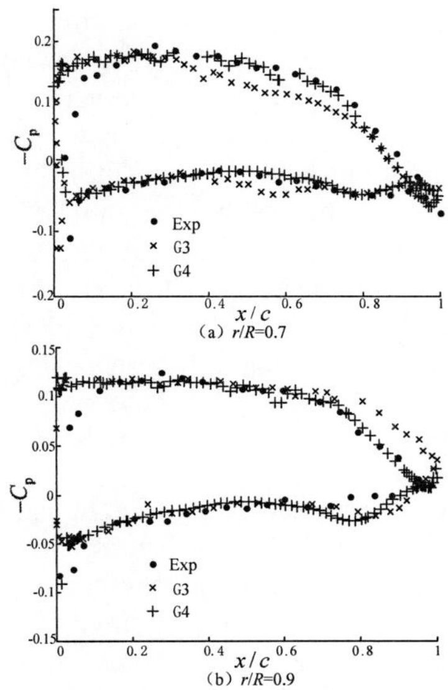

Figure 7: Pressure coefficient distribution at different sections.

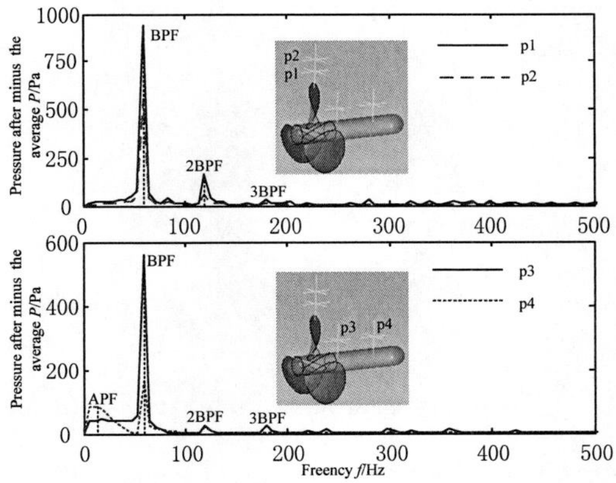

Figure 8: Fluctuated pressure spectrum for unsteady flow field.

After a steady solution has been computed, to be sure that the flow is really steady, an unsteady simulation should be carried out with the existing steady flow field as the initial condition. What's more important, the subsequent transient simulation is used to get the fluctuated source. The time step corresponding to 2 degrees of propeller rotation is $\Delta t = 2.78 \times 10^{-4}$ s, and the total time is 0.15s corresponding to 3 circles. The last week's transient results are used to extract the pressure and normal velocity used in equation (4). In the process of unsteady simulation, special monitor points are set to get the time series pressure which can determine the validity of the calculation indirectly in terms of the fluctuating information. Four points are seen in Fig. 8. The calculated the pressure amplitude with the FFT transformation after minus the average value is shown in Fig. 8. The monitors are placed on radial and axial direction of the blades. The results reveal that the fluctuated pressure of both of the point p1 and p2 can represent the blade passing frequency (BPF) and its harmonics information due to the rotating of blades, and the amplitude will decrease with the radial distance increase. As for the axial point p3 which is relative near to the blades, it is still sensitive to the rotating effect. And both of its BPF and harmonics appear again. When the point shifting downstream to the p4 point, the propeller wake effect weakens, the amplitude of BPF decreases rapidly, and the axial passing frequency (APF) information presents at this time.

3.2 Acoustic prediction

After the load noise source term according to equation (2) has been solved, the fluctuated pressure in time domain on the blades is transformed to the acoustic nodes using one to one model within frequency domain. Fig. 9 shows the blades pressure distribution in frequency domain obtained by the DBEM equation. The tip area and trailing edge near the root section contribute mainly to the noise under the low frequency. Specifically, the tip vortex core resulting low pressure and the trailing edge vortex shedding as the noise source take a dominating effect. Wherein, for the relatively high frequency, the noise highlight area shifts to the load centered region on the blades wall, and the load noise is the main noise source at this time.

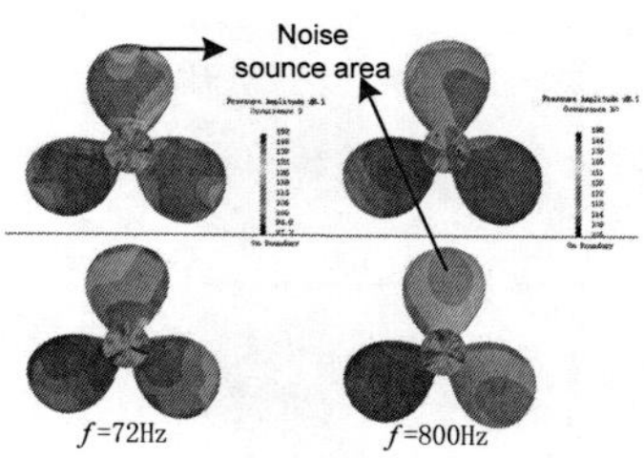

Figure 9: Blades pressure distribution in frequency domain.

To be consistent with the prediction in references [4] and [5], the noise directivity characteristics on the vertical plane along the axis and the sound pressure level spectrum in the same broadband frequency on given field point in the wake flow of 4119 propeller within spatial non-uniform inflow are analyzed. The calculated sound pressure distribution on axial vertical circinal-plane with the radius of $10R$ is seen in Fig. 10. In the low frequency broadband, the load noise presents obvious axial dipole information, the acoustic energy concentrated with a strong radiation tendency towards the observer on the hub axis. Within the high frequency, the dipole nature of the acoustic source also presents, but the source concentrated area enlarging with a deflection to the axial direction, and the slope angle depends on the frequency and inflow velocity. If the field points series consist of a three-dimensional spherical surface, on where the sound pressure distribution in the low frequency is seen in Fig. 11 with a qualitative agreement with and the calculated results of reference [5] because of the unawareness of exact frequency under 200Hz in that. Fig. 12 shows the sound pressure spectrum on the axial point downstream with a distance of 10 times the propeller radius. Because the ship wake used in this paper is not the same as that in references, the tendency and quantitative agreement with the reference results just can be obtained. From Figs. 11 and 12, it can interpret indirectly the availability of the coupling method given in this paper to predict the underwater radiated noise of ship propellers.

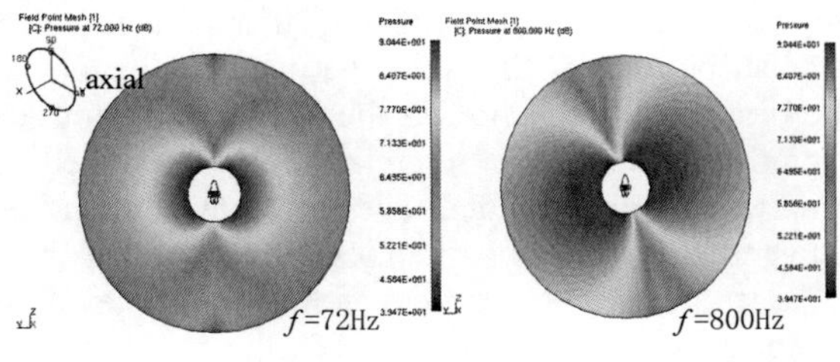

Figure 10: Sound pressure distribution on the plane.

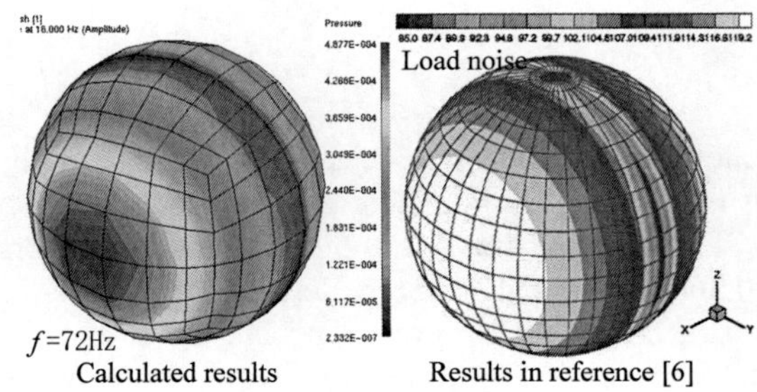

Figure 11: Sound pressure distribution on spherical surface.

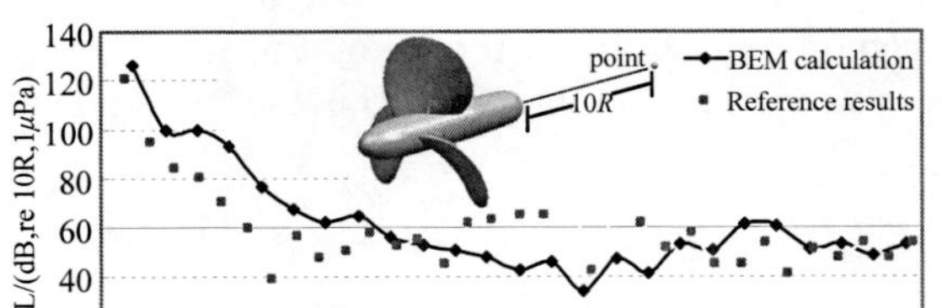

Figure 12: Sound pressure spectra on visual measured point.

The overall sound pressure level of the visual measured point calculation according to the equation

$$SPL_{total} = 10\log\left(\sum_{i=1}^{i=n} 10^{0.1SPL(i)}\right) \tag{5}$$

is 126.4dB, where the *SPL(i)* is the *SPL* in each one-third Octave band to the equivalent 1Hz bandwidth by means of the correction formula accounting for the frequency broadband.

4 Conclusions

This paper presents the weak coupling LES and BEM numerical acoustics method in frequency domain to predict the underwater radiated load noise of DTRC4119 propeller. The sound strength distribution on the blades surface, sound pressure distribution on the axial vertical plane and spatial spherical surface and the broadband spectra of the measured point downstream are presented and compared with that in references. The calculated steady flow field is fit well with the experiment data. The unsteady simulation can precisely present the fluctuated BPF and its harmonics information. The predicted spatial sound pressure distribution and the field point's noise level spectra are both reasonable with the tendency and quantity are the same as that in references, which can indicate the availability of this coupling method to predict the propeller's underwater radiated noise in engineering.

5 Comments

The analyzed condition given in references [4] and [5] is n=120rpm, v=1.6m/s. But the diameter D is unknown. For the generally propeller model's open water experiment and CFD analysis, the D is about 0.25 to 0.35 meters, and the ITTC report gives D=0.305m for the 4119 propeller. So the advance ratio $J=v/(nD)$ is 2.29 to 3.2, lying in the fourth quadrant, and the thrust coefficient is negative which is impossible for analysis. On the other hand, as the design point of 4119 propeller is J=0.833, if the rotating speed is right, the D will be 0.96m, which is nearly to the full scale. Nowadays, even the successful viscous CFD calculation, the prediction error will be about 4~5% for the full scale, and let alone the non-viscous panel method. Hence, for the promising fluctuated source prediction, the n will be 20rps as same as the commonly used in the hydrodynamics.

References

[1] Carlton J S. *Marine propellers and propulsion*. Second Edition. Elsevier Ltd.: Netherlands, pp. 240-283, 2007.

[2] Caridi D. *Industrial CFD simulation of aerodynamic noise*. Ph.D dissertation, University Degli Studi Di Napoli Federico II, 2008.

[3] Ross D. *Mechanics of underwater noise*. Pergamon Press: New York, 1976.

[4] Seol H, Jung B, Suh J C, *et al*. Prediction of non-cavitation underwater propeller noise. *Journal of Sound and Vibration*, **257(1)**, pp. 131-156, 2002.

[5] Seol H, Suh J C, Lee S. Development of hybrid method for the prediction of underwater propeller noise. *Journal of Sound and Vibration*, **288**, pp. 345-360, 2005.

[6] Carlton J S. *Marine propellers and propulsion*. Second Edition. Elsevier Ltd.: Netherlands, pp. 435-438, 2007.

[7] Hu Jiang. *Research on propeller cavitation characteristics and low noise propeller design*. Ph.D thesis, University of Harbin Engineering, 2006.

[8] Kato C, Yamade Y, Wang H, *et al*. Numerical prediction of sound generated from flows with a low mach number. *Computers & Fluids*, **36**, pp. 53-68, 2007.

[9] Testa C, *Acoustic formulations for aeronautical and naval rotorcraft noise prediction based on the Ffowcs Williams and Hawkings equation*. Ph.D dissertation, Delft University of Technology, 2008.

[10] Carley M, *Time domain calculation of noise generated by a propeller in a flow*. Ph.D dissertation, Department of Mechanical Engineering, Trinity College, Ireland, 1996.

[11] Wang M, Freund J B, Lele S K. Computation prediction of flow-generated sound. *Annual Review of Fluid Mech*, **38**, pp. 483-512, 2006.

[12] LMS Inc. *BEM acoustics*. LMS Virtual Lab online Help, 2006.

[13] LMS International. *Numerical Acoustics*. LMS Virtual Lab Theoretical Manual, 2006.

[14] Kobayashi T. Large eddy simulation for engineering applications. *Fluid Dynamics Research*, **38**, pp. 84-107, 2006.

[15] Holm D. D., Geurts B. Commutator errors in large-eddy simulation. *Journal of Physics A*: *Mathematical and General*, **39**, pp. 2213-2229, 2006.

[16] Vysohlid M. *Large eddy simulation of crashback in marine propellers*. Ph.D dissertation, University of Minnesota, 2007.

[17] Escobar M. Finite element simulation of flow-induced noise using Lighthill's acoustic analogy. Ph.D thesis, University Erlangen-Nurnberg, 2007.

[18] Yang Qiongfang, Wang Yongsheng, GUO Wei et al. Procedural realization of pre-operation in CFD prediction of propeller hydrodynamics. Accepted by *Journal of Ship Mechanics*, 2010 (in Chinese).

[19] Yang Qiongfang, Wang Yongsheng, Huang Bin et al. Integrated lifting line theory and RANS simulation for propeller design and non-cavitation and cavitation hydrodynamics prediction. Accepted by *Journal of Shanghai Jiaotong University*, 2010 (in Chinese).

[20] Yang Qiongfang, Wang Yongsheng, Fan Lu. Hydrodynamic performance analysis and optimization design of mechatronic ducted propeller with CFD. *Chinese Journal of Mechanical Engineering*, **46(1)**, pp. 162-168, 2010.

URANS and LES methodology for two-dimensional natural convection in a differentially heated cavity by BEM

L. Škerget & J. Ravnik
University of Maribor, Faculty of Mechanical Engineering, Slovenia

Abstract

The paper considers the fundamental aspects of turbulence modelling for incompressible Boussinesq fluid flow and corresponding numerical models based on the boundary-domain integral equations. Two different solution methodologies of turbulent flow circumstances are considered, e.g. the classical statistical methodology known as Unsteady Reynolds-Averaged Navier–Stokes (URANS) equations and deterministic Large-Eddy-Simulation (LES) formulation. The velocity-vorticity formulation of the mean/filtered equations is applied, while the averaged/filtered pressure field is determined by solving the Poisson velocity equation. Chaotic natural convection in a differentially heated cavity of aspect ratio 4 with adiabatic horizontal walls is studied by both mentioned methodologies of the unsteady two-dimensional governing equations.
Keywords: turbulence, large eddy simulation, unsteady RANS, boundary element method.

1 Introduction

The set of partial differential equations governing the motion of viscous fluid is known as nonlinear Navier–Stokes equations. This equation system is generally considered to be the fundamental description for all laminar as well as turbulent flows, although some statistical averaging or deterministic filtering procedure is required in practice to predict the turbulence and simulate numerically the flow at higher Reynolds or Rayleigh number values due to the enormous computational effort needed. In the LES methodology the classical Smagorinsky model with Van Driest damping closed to cavity walls is considered, while in the URANS

WIT Transactions on Modelling and Simulation, Vol 52, © 2011 WIT Press
www.witpress.com, ISSN 1743-355X (on-line)
doi:10.2495/BE110091

methodology the low Reynolds two equation $k-\epsilon$ Launder and Sharma turbulent model is applied.

The present investigation is concerned with URANS and LES of complex buoyancy-driven Boussinesq fluid flow in two dimensions in a 1:4 square differentially heated enclosure. For aspect ratio larger than or equal to 4, the onset of unsteadiness is due to the instability of vertical boundary layers. Although the turbulent flow in the rectangular cavity is basically three-dimensional, two-dimensional model can be applied in this case due to minor differences between the 3D and 2D flow field results.

2 Governing mean/filtered flow equations

2.1 Primitive variables formulation

The governing equations for the mean/filtered flow can be written in terms of effective momentum diffusivity ν_{ef} and thermal diffusivity a_{ef}, respectively, as follows

$$\frac{\partial \overline{v_j}}{\partial x_j} = 0, \tag{1}$$

$$\rho_o \frac{D\overline{v_i}}{Dt} = -\frac{\partial \overline{p}^\star}{\partial x_i} + \frac{\partial}{\partial x_j}\left(\rho_o 2\nu_{ef}\overline{\dot{\varepsilon}_{ij}}\right) + \overline{\rho} g_i, \tag{2}$$

$$c_o \frac{D\overline{T}}{Dt} = \frac{\partial}{\partial x_j}\left(c_o a_{ef}\frac{\partial \overline{T}}{\partial x_j}\right) + \overline{S_T}, \tag{3}$$

where the effective transport coefficient for the mean/filtered flow equations are given by the definitions, e.g. $\nu_{ef} = \nu + \nu_t$, $a_{ef} = a + a_t$, or similarly $\nu_{ef} = \nu + \nu_s$ and $a_{ef} = a + a_s$, respectively. The modified mean pressure term $\overline{p}^\star$ represents the sum of the static pressure and complementary volumetric part of Reynolds stress, such as

$$\overline{p}^\star = \overline{p} + \frac{2}{3}\rho_o k, \tag{4}$$

while in the modified filtered pressure term $\overline{p}^\star$ the complementary spherical tensor part or the trace of subgrid-scale stress tensor has been lumped into the pressure by defining

$$\overline{p}^\star = \overline{p} + \frac{1}{3}\rho_o \tau_{kk}^R. \tag{5}$$

The momentum equation Eqn. (2) can be written in a form suitable for velocity-vorticity formulation, e.g. in a vector form

$$\rho_o \frac{D\overline{\vec{v}}}{Dt} = -\mathrm{rot}(\eta_{ef}\overline{\vec{\omega}}) + 2\,\mathrm{grad}\,\overline{\vec{v}}\cdot \mathrm{grad}\,\eta_{ef} + 2\,\mathrm{grad}\,\eta_{ef} \times \overline{\vec{\omega}}$$
$$-\mathrm{grad}\,\overline{p}^\star + \overline{\rho}\overline{\vec{g}}, \tag{6}$$

The effective dynamic viscosity η_{ef} and effective heat conductivity k_{ef} can be given as a sum of a constant and variable part

$$\eta_{ef} = \eta_{efo} + \widetilde{\eta_{ef}}\,, \quad k_{ef} = k_{efo} + \widetilde{k_{ef}}\,, \tag{7}$$

therefore the momentum and energy Eqns. (6) and (3) can be written in analogy to the basic conservation equations formulated for the constant material properties

$$\rho_o \frac{D\overline{\vec{v}}}{Dt} = -\eta_{efo}\,\mathrm{rot}\,\overline{\vec{\omega}} - \mathrm{grad}\,\overline{p}^{\star} + \overline{\rho}\vec{g} + \overline{\vec{f}^{m}}\,, \tag{8}$$

$$c_o \frac{D\overline{T}}{Dt} = k_{efo}\Delta\overline{T} + \overline{S_T} + \overline{S_T^m}\,, \tag{9}$$

where the pseudo body force term $\overline{\vec{f}^m}$ and pseudo heat source term $\overline{S_T^m}$, are introduced into the momentum Eqn. (8) and energy Eqn. (9), respectively, capturing the variable transport property effects, and given by expressions

$$\overline{f_i^m} = -e_{ijk}\frac{\partial\overline{\omega_k}}{\partial x_j}\widetilde{\eta_{ef}} + e_{ijk}\frac{\partial\eta_{ef}}{\partial x_j}\overline{\omega_k} + 2\frac{\partial\eta_{ef}}{\partial x_j}\frac{\partial\overline{v_i}}{\partial x_j}\,, \tag{10}$$

while the pseudo heat source term is given by expression

$$\overline{S_T^m} = \frac{\partial}{\partial x_j}\left(\widetilde{k_{ef}}\frac{\partial\overline{T}}{\partial x_j}\right). \tag{11}$$

Once the form of the eddy diffusivity coefficients are specified then the mean/filtered governing transport equations can be solved in the same manner as a laminar flow since the equations are the same except for augmented diffusivity coefficients. Though the turbulent flow problem has been reduced to a familiar system of partial differential transport equations, there remains the nontrivial task of determining how the eddy diffusivity coefficients vary with the flow field.

2.2 Velocity-vorticity mean/filtered flow formulation

With the mean/filtered vorticity vector $\overline{\omega_i}$ representing the curl of the velocity field $\overline{v_i}$

$$\overline{\omega_i} = e_{ijk}\frac{\partial\overline{v_k}}{\partial x_j} \quad \text{and} \quad \frac{\partial\overline{\omega_j}}{\partial x_j} = 0\,, \tag{12}$$

the fluid motion computation scheme is partitioned into its kinematics, given by the elliptic mean/filtered velocity vector equation

$$\frac{\partial^2\overline{v_i}}{\partial x_j\partial x_j} + e_{ijk}\frac{\partial\overline{\omega_k}}{\partial x_j} = 0\,, \tag{13}$$

and kinetics given by mean/filtered vorticity transport equation, obtained as a curl of the mean/filtered momentum Eqn. (8), e.g., written in Cartesian tensor notation formulation

$$\frac{\partial \overline{\omega_i}}{\partial t} + \frac{\partial \overline{v_j}\,\overline{\omega_i}}{\partial x_j} = \nu_{efo}\frac{\partial^2 \overline{\omega_i}}{\partial x_j \partial x_j} + \frac{\partial \overline{\omega_j}\,\overline{v_i}}{\partial x_j}$$
$$+\frac{1}{\rho_o} e_{ijk}\frac{\partial \overline{\rho} g_k}{\partial x_j} + \frac{1}{\rho_o} e_{ijk}\frac{\partial \overline{f_k^m}}{\partial x_j}, \quad (14)$$

which reduces for two-dimensional plane flow case, to the following scalar mean/filtered vorticity statement

$$\frac{\partial \overline{\omega}}{\partial t} + \frac{\partial \overline{v_j}\,\overline{\omega}}{\partial x_j} = \nu_{efo}\frac{\partial^2 \overline{\omega}}{\partial x_j \partial x_j} - \frac{1}{\rho_o} e_{ij}\frac{\partial \overline{\rho} g_i}{\partial x_j} - \frac{1}{\rho_o} e_{ij}\frac{\partial \overline{f_i^m}}{\partial x_j}. \quad (15)$$

The vorticity transport Eqn. (14) is highly nonlinear partial differential equation. Due to the buoyancy force and variable effective transport properties, acting as additional nonlinear vorticity source terms, the vorticity transport equation is coupled to the energy/mass and transport equations for the turbulence quantities, making the numerical computation procedure very challenging.

3 Eddy-viscosity turbulence models

3.1 Two-equation LRN $k-\epsilon$ turbulence models

In the $k-\epsilon$ turbulence models, the turbulent motion is characterized by two quantities, namely the turbulent kinetic energy k and the turbulent energy dissipation rate ϵ, e.g. given by relations

$$k = \frac{1}{2}\overline{v_i' v_i'}, \qquad \epsilon = \nu \overline{\frac{\partial v_i'}{\partial x_j}\frac{\partial v_i'}{\partial x_j}}, \quad (16)$$

while the turbulent viscosity is given by the Kolmogorov–Prandtl relation

$$\eta_t = C_\eta \rho_o f_\eta \frac{k^2}{\epsilon}, \quad (17)$$

which relates the eddy viscosity directly to the turbulence variables, k and ϵ, and where $C_\eta = 0.09$ is an empirical constant. The values of k and ϵ come directly from the differential transport equations for the turbulent kinetic energy and the eddy dissipation rate

$$\rho_o \frac{Dk}{Dt} = \frac{\partial}{\partial x_j}\left[\left(\eta_o + \frac{\eta_t}{\sigma_k}\right)\frac{\partial k}{\partial x_j}\right] + P_k - \rho_o\widetilde{\epsilon} - \rho_o D, \quad (18)$$

$$\rho_o \frac{D\widetilde{\epsilon}}{Dt} = \frac{\partial}{\partial x_j}\left[\left(\eta_o + \frac{\eta_t}{\sigma_\epsilon}\right)\frac{\partial \widetilde{\epsilon}}{\partial x_j}\right] + C_{\epsilon 1} f_{\epsilon 1}\frac{\widetilde{\epsilon}}{k}P_k - C_{\epsilon 2} f_{\epsilon 2}\rho_o \frac{\widetilde{\epsilon}^2}{k} + \rho_o E, \quad (19)$$

where the model variable $\widetilde{\epsilon}$ is defined as

$$\epsilon \equiv \widetilde{\epsilon} + D \qquad \text{with} \qquad D = 2.0\nu_o \left(\frac{\partial k^{1/2}}{\partial x_j} \right)^2 . \tag{20}$$

Since the dissipation rate at the wall is equal to

$$\epsilon|_{wall} \equiv D|_{wall} = 2.0\nu_o \left(\frac{\partial k^{1/2}}{\partial x_j} \right)^2 \Bigg|_{wall} , \tag{21}$$

the variable $\widetilde{\epsilon}$ is zero at the wall which simplifies the specification of wall boundary conditions, i.e.

$$k = \widetilde{\epsilon} = 0 . \tag{22}$$

Further, such definition of new variable $\widetilde{\epsilon}$ also gives rise to an extra term E in the $\widetilde{\epsilon}$ transport equation

$$E = 2.0\nu_o\nu_t \left(\frac{\partial^2 \overline{v_i}}{\partial x_j \partial x_k} \right)^2 . \tag{23}$$

The turbulent kinetic energy production term P_k is due to viscous and buoyancy forces and is modelled, e.g. by the following relation

$$P_k = 2\eta_t \overline{\dot{\varepsilon}_{ij}} \frac{\partial \overline{v_i}}{\partial x_j} - \eta_t g_i \frac{\partial \rho / \partial x_i}{\sigma_\rho \rho} . \tag{24}$$

The damping functions are expressed as functions of the local turbulence Reynolds number Re_t as follows:

$$f_\eta = \exp \left[- \frac{3.4}{(1 + 0.02 Re_t)^2} \right] \qquad \text{and} \qquad Re_t = \frac{\rho k^2}{\eta \epsilon} , \tag{25}$$

$$f_{\epsilon 1} = 1 \qquad \text{and} \qquad f_{\epsilon 2} = 1.00 - 0.3 \exp\left(-Re_t^2 \right) , \tag{26}$$

with $C_{\epsilon 1} = 1.44$, $C_{\epsilon 2} = 1.92$, $\sigma_k = 1.0$, and $\sigma_\epsilon = 1.3$, while $\sigma_\rho = 0.9$, respectively.

3.2 Subgrid-scale closure/modeling

One of the most popular Boussinesq eddy-viscosity subgrid closure model is due to Smagorinsky, e.g. which correlates τ_{ij}^R to the large-scale strain-rate tensor $\overline{\dot{\varepsilon}_{ij}}$

$$\tau_{ij}^R = -2\eta_s \overline{\dot{\varepsilon}_{ij}} + \frac{1}{3} \rho_o \tau_{kk}^R \delta_{ij} . \tag{27}$$

The subgrid viscosity η_s can be expressed as

$$\nu_s \propto l_s \, v_s , \tag{28}$$

where l_s is the length scale of the unresolved flow and v_s its velocity scale. The subgrid viscosity can be expressed as

$$l_s = \triangle = (\triangle\Omega)^{1/3}, \tag{29}$$

where $\triangle\Omega$ is the volume of the computational internal cell. The velocity scale is related to the gradients of $\overline{v_i}$ and it is defined as

$$v_s = l_s \bar{\dot{\gamma}}, \tag{30}$$

where $\bar{\dot{\gamma}}$ is the deformation velocity of the resolved flow or the magnitude of the large-scale strain-rate tensor $\overline{\dot{\varepsilon}_{ij}}$ defined as

$$\bar{\dot{\gamma}} = (2\overline{\dot{\varepsilon}_{ij}}\;\overline{\dot{\varepsilon}_{ij}})^{1/2} \qquad \text{and} \qquad \overline{\dot{\varepsilon}_{ij}} = \frac{1}{2}\left(\frac{\partial \overline{v_i}}{\partial x_j} + \frac{\partial \overline{v_j}}{\partial x_i}\right). \tag{31}$$

Finally the subgrid-scale viscosity is defined as

$$\nu_s = (C_s \triangle)^2 \bar{\dot{\gamma}}, \tag{32}$$

where C_s is the Smagorinsky constant. It is found that values of C_s vary from 0.065 to 0.2 for isotropic turbulence. Furthermore, C_s has to be reduced in the near wall region to account for the turbulence anisotropy. Thus, the correct distribution of C_s in the near wall region is obtained by using so-called damping functions, e.g. the most often used is the van Driest damping function

$$C_s = C_{so}\left[1 - \exp(-Re_\tau/25)\right]^2. \tag{33}$$

The subgrid-scale heat flux $\overline{v_j T}$ can be modelled as simple gradient diffusion hypothesis

$$\overline{v_j T} = -\frac{\nu_s}{Pr_t}\frac{\partial \overline{T}}{\partial x_j}. \tag{34}$$

4 Boundary-domain integral equations

The kinematics of plane motion is given by two scalar equations as follows:

$$c(\xi)\,\overline{v_i}(\xi) + \int_\Gamma \overline{v_i} q d\Gamma = \int_\Gamma \frac{\partial v_i}{\partial n} u^\star d\Gamma + e_{ij}\int_\Gamma \overline{\omega} n_j u^\star d\Gamma - e_{ij}\int_\Omega \overline{\omega} q_j^\star d\Omega. \tag{35}$$

where $u^\star$ stands for the elliptic Laplace fundamental solution and $q^\star$ is its normal derivative, e.g. $q^\star = \partial u^\star/\partial n = \vec{q^\star} \cdot \vec{n}$, while the vector flux variable is defined as $q_i^\star = \partial u^\star/\partial x_i$. The most important issues in numerical modelling of incompressible fluid flow is to obtain a divergence free final solution for the velocity and vorticity vector field functions. Thus, the proper kinematic integral representation should preserve the compatibility and restriction conditions for the velocity and vorticity field functions. Accounting for the additional compatibility and restriction conditions for velocity and vorticity fields, e.g. $\vec{\omega} = \mathrm{rot}\,\vec{v}$ and $\mathrm{div}\,\vec{v} = 0$, the following boundary integral representation for the general flow situation can be stated for the two-dimensional plane flow kinematic case as follows

$$c(\xi)\,\overline{v_i}\,(\xi) + \int_\Gamma \overline{v_i} q^\star d\Gamma = e_{ij} \int_\Gamma \overline{v_j} q_t^\star d\Gamma - e_{ij} \int_\Omega \overline{\omega} q_j^\star d\Omega\,. \qquad (36)$$

Using unique feature of global integral representation for boundary vorticity values, the vector eq.(36) has to be written in its tangential form in order to obtain an appropriate non-singular implicit system of equations for unknown boundary vorticity values.

Considering the kinetics in an integral representation one has to take into account the parabolic diffusion-convection character of the vorticity transport equation. With the use of the linear parabolic diffusion operator the vorticity equation can be formulated as a scalar inhomogeneous parabolic diffusion equation as follows:

$$c(\xi)\,\overline{\omega}\,(\xi, t_F) + \int_\Gamma \overline{\omega} Q^\star d\Gamma = \frac{1}{\eta_o} \int_\Gamma (\eta_o \frac{\partial \overline{\omega}}{\partial n} - \rho_o v_n \overline{\omega} + \rho g_t) U^\star d\Gamma$$

$$+ \frac{1}{\eta_o} \int_\Omega (\rho_o \overline{v_j} \overline{\omega} + e_{ij}(\rho g_i + \overline{f_i^m})) Q_j^\star d\Omega + \int_\Omega \overline{\omega}_{F-1} u_{F-1}^\star d\Omega, \qquad (37)$$

where a constant variation of all field functions within the individual time increment $\Delta t = t_F - t_{F-1}$ is assumed, e.g. the values at $t = t_F$ are considered for each time step, where v_n and g_t are the normal velocity, and the tangential gravity, respectively, e.g. $v_n = \vec{v} \cdot \vec{n}$, $g_t = \vec{g} \cdot \vec{t} = -e_{ij} g_i n_j$.

The integral representation of the heat energy diffusion-convection transport equation is derived considering the linear parabolic diffusion differential operator and therefore the equation may be rewritten in the form

$$c(\xi)\,\overline{T}\,(\xi, t_F) + \int_\Gamma \overline{T} Q^\star d\Gamma = \frac{1}{k_o} \int_\Gamma \left(k_o \frac{\partial \overline{T}}{\partial n} - c_o \overline{v_n} \overline{T} \right) U^\star d\Gamma$$

$$+ \frac{1}{k_o} \int_\Omega (c_o \overline{v_j} \overline{T} + \overline{S_T^m}) Q_j^\star d\Omega + \int_\Omega \overline{T}_{F-1} u_{F-1}^\star d\Omega, \qquad (38)$$

where a constant variation of all field functions within the individual time increment $\Delta t = t_F - t_{F-1}$ is assumed.

5 Numerical aspects/iterative strategy

The boundary element implementation of the two equation $k - \epsilon$ low-Reynolds-number model is basically straightforward since the transport equations for the turbulent quantities are of the familiar diffusion-convection type. The details of the iterative scheme can be formulated as follows. For the given velocity field $\vec{v}$ the k and ϵ equations have to be solved. The first point of the iterative scheme is that the equations for k and ϵ are coupled iteratively. Therefore, the nonlinear transport equation is solved for k first assuming that ϵ is known and then the computed value of k is used to solve the nonlinear transport equation for ϵ. Since it is not desired to deal with problems with a negative production term, the term P_k is kept constant in Eqn. (19) and also production term is kept constant in Eqn. (18). In the innermost iterative loop the linearized k equation is solved.

Next the linearized ϵ equation is solved. Now, the iterative scheme can be written in detail, as follows:

1. Compute P_k and P_ϵ
2. Repeat until convergence for ν_t
   - 2.1 Solve for k:
     - 2.1.1 Solve linearized k equation
     - 2.1.2 $k_i = ur \cdot k_i + (1 - ur) \cdot k_{i-1}$
   - 2.2 Check convergence for k. If not, go to 2.1.
   - 2.3 Update ν_t using Eqn. (17)
   - 2.4 Solve for ϵ:
     - 2.4.1 Solve linearized ϵ equation
     - 2.4.2 $\epsilon_i = ur \cdot \epsilon_i + (1 - ur) \cdot \epsilon_{i-1}$
   - 2.4 Check convergence for ϵ. If not, go to 2.4.
   - 2.5 Update ν_t using Eqn. (17)
3. Check convergence for ν_t. If not, go to 2.

The main advantages of the Smagorinsky model are its simplicity and its stability. Whether filtering is introduced or not, the LES equations with subgrid-scale eddy viscosity model are solved numerically for the time evolution of the LES field functions. This involves discretization in space and time, which introduces differences between the differential equations and their numerical equivalent. The solution iterative strategy is to solve for large scale velocity $\vec{v}$ and vorticity $\vec{\omega}$ field functions and then to compute subgrid-scale eddy viscosity until convergence, repeating the iterative process if needed. The solution scheme is as follows:

1. Solve the filtered Navier–Stokes LES equations
   - 1.1 Update subgrid-scale eddy viscosity ν_s
2. Check convergence for $\vec{\omega}$. If not, go to 1.

6 Two-dimensional natural convection in a differently heated cavity of aspect ratio 4

We consider a cavity of height $H = 4$ and width $W = 1$, aspect ratio $A = H/W = 4$, filled with a Newtonian viscous fluid. It is submitted to a temperature difference $\triangle T = T_h - T_c > 0$ at the vertical walls, with uniform temperatures $T_h = 0.5$ and $T_c = -0.5$, respectively, while the top and bottom

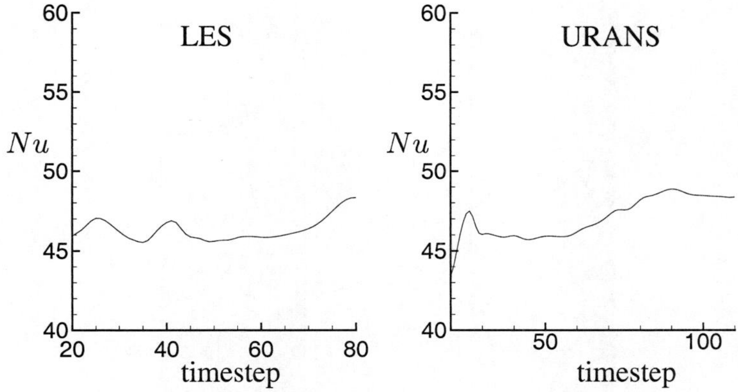

Figure 1: Development of heat flux expressed by Nusselt number.

walls are adiabatic.Two non-uniform numerical models are considered consisting of $M = 40 \times 160$ and $M = 60 \times 240$ macro elements with the aspect ratio of $1 : 4$ in the $x-$ and $y-$direction. Three-node quadratic boundary elements and nine-node quadratic internal cells are applied. The time dependent numerical simulation is performed with the dimensionless time step $\triangle t = 5 \cdot 10^{-4}$.

Turbulent natural convection in a differently heated air-filled cavity of aspect ratio 4 with adiabatic horizontal walls is investigated by large eddy simulation LES and unsteady mean flow URANS numerical integration of the unsteady two-dimensional governing equations. In order to approach chaotic flows which exhibits randomness in space as well as in time, simulations for different Rayleigh number values, e.g. $Ra = 6.4 \cdot 10^8$, and for Prandtl number value $Pr = 0.71$ are performed.

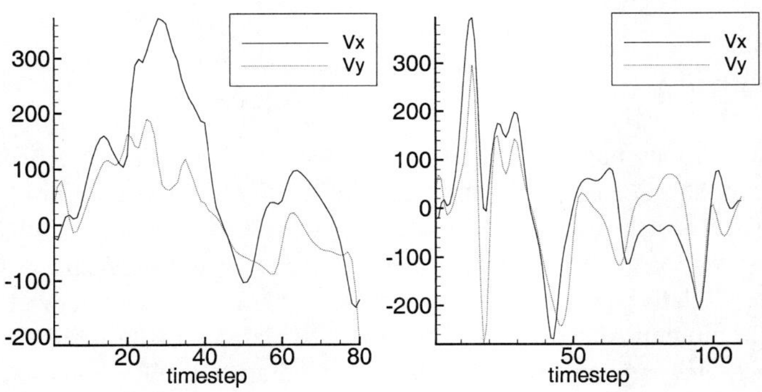

Figure 2: Time traces of horizontal and vertical velocity components recorded at location (0.75,0.25). LES turbulence model (left), URANS turbulence model (right). $Ra = 6.4 \cdot 10^8$

www.witpress.com, ISSN 1743-355X (on-line)

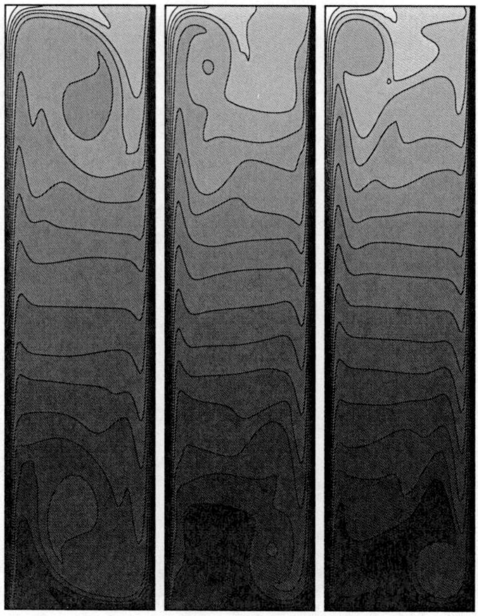

Figure 3: The temperature field at $Ra = 6.4 \cdot 10^8$. Three instantaneous timesteps are shown. Turbulence is modelled by URANS approach.

Figure 1 shows heat flux expressed by Nusselt number. The value of 48.3 was recorded at the final time step. This compares well with $Nu = 49.2$, which was reported by Xin and Le Quéré [9] and with $Nu = 49.08$ reported by Ravnik et al. [8]. Figure 2 shows the time traces of horizontal and vertical velocity components recorded at location (0.75,0.25) for both turbulence models. The location is inside the vortex, which is located in the corner of the enclosure. We observe rapid changes in the velocity field. Finally, Figures 3 and 4 show instantaneous temperature contours for several time instants. One can readily observe the unsteady nature of the flow.

7 Conclusion

In this work a numerical procedure based on the boundary element method for the simulation of turbulent buoyancy-driven two-dimensional fluid flow in a differentially heated air-filled cavity of aspect ratio 4 is investigated. The flow circumstances for the Rayleigh number value $Ra = 6.4 \cdot 10^8$ are presented in this paper. Relatively course mesh consisting of 40×160 macro elements is used in the numerical model. The periodic oscillations of the field functions in the downstream parts of the boundary layers in the form of Tollmien–Schlichting waves are detected. Rather large fluctuations are observed in the cavity corners where the flow is very chaotic. The main cavity core is still well stratified and basically motionless, therefore the flow is still far from being turbulent. With increasing Ra the cavity core becomes deorganised and chaotic. Finer mesh, e.g.

Figure 4: The temperature field at $Ra = 6.4 \cdot 10^8$. Three instantaneous timesteps are shown. Turbulence is modelled by LES approach.

consisting of 60×240 macro elements, have to be used to accurately simulate turbulent flow structures at higher Ra number values.

Low Reynolds number $k - \epsilon$ and LES turbulence modelling methodologies have been considered. Iterative strategy of highly nonlinear and coupled governing equations is discussed. The main goal of the paper is to increase the applicability of BEM technique to solve real turbulent fluid flow problems.

References

[1] G. Biswas and V. Eswaran (2002) *Turbulent Flows: Fundamentals, Experiments and Modeling*. Alpha Science International Ltd., Pangbourne, UK.

[2] Tennekes, H. and Lumley, J.L.: *A First Course in Turbulence*. MIT Press, (1972).

[3] Abe, K., Kondoh, T. and Nagano, Y.: *A New Turbulence Model for Predicting Fluid Flow and Heat Transfer in Separating and Reattaching Flows - I. Flow Field Calculations*. Int. Journal of Heat and Mass Transfer, Vol. 37, pp. 139-151, (1994).

[4] L. Škerget and J. Ravnik (2009) BEM simulation of compressible fluid flow in an enclosure induced by thermoacoustic waves. *Eng. Anal. Boundary Elements*, **33**, 561–571.

[5] L. Škerget and J. Ravnik (2010) Simulation of fluid flow by BEM, in *Recent Developments in Boundary Element Methods : a Volume to Honour Professor John T. Katsikadelis*, WIT Press, Southampton, pp. 213–226.

[6] L. Škerget and J. Ravnik (2010) Solution of velocity-vorticity URANS by BEM, in *Boundary elements and other mesh reduction methods XXXII, (WIT transactions on modelling and simulation)*, WIT Press, Southampton, pp. 29–40.

[7] L. Škerget and J. Ravnik (2009) Boundary element analysis of general laminar and turbulent fluid flow problems, in *2nd South-East European Conference on Computational Mechanics*. Institute of Structural Analysis & Seismic Research, National Technical University of Athens.

[8] J. Ravnik, L. Škerget and M. Hriberšek (2006) 2D velocity vorticity based LES for the solution of natural convection in a differentially heated enclosure by wavelet transform based BEM and FEM, *Eng. Anal. Boundary Elements*, **30**, 671–686.

[9] S. Xin and P. Le Quéré (1995) Direct numerical simulations of two-dimensional chaotic natural convection in a differentially heated cavity of aspect ratio 4, *J. Fluid Mech.*, **304**, 87–118.

BEM simulations of diffraction-optimized noise barriers

S. Gasparoni, M. Haider, M. Conter & R. Wehr
AIT Austrian Institute of Technology, Austria

Abstract

Traffic-related noise has been increasing steadily. Noise barriers are one of the main tools used for noise abatement, and there is still potential for optimization and improvement of the acoustic performance by employing non-standard designs. Simulations are a cost-efficient tool for predicting and planning new noise barrier solutions.

The following paper studies some non-standard barrier shapes with particular focus on the formation of a virtual soft-plane for some frequencies. Destructive diffraction from the top edge of the barrier is used in order to optimize the shielding effect of the barrier.

Through the use of 2D-BEM simulations different barrier profiles and their effect of shielding are studied. The focus is to obtain useful shielding in the far-field region with intelligent shapes thus permitting a reduction of the barrier height.
Keywords: noise barriers, noise abatement, diffraction, 2D-BEM simulations.

1 Introduction

Traffic noise has become an important problem with the increase of traffic volume. To counteract this, noise barriers are the most used traffic noise abatement tool; it is in the public interest to reduce the height of the barriers, that being a natural way to reduce the material, and thus the costs.

Even if good results can be achieved with the use of appropriate absorbing materials on the surfaces or on part of the surfaces (for example on the top of a T-shape barrier (Fig 1)), the porosity of this kind of materials makes them very sensitive to clogging by dirt and changes their absorbance with time.

WIT Transactions on Modelling and Simulation, Vol 52, © 2011 WIT Press
www.witpress.com, ISSN 1743-355X (on-line)
doi:10.2495/BE110101

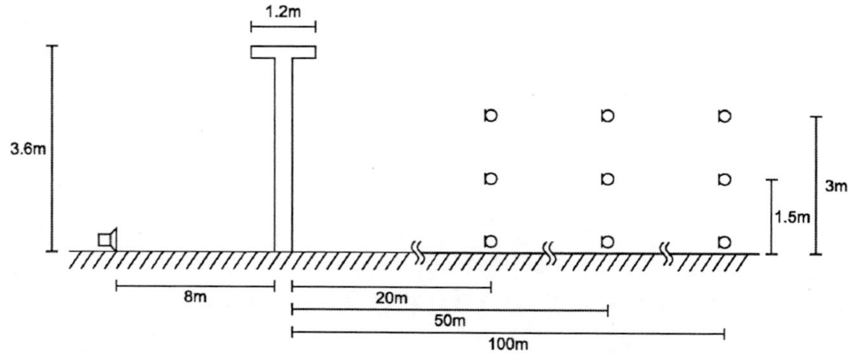

Figure 1: The simplified set-up of the simulations. The source is put on the ground at 8 m from the barrier. On the other side, a grid of 9 microphones is used to explore the far field.

In this paper two non-standard geometrical forms of noise barriers are investigated in order to overcome this problem.

2 Investigation

Perfectly reflecting (acoustically hard) materials are not considered the best choice for noise barriers, as they generate many unwanted reflections. Absorbing materials seem much more appropriate, but their impedance changes rapidly with time as mentioned above. One interesting idea is to use geometrical shapes to obtain a specific input impedance, as suggested in [1–3], where different shapes are analysed.

Let us consider the barrier b) depicted in Fig 2, which will be called "fork" barrier below. With rigid surfaces the specific input impedance at the open side can be approximated by

$$Z_{in} = i\cot(kd) \tag{1}$$

where d is the depth of the fork and k the wave vector, as can be seen from (1). According to this equation, at frequencies f_n, with $k_n d = (2n+1)\pi/2$, the impedance is zero. This means that the fork element, for the range close to those frequencies, plays the role of a soft plane, with complete absorbance. The condition of soft plane can never be realized 100% with the use of absorbent materials, which makes the geometric solution a useful alternative. As this impedance is only dependent on the geometry, the problem of the time-variance of the absorbent materials is practically solved. On the other hand, this solution is efficient only for some frequencies.

This problem can be coped with using a barrier whose channels have different lengths, as depicted in Fig 2 c), the "fork gradient". This corresponds to using a strip of absorbent material whose impedance changes gradually along the length of the material. The idea is similar to chirped mirrors in optics, made out of different layers that can filter different wavelengths.

For our investigation we performed a 2D BEM analysis, assuming the invariance of the system on the y axis. We used OpenBEM, an open source software developed in Matlab environment by the University of Southern Denmark [4]. OpenBEM solves the Helmholtz equation with a direct collocation approach.

The set-up for the simulation is shown in Fig 1.

In the following, the ground is assumed to be perfectly reflecting. The source is placed on the ground, in order to prevent unwanted reflections, and at 8m distance from the barrier. On the other side of the barrier, 9 microphones are placed in a regular grid structure, at the different heights of 0, 1.5 m and 3 m from the ground, and at the distances 20, 50, 100m from the barrier. Simulations are performed at the middle frequencies of the third octave bands. First simulations are performed without barrier, then with the three different barriers depicted in Fig 2.

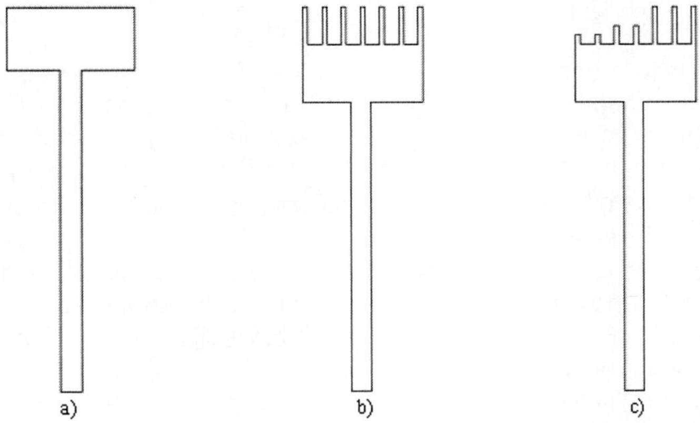

Figure 2: Different barriers that have been used for the simulation. a) will be referred to as T-shape, b) as fork shape, c) as fork gradient.

The insertion loss (IL) is calculated, according to the formula:

$$IL(dB) = -10\log_{10}(R) \tag{2}$$

$$R = \frac{1}{n}\sum_{i=1}^{n}\left(\frac{p_i}{\overline{p_i}}\right)^2$$

where p_i represents the pressure on the i-th microphone and $\overline{p_i}$ the pressure on the i-th microphone position of the configuration without barrier. The results can be seen in Fig 3.

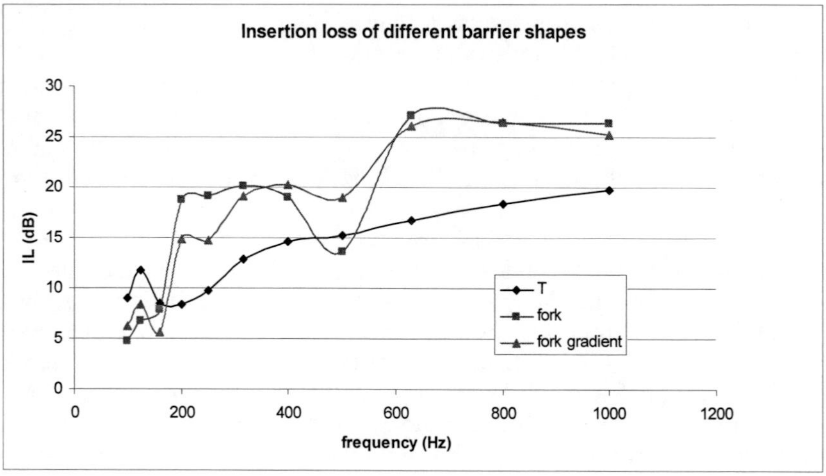

Figure 3: Insertion loss relative to the barriers of shape T, fork and fork gradient.

The expected improvement of the IL can be seen in Fig 3, for some frequencies an improvement of about 10dB for the "fork" shape can be found. On the other hand we also see a dip at around 500 Hz that corresponds to the maximum of the impedance. Following equation (1) the maximum IL is expected when $kd = \pi / 2$, which in our case (d = 400 mm) occurs at the frequency f = 210 Hz.

The graph of the fork barrier also shows a peak between 600 and 800 Hz, which can be seen again in the field lines in Fig.4 as the building of a soft plane.

The use of the "fork gradient", where different channels of different depths are used, presents an input impedance that changes along the length. It represents a considerable improvement in the insertion loss as the attenuation is better distributed along the range of the considered frequencies, as can be seen from the graph. The minimum at 500Hz is now still visible but has an increase of about 5dB compared to the previous shape.

It is interesting to have a closer look at the profile at the frequencies between 400Hz and 600Hz, to see how narrow the dip is. In order to achieve a better accuracy, simulations have been performed at additional frequencies. From Fig 4 it is clear that the profile relative to the fork barrier is smooth and the dip is a broad one. The same applies to a higher degree for the fork gradient barrier.

An average between the 9 microphones positions was calculated in order to represent an arbitrary point in the far field. In Fig 5 two graphs are shown that correspond to the points on the ground, one at 50 m and one at 100 m from the barrier. The behaviour is indeed very similar to that of the chosen average (Fig 3).

For a better quantification of the results it is useful to calculate a single number value and weight it with traffic spectra. The traffic spectrum Lj of [5] has been used to define the quantity

$$D_{1000} = -10 Log\left(\frac{\sum_{j=1}^{fre} 10^{0.1L_j} R_j}{\sum_{j}^{fre} 10^{0.1L_j}} \right)$$

where the index j counts up to the frequency 1000Hz. For the three shapes analysed the single-number values are given in the following table

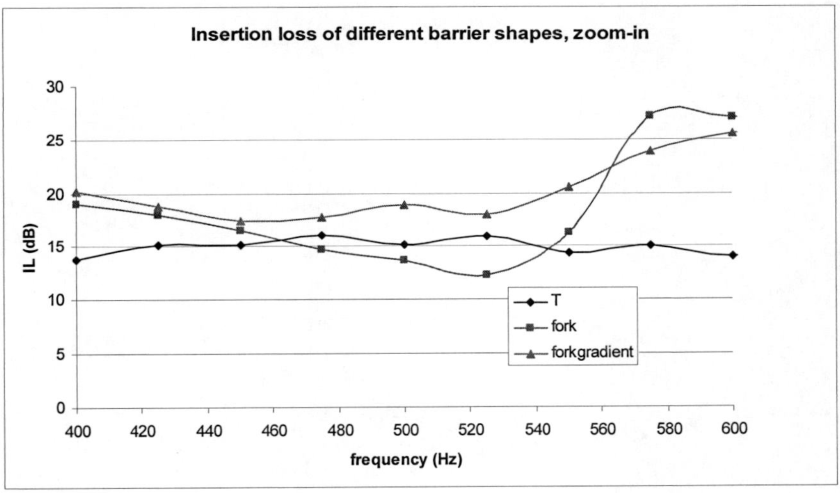

Figure 4: A zoom in of Fig. 3 with higher accuracy for frequencies from 400 to 630 Hz.

	D1000
T-shape	14,4
fork	16,6
fork gradient	16,8

It is interesting to have a look at Fig 5. Here a graphical representation of the "fork" (left) and of the "fork gradient" (right) barrier is given, where field lines at 500 and 630Hz can be seen. On the edge of the barrier, the formation of the soft-plane is clearly visible. According to equation (1) the soft plane is expected at about 630Hz.

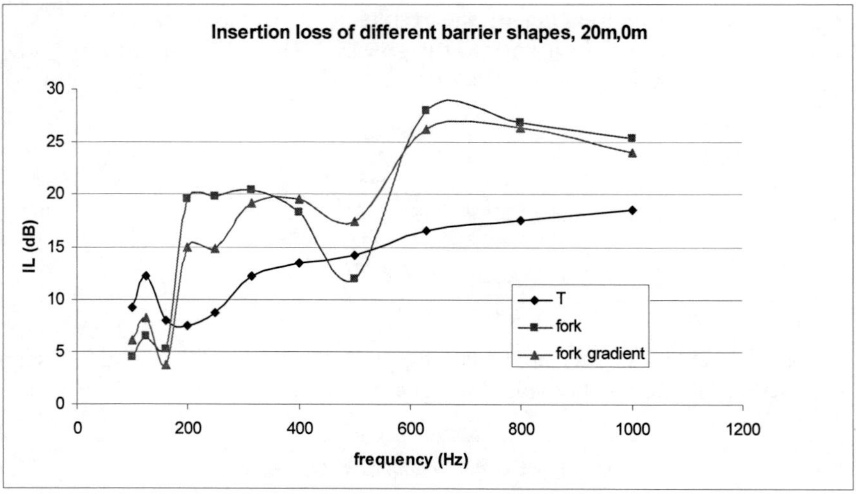

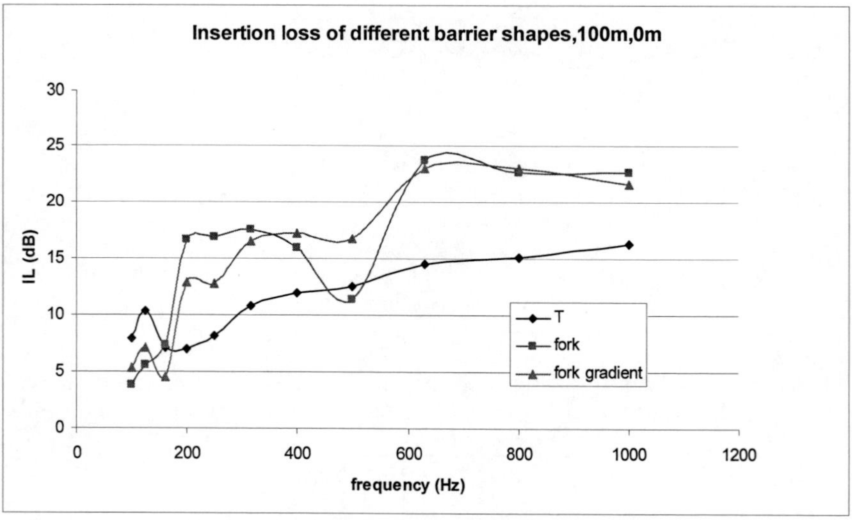

Figure 5: The insertion losses at two different points (20,0; 100,0) have graphs that are very similar to the graph of the average (fig 3).

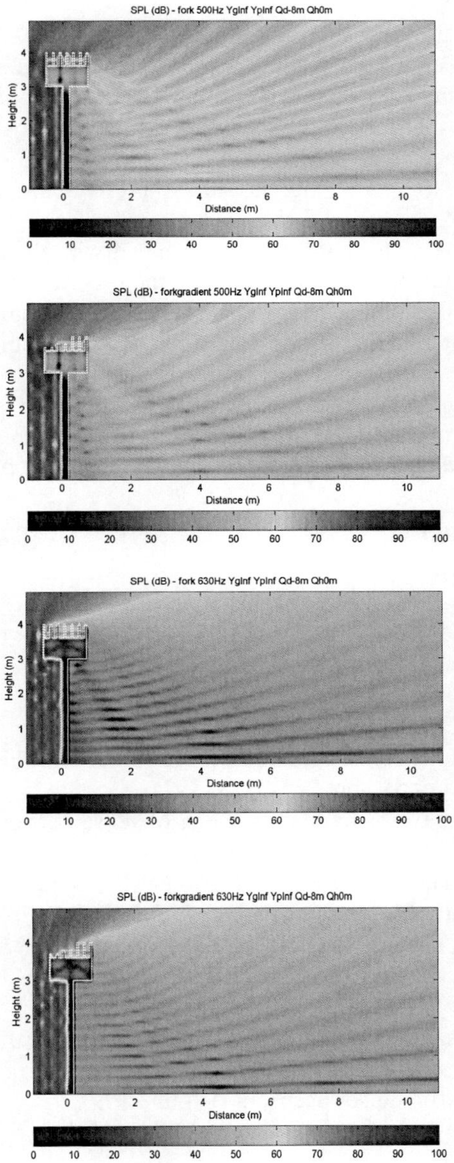

Figure 6: Graphical representation of the simulations for the frequencies of 500 and 630Hz for the fork barrier (left) and the fork gradient (right). The formation of the soft plane at these frequencies is visible.

3 Conclusions

Non-standard noise barriers for optimal far-field shielding have been investigated. A purely geometrical solution is not as prone to deterioration as absorbent barriers that tend to change their acoustic properties with time. The results of two-dimensional BEM analysis for the "fork" and "fork gradient" configurations are encouraging, as they present an improved shielding performance.

Moreover, as the performance can be changed with the change of the shape, in particular with the change of the depth of the used channels, these results can be used for a possible tuning of the barrier. If a mechanism is included into the barrier so that the height of the channels can be changed, this gives the possibility to change the spectral profile of the insertion loss. This could be useful in building sustainable barriers, in view of expected but not yet quantifiable shifts of the traffic noise spectrum in the future, as for example due to the increasing number of e-cars on the main transportation routes.

Acknowledgement

We would like to acknowledge Vicente Cutanda Henríquez for the help and support in working with OpenBEM.

References

[1] Fujiwara K., Hothersall D., Kim C., Noise barriers with reactive surfaces, *Applied Acoustics* 53 , pp. 255-272 (1997)

[2] Ishizuka T., Fujiwara K., Performance of noise barriers with various edge shapes and acoustical conditions, *Applied Acoustics* 65 , pp. 125-141(2004)

[3] Gasparoni S., Haider M., Conter M., Wehr R., Breuss S., BEM simulations of noise barriers. *Proceedings of the 39th Internoise*, SPA (2010)

[4] Henríquez V.C., Juhl P. M., OpenBEM - An open source Boundary Element Method software in Acoustics, *Proceedings of the 39th Internoise,* SPA (2010)

[5] CEN, European Standard EN 1793-3 Road traffic noise reducing devices – Test method for determining the acoustic performance – Part 3: Normalised traffic noise spectrum, Belgium 1997

Section 3
Computational methods

A shape sensitivity analysis approach based on the boundary element method

T. Matsumoto[1], T. Takahashi[1], K. Shibata[2] & T. Yamada[1]
[1] *Nagoya University, Japan*
[2] *Kyushu Electric Power Co. Inc., Japan*

Abstract

A revised shape design sensitivity formulation is presented for elastostatic problems based on the adjoint method and the boundary element method. The objective function is assumed as a functional consisting of the boundary quantities and those given at some finite number of points in the domain of the solid. The gradient of the objective function is derived and an adjoint state is introduced so that the unknown sensitivity coefficients of the displacement and traction on the boundary and in the domain are eliminated from the gradient expression. Since the original boundary value problem and the adjoint problem are governed by the same differential equations and the boundary condition types, and also the derived sensitivity formulation is expressed with only the boundary integrals and the quantities at some discrete points in the domain, the boundary element method can be used as the effective computational tool. Also, the recent development of the fast-multipole boundary element method enables a large-scale shape optimization analysis of complicated structures. The validity of the derived formulation is tested through some numerical example problems.
Keywords: elastostatics, shape sensitivity, adjoint method, boundary element method, topology optimization.

1 Introduction

Although the boundary element method (BEM) has an advantage of boundary only modeling, it has not become a popular simulation tool in engineering applications because it required high storage and computation costs. For shape and topology optimization problems, the finite element method (FEM) has widely been used as the computational method. However, following the development of

WIT Transactions on Modelling and Simulation, Vol 52, © 2011 WIT Press
www.witpress.com, ISSN 1743-355X (on-line)
doi:10.2495/BE110111

fast computation algorithm [1,2] in recent decades, BEM may become a powerful alternative to FEM in shape optimization problems.

Shape optimization as well as topology optimizations problems define objective functions of the quantities defined on the boundary and in the domain. All such quantities are also functions of shape parameters, and the optimum values of the shape parameters minimizing the objective functions are calculated. The gradients, or sensitivities, of the objective function with respect to the design variables are also needed in most of the solution procedures to find the direction to the minimum point of the objective function. The gradient is related to a variation of the objective function, and consists of the sensitivities of the boundary and internal quantities. To calculate these sensitivities, direct differentiation method [3–5] and adjoint variable method [6, 7] has been proposed. When using BEM, the direct differentiation method uses an additional boundary integral equation obtained by differentiating the original boundary integral equation with respect to an arbitrary shape design variable. The direct differentiation is, however, costly for problems with a large number of design variables because the additional boundary integral equation for the sensitivity must be solved for every design variable. The adjoint variable method defines an additional system that eliminates the unknown sensitivities on the boundary and in the domain. Therefore this method is more efficient because we have to repeat the boundary element calculation only for the original problem and the adjoint problem to calculate the gradient of the objective function.

In this paper, we consider an objective function that is appropriate to evaluate by means of BEM. The objective function is assumed to consist only of the quantities on the boundary and at some discrete points in the domain. Adjoint variable method is applied to the defined objective function and some numerical examples are shown to demonstrate the effectiveness of the approach.

2 Formulations

2.1 Boundary element method for elastostatics

The governing differential equation for linear isotropic elastic solids is the following Navier's equation:

$$C_{ijkl}u_{k,li} + b_j = Gu_{j,kk} + \frac{G}{(1-2\nu)}u_{k,kj} + b_j = 0 \quad \text{in } \Omega \tag{1}$$

where C_{ijkl} denotes the elastic tensor, u_i and b_i are the displacement and body force vectors, respectively, G is the shear modulus, ν is Poisson's ratio, and Ω is the domain under consideration. The index is assumed to change from 1 to 3 for three-dimensional problems, and from 1 to 2 for two-dimensional case. For terms with repeated indices, summation convention is assumed. The indices after a comma denote differentiations with respect to the coordinate axes.

www.witpress.com, ISSN 1743-355X (on-line)

The boundary conditions are as follows:

$$u_i = \bar{u}_i \quad \text{on } \Gamma_u \tag{2}$$

$$t_i = \bar{t}_i \quad \text{on } \Gamma_t \tag{3}$$

where t_i is traction, $\bar{u}_i$ and $\bar{t}_i$ are the prescribed known functions of u_i and t_i on some parts of the boundary Γ_u and Γ_t, respectively.

The boundary integral equation derived corresponding to Eq. (1) becomes [8]

$$c_{ij}u_i(y) + \int_\Gamma t^*_{ij}(x,y)u_i(x)d\Gamma(x) = \int_\Gamma u^*_{ij}(x,y)t_i(x)d\Gamma(x) + \int_\Omega u^*_{ij}(x,y)b_i(x)d\Omega(x), \quad y \in \Gamma \tag{4}$$

where x and y are points on the boundary, c_{ij} is a constant tensor, becoming $1/2\delta_{ij}$ when y lies at a smooth part of the boundary, u^*_{ij} is the fundamental solution, and t^*_{ij} is the traction related to u^*_{ij}. For two-dimensional, plane strain, case, u^*_{ij} and t^*_{ij} are given as

$$u^*_{ij}(x,y) = \frac{1}{8\pi G(1-\nu)}\left\{(3-4\nu)\delta_{ij}\ln\left(\frac{1}{r}\right) + r_{,i}r_{,j}\right\} \tag{5}$$

$$t^*_{ij}(x,y) = \frac{-1}{4\pi(1-\nu)r}\left[\frac{\partial r}{\partial n}\{(1-2\nu)\delta_{ij} + 2r_{,i}r_{,j}\} + (1-2\nu)(r_{,i}n_j - r_{,j}n_i)\right] \tag{6}$$

where r is the distance between x and y, n_i is the unit outward normal vector at x, and $\partial r/\partial n$ is the derivative of r in n_i direction at x.

In what follows, we assume that no body force exists in the domain for simplicity. Discretizing Eq. (4), we have the following system of algebraic equations:

$$[H]\{u\} = [G]\{t\} \tag{7}$$

and rearranging this so that all the unknowns come to the left-hand side and all the others to the right-hand side result in

$$[A]\{X\} = \{Y\} \tag{8}$$

where $\{X\}$ is the vector consisting only of unknown nodal values, while $\{Y\}$ is the vector obtained by multiplying the known nodal values with corresponding parts of the coefficient matrix.

Once Eq. (8) is solved we obtain all the displacement and tractions on the boundary. The stress component on the boundary can be calculated using them by

$$\sigma_{ij} = \frac{1}{1-\nu}\left(\nu\delta_{ij} - n_i n_j\right) t_k n_k + t_i n_j + t_j n_i$$
$$+ G\left\{\frac{2\nu}{1-\nu}\left(2\delta_{ij} - n_i n_j\right)\alpha_k + \left(\delta_{ik} - n_i n_k\right)\alpha_j + \left(\delta_{jk} - n_j n_k\right)\alpha_i\right\}\frac{\partial u_k}{\partial \alpha}$$
$$+ G\left\{\frac{2\nu}{1-\nu}\left(2\delta_{ij} - n_i n_j\right)\beta_k + \left(\delta_{ik} - n_i n_k\right)\beta_j + \left(\delta_{jk} - n_j n_k\right)\beta_i\right\}\frac{\partial u_k}{\partial \beta} \tag{9}$$

where α_k and β_k are tangential vectors on the boundary, and $\partial u_k/\partial\alpha$ and $\partial u_k/\partial\beta$ are corresponding tangential derivatives of u_k.

The internal displacement can be calculated by using Eq. (4) with $c_{ij} = \delta_{ij}$. The formula for calculating internal stresses is also obtained by differentiating Eq. (4) with $c_{ij} = \delta_{ij}$ at an internal point and substituting it into Hooke's law [8].

2.2 Objective function of shape optimization problem and its sensitivity

We consider the following objective function

$$J = \int_\Gamma g(u_i, t_i) d\Gamma(x) + \sum_s \int_\Omega h(u_i, \sigma_{ij})\delta(x - z^s)\, d\Omega, \quad z^s \in \Omega(x) \tag{10}$$

where Γ denotes the boundary, Ω the domain, u_i and t_i the displacement and traction, respectively, σ_{ij} the stress components, z_s, $(s = 1, 2, \ldots)$ the discrete points in the domain, and $\delta(x - z_s)$ the Dirac delta function. $g(u_i, t_i)$ is a function defined with u_i and t_i on the boundary, while $h(u_i, \sigma_{ij})$ is that defined in the domain.

Note that the second domain integral is not in fact an integral because Dirac's delta functions exist in the integrands, therefore, this type of objective function is quite appropriate to treat with BEM.

The gradient of J with respect to an arbitrary shape design variable becomes

$$J' = \int_\Gamma \left(\frac{\partial g}{\partial u_i}\dot{u}_i + \frac{\partial g}{\partial t_i}\dot{t}_i\right) d\Gamma + \int_\Gamma g\, d\dot{\Gamma}$$
$$+ \sum_s \int_\Omega \left(\frac{\partial h}{\partial u_i}\dot{u}_i + \frac{\partial h}{\partial \sigma_{ij}}\dot{\sigma}_{ij}\right)\delta(x - z^s)\, d\Omega$$
$$+ \sum_s \int_\Omega h(u_i, \sigma_{ij})\delta(x - z^s)\, d\dot{\Omega} + \sum_s \int_\Omega h((u_i, \sigma_{ij})\dot{\delta}(x - z^s)\, d\Omega \tag{11}$$

An overscribed dot $(\dot{\ })$ in the above expression denotes a material derivative [9], $\dot{u}_i$, $\dot{t}_i$, $\dot{\sigma}_{ij}$ are understood in the following sense:

$$\dot{u}_i = u'_i + u_{i,j}\dot{x}_j \tag{12}$$

$$\dot{t}_i = t'_i + t_{i,j}\dot{x}_j \tag{13}$$

$$\dot{\sigma}_{ij} = \sigma'_{ij} + \sigma_{ij,k}\dot{x}_k \tag{14}$$

where a prime $(')$ denotes a differentiation with respect to the design variable before the shape change is applied. Also, $\dot{d\Gamma}$ and $\dot{d\Omega}$ are written as follows [9]:

$$\dot{d\Gamma} = \left(\dot{x}_{m,m} - \dot{x}_{i,j}n_i n_j\right)\, d\Gamma \tag{15}$$

$$\dot{d\Omega} = \dot{x}_{m,m}\, d\Omega \tag{16}$$

We find from Eq. (11) that in order to evaluate Eq. (10) we have to calculate the sensitivities of the $\dot{u}_i$, $\dot{t}_i$, $\dot{\sigma}_{ij}$ as many as the number of the design variable. Although these quantities can be calculated by using the direct differentiation method based on the differentiation of the boundary integral equation with respect to the design variable, it is not efficient when the number of the design variables is large. Therefore, we apply the adjoint variable method to eliminate these unknown sensitivities from Eq. (11). We now observe

$$\dot{(u_{k,l})} = \dot{(u_k)}_{,l} - u_{k,m}\dot{x}_{m,l} \tag{17}$$

where $\dot{(\ )}$ denotes a material derivative of the quantity enclosed with parentheses. Using Eq. (17) we have the material derivative of the stress components as follows:

$$\begin{aligned}\dot{\sigma}_{ij} &= C_{ijkl}\dot{(u_{k,l})} \\ &= C_{ijkl}\dot{(u_k)}_{,l} - C_{ijkl}u_{k,m}\dot{x}_{m,l}\end{aligned} \tag{18}$$

Using Eq. (18), we can modify the integral of $\dot{\sigma}_{ij}$ in Eq. (11) as

$$\begin{aligned}\int_\Omega \frac{\partial h}{\partial \sigma_{ij}}\dot{\sigma}_{ij}\delta(x - z^s)\, d\Omega &= \int_\Omega \frac{\partial h}{\partial \sigma_{ij}} C_{ijkl}\dot{(u_k)}_{,l}\delta(x - z^s)\, d\Omega \\ &\quad - \int_\Omega \frac{\partial h}{\partial \sigma_{ij}} C_{ijkl}u_{k,m}\dot{x}_{m,l}\delta(x - z^s)\, d\Omega \\ &= -\int_\Omega \left[\frac{\partial h}{\partial \sigma_{ij}}\delta(x - z^s)\right]_{,l} C_{ijkl}\dot{u}_k\, d\Omega \\ &\quad - \int_\Omega \frac{\partial h}{\partial \sigma_{ij}} C_{ijkl}u_{k,m}\dot{x}_{m,l}\delta(x - z^s)\, d\Omega\end{aligned} \tag{19}$$

Then, Eq. (11) becomes

$$J' = \int_\Gamma \left(\frac{\partial g}{\partial u_i} \dot{u}_i + \frac{\partial g}{\partial t_i} \dot{t}_i \right) d\Gamma + \int_\Gamma g \, d\dot{\Gamma} + \sum_s \int_\Omega \frac{\partial h}{\partial u_i} \dot{u}_i \delta(x - z^s) \, d\Omega$$
$$- \sum_s \int_\Omega \left(\frac{\partial h}{\partial \sigma_{ij}} \right)_{,l} C_{ijkl} \dot{u}_k \delta(x - z^s) \, d\Omega$$
$$- \sum_s \int_\Omega \frac{\partial h}{\partial \sigma_{ij}} C_{ijkl} \dot{u}_k \delta_{,l}(x - z^s) \, d\Omega + \sum_s \int_\Omega h \dot{\delta}(x - z^s) \, d\Omega$$
$$- \sum_s \int_\Omega \frac{\partial h}{\partial \sigma_{ij}} C_{ijkl} u_{k,m} \dot{x}_{m,l} \delta(x - z^s) \, d\Omega + \sum_s \int_\Omega h \delta(x - z^s) \, d\dot{\Omega} \quad (20)$$

We consider the following augmented objective function instead of Eq. (10):

$$P = J + I \quad (21)$$

where I is the integral of the left-hand side of Navier's equation times Lagrange multipliers λ_j, $(j = 1, 2, 3)$ over the domain, as follows:

$$I = \int_\Omega \lambda_j C_{ijkl} u_{k,li} \, d\Omega \quad (22)$$

where C_{ijkl} is the elastic constant tensor. Integrating I by parts gives the following weak form:

$$I = \int_\Gamma \lambda_j t_j \, d\Gamma - \int_\Omega \lambda_{j,i} C_{ijkl} u_{k,l} \, d\Omega \quad (23)$$

From Eq. (23), we have

$$I' = \int_\Gamma \dot{\lambda}_j t_j \, d\Gamma + \int_\Gamma \lambda_j \dot{t}_j \, d\Gamma + \int_\Gamma \lambda_j t_j \, d\dot{\Gamma}$$
$$- \int_\Omega (\dot{\lambda_{j,i}}) C_{ijkl} u_{k,l} \, d\Omega - \int_\Omega \lambda_{j,i} C_{ijkl} (\dot{u_{k,l}}) \, d\Omega - \int_\Omega \lambda_{j,i} C_{ijkl} u_{k,l} \, d\dot{\Omega} \quad (24)$$

Again, we observe some relationships for the material derivative of the Lagrange multiplier as

$$\dot{\lambda}_j = \lambda_{j,m} \dot{x}_m \quad (25)$$

$$(\dot{\lambda_{j,i}}) = \lambda_{j,im} \dot{x}_m \quad (26)$$

www.witpress.com, ISSN 1743-355X (on-line)

Using Eqs. (17), (25) and (26) into Eq. (24) and applying integration by parts give

$$
\begin{aligned}
I' = & \int_\Gamma \lambda_{j,m}\dot{x}_m t_j \, d\Gamma + \int_\Gamma \lambda_j \dot{t}_j \, d\Gamma + \int_\Gamma \lambda_j t_j \, d\dot{\Gamma} \\
& - \int_\Gamma \lambda_{j,i} C_{ijkl} \dot{u}_k n_l \, d\Gamma + \int_\Gamma \lambda_{j,i} C_{ijkl} u_{k,m} \dot{x}_m n_l \, d\Gamma \\
& + \int_\Omega \lambda_{j,il} C_{ijkl} \dot{u}_k d\Omega - \int_\Omega \lambda_{j,il} C_{ijkl} u_{k,m} \dot{x}_m \, d\Omega \\
& - \int_\Gamma \lambda_{j,i} C_{ijkl} u_{k,l} \dot{x}_m n_m \, d\Gamma
\end{aligned} \tag{27}
$$

Thus, we obtain the gradient of the augmented objective function P, as follows:

$$
\begin{aligned}
P' = & \int_{\Gamma_t} \left(\frac{\partial g}{\partial u_k} - \tau_k \right) \dot{u}_k \, d\Gamma + \int_{\Gamma_u} \left(\frac{\partial g}{\partial u_k} - \tau_k \right) \dot{u}_k \, d\Gamma \\
& + \int_{\Gamma_t} \left(\frac{\partial g}{\partial t_i} + \lambda_i \right) \dot{t}_i \, d\Gamma + \int_{\Gamma_u} \left(\frac{\partial g}{\partial t_i} + \lambda_i \right) \dot{t}_i \, d\Gamma \\
& + \int_\Omega \left\{ C_{ijkl}\lambda_{j,il} + \sum_s \left[\frac{\partial h}{\partial u_k} \delta(x - z^s) - \left(\frac{\partial h}{\partial \sigma_{ij}} \right)_{,l} C_{ijkl} \delta(x - z^s) \right. \right. \\
& \qquad \left. \left. - \frac{\partial h}{\partial \sigma_{ij}} C_{ijkl} \delta_{,l}(x - z^s) \right] \right\} \dot{u}_k d\,\Omega \\
& + \int_\Gamma g \, d\dot{\Gamma} + \int_\Gamma \lambda_{j,m} \dot{x}_m t_j \, d\Gamma \\
& + \int_\Gamma \lambda_j t_j \, d\dot{\Gamma} + \int_\Gamma \tau_k u_{k,m} \dot{x}_m \, d\Gamma \\
& - \int_\Gamma \lambda_{j,i} C_{ijkl} u_{k,l} \dot{x}_m n_m \, d\Gamma - \int_\Omega \lambda_{j,il} C_{ijkl} u_{k,m} \dot{x}_m \, d\Omega \\
& + \sum_s \int_\Omega h \dot{\delta}(x - z^s) \, d\Omega \\
& - \sum_s \int_\Omega \frac{\partial h}{\partial \sigma_{ij}} C_{ijkl} u_{k,m} \dot{x}_{m,l} \delta(x - z^s) \, d\Omega \\
& + \sum_s \int_\Omega h \delta(x - z^s) \, d\dot{\Omega}
\end{aligned} \tag{28}
$$

where τ_k is defined as the traction corresponding to λ_j as

$$
\tau_k = C_{ijkl} \lambda_{j,i} n_l \tag{29}
$$

www.witpress.com, ISSN 1743-355X (on-line)

Notice that $\dot{u}_i$ and $\dot{t}_i$ are also known on Γ_u and Γ_t, respectively, whereas $\dot{u}_i$ on Γ_t, $\dot{u}_i$ on Γ_u, and $\dot{u}_i$ in Ω are all unknown. Therefore, in order to eliminate these unknown sensitivities from Eq. (28), we consider the following adjoint problem for λ_i:

$$C_{ijkl}\lambda_{j,il}(x) + \sum_s \left\{ \frac{\partial h}{\partial u_k}\delta(x - z^s) - \left(\frac{\partial h}{\partial \sigma_{ij}}\right)_{,l} C_{ijkl}\delta(x - z^s) \right.$$
$$\left. - \frac{\partial h}{\partial \sigma_{ij}} C_{ijkl}\delta_{,l}(x - z^s) \right\} = 0, \qquad x \in \Omega \tag{30}$$

$$\tau_k(x) = \frac{\partial g}{\partial u_k}(x) \qquad x \in \Gamma_t \tag{31}$$

$$\lambda_i(x) = -\frac{\partial g}{\partial t_i}(x) \qquad x \in \Gamma_u \tag{32}$$

Equation (30) is the same as Navier's equation with body force terms at discrete points. The boundary conditions given by Eqs. (31) and (32) have the same type of the original boundary condition given by Eqs. (2) and (3). Therefore, we can solve the adjoint problem very efficiently using the same coefficient matrices of BEM.

By using λ_i as the solution of Eqs. (30), (31), and (32), the gradient of P can be calculated by using the following expression.

$$P' = \int_{\Gamma_u} \left(\frac{\partial g}{\partial u_i} - \tau_i\right) \dot{u}_i \, d\Gamma + \int_{\Gamma_t} \left(\frac{\partial g}{\partial t_i} + \lambda_i\right) \dot{t}_i \, d\Gamma$$
$$+ \int_\Gamma g \, d\dot{\Gamma} + \int_\Gamma \lambda_{j,m}\dot{x}_m t_j \, d\Gamma + \int_\Gamma \lambda_j t_j \, d\dot{\Gamma}$$
$$+ \int_\Gamma \tau_k u_{k,m}\dot{x}_m \, d\Gamma - \int_\Gamma \lambda_{i,j}\sigma_{ij}\dot{x}_m n_m \, d\Gamma$$
$$- \sum_s \left\{ \frac{\partial h}{\partial \sigma_{ij}}\sigma_{ij,m}\dot{x}_m + \frac{\partial h}{\partial u_i} u_{i,m}\dot{x}_m + h\dot{x}_{m,m} \right\} \tag{33}$$

where $u_{i,j}$ can be calculated by using

$$u_{i,j} = \frac{\partial u_i}{\partial n} n_j + \frac{\partial u_i}{\partial \alpha}\alpha_j + \frac{\partial u_i}{\partial \beta}\beta_j \tag{34}$$

with

$$\frac{\partial u_i}{\partial n} = \frac{1}{G}\left(t_i - \frac{1}{2(1-\nu)} t_k n_k n_i\right) - \left(\frac{\nu}{1-\nu} n_i \alpha_k + n_k \alpha_i\right)\frac{\partial u_k}{\partial \alpha}$$
$$- \left(\frac{\nu}{1-\nu} n_i \beta_k + n_k \beta_i\right)\frac{\partial u_k}{\partial \beta} \tag{35}$$

The formula to calculate $\lambda_{i,j}$ is obtained by simply replacing u in Eqs. (34) and (35) with λ.

www.witpress.com, ISSN 1743-355X (on-line)

3 Numerical examples

We consider a rectangular plate, subjected to a roller support on one end and a uniform tensile stress $\bar{t} = 200\,[\mathrm{MPa}]$ on the other end, as shown in Figure 1. Young's modulus and Poisson's ratio are assumed as 216 GPa and 0.3, respectively. The size of the plate is initially set as $L_1 = 0.1\,[\mathrm{m}]$ and $L_2 = 0.02\,[\mathrm{m}]$.

The functions g and h in Eq. (10) are given as follows:

$$g = 0 \tag{36}$$

$$h = \frac{1}{2}\left(u_1(\mathrm{P}) - 0.00004\right)^2 \tag{37}$$

In this example, we observe that changing L_1 yields a different value of u_1, thus, L_1 is chosen to be a design variable to validate the derived formulas.

The boundaries of the rectangular plate is discretized into quadratic elements uniformly. We show in Table 1 the sensitivities of the objective function and errors obtained for different number of elements using the present approach.

Next we consider another rectangular plate model, as shown in Figure 2. Both sides of the plate are fixed and a concentrated force $F = 2.67 \times 10^6\,[\mathrm{N}]$ is applied at the center of the lower boundary. We intend to change the von Mises stress at internal point P to the target value $\bar{\sigma} = 30\,[\mathrm{MPa}]$ by changing the x_2 coordinate of the nodes **a**, **b**, and **c** on a part of the boundary Γ_D. The related functions of the

Table 1: Sensitivities and their errors obtained for various number of element discretization of the rectangular plate model.

Number of nodes	Sensitivity P'	Error [%]
24	-4.9434×10^{-9}	3.37×10^{-1}
48	-4.9397×10^{-9}	4.49×10^{-1}
120	-4.9622×10^{-9}	5.02×10^{-3}
240	-4.9619×10^{-9}	2.82×10^{-4}

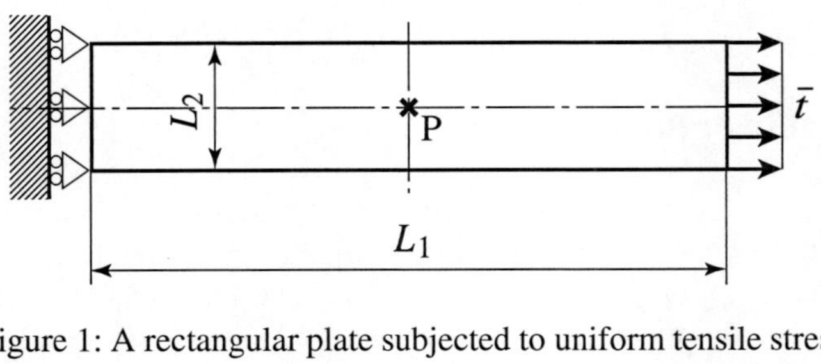

Figure 1: A rectangular plate subjected to uniform tensile stress.

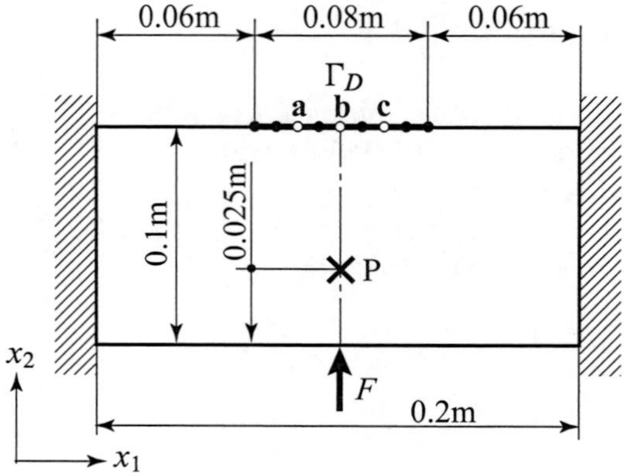

Figure 2: A rectangular plate subjected to a concentrated force.

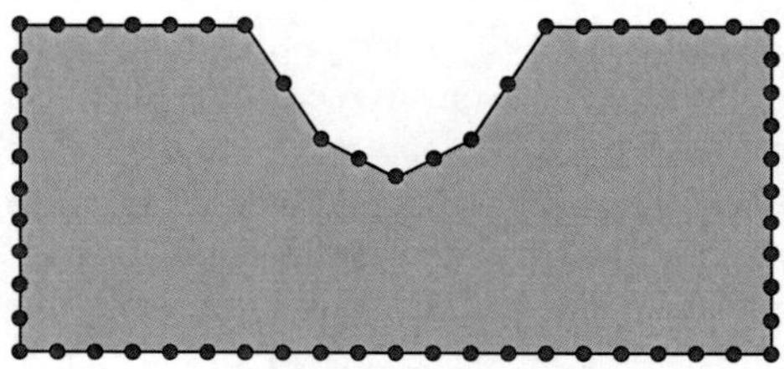

Figure 3: Final shape of the rectangular plate subjected to a concentrated force.

objective function are given as

$$g = 0 \tag{38}$$

$$h = \frac{1}{2}\left(\sigma_{\mathrm{M}}(P) - \bar{\sigma}\right)^2 \tag{39}$$

where $\sigma_{\mathrm{M}}(P)$ is the von Mises stress at the internal point P. The entire boundary is discretized into 30 quadratic elements uniformly. The concentrated force is given as an equivalent traction at the node placed at the point where the concentrated force is acting. In Figure 3 is shown the final shape of the plate obtained by the present procedure for calculating the shape sensitivities. The von Mises stress at point P in the optimum shape is found to be 3.0034128×10^7 Pa, which is turned out to be very close to the target stress value.

4 Concluding remarks

A shape sensitivity expression for objective functions appropriate to evaluate using BEM, has been derived based on the adjoint variable method. The objective function consists of only boundary integrals of the displacement and traction, and values defined with the internal displacement and stress at discrete points in the domain. The differential equation of the adjoint system has become a Navier's equation with body forces at the discrete points in the domain, and the boundary conditions are of the same type as those of the original problem. Therefore, the same coefficient matrices can be consistently used also for the adjoint problem. The derived adjoint variable approach was applied to some numerical examples to validate its effectiveness.

References

[1] Rokhlin, V., Rapid solution of integral equations of classical potential theory. *Journal of Computational Physics*, **60**(2), pp. 187-207, 1985.
[2] Yoshida, K., PhD Thesis, Kyoto University, Japan, 2001.
[3] Barone, M.R. & Yang, R.J., A boundary element approach for recovery of shape sensitivities in three dimensional elastic solids. *Comp. Meth. in Appl. Mech. Engng.*, **74**, pp. 69-82, 1989.
[4] Matsumoto, T., Tanaka, M., Miyagawa, M. & Ishii, N., Optimum design of cooling lines in injection moulds by using boundary element design sensitivity analysis. *Finite Elements in Analysis and Design*, **14**, pp. 177-185, 1993.
[5] Matsumoto, T., Tanaka, M. & Yamada, Y., Design Sensitivity Analysis of Steady-State Acoustic Problems using Boundary Integral Equation Formulation. *JSME International Journal, Series C*, **38**(1), pp. 9-16, 1995.
[6] Haug, E.J., Choi, K.K. & Komkov, V., *Design Sensitivity Analysis of Structural Systems*, Academic Press, 1986.
[7] Choi, J.H. & Kwak, B.M., Boundary integral equation method for shape optimization of elastic structures, *Int. J. Numer. Meth. Engng.*, **26**, pp. 1579-1595, 1988.
[8] Brebbia, C.A., Telles, J.C.F. & Wrobel, L.C., *Boundary Element Techniques: Theory and Applications in Engineering*, Springer, 1984,
[9] Arora, J.S., Lee, T.H. & Cardoso, J.B., Structural shape sensitivity analysis: relationship between material derivative and control volume approach, *AIAA J.*, **30**(6), pp. 1638-1648, 1992.

An efficient boundary element modeling of the time domain integral equations for thin wires radiating in a presence of a lossy media

D. Poljak, S. Antonijevic & V. Doric
University of Split, Croatia

Abstract

The paper deals with the space-time Galerkin-Bubnov scheme of the Indirect Boundary Element Method for the solution of time domain integral equations for thin wires in the presence of lossy media of the Pocklington and Hallen type, respectively. The presence of a dissipative half-space is taken into account via the corresponding space-time reflection coefficients. Some illustrative computational examples related to the overhead wires and grounding electrodes are presented.
Keywords: boundary elements, time domain modeling, Hallen equation, Pocklington equation, thin wires.

1 Introduction

Generally, a direct time-domain analysis of thin wire configurations in the presence of lossy media can be carried out by using the appropriate space–time integral equations of either Pocklington or Hallen type [1–4]. One of the most efficient numerical solution approaches to both equation types is related to the Galerkin-Bubnov Indirect Boundary Element Method (GB-IBEM) [2]. When applied to the solution of the Hallen integral equation the method appears to be relatively complex comparing to various procedures for the solution of Pocklington equations, but, at the same time, it is proven to be highly efficient and accurate and unconditionally stable [2, 4]. On the other hand, the implementation of GB-IBEM to the solution of the Pocklington type equation is relatively simple, but suffers from numerical instabilities. The origin of these instabilities is the existence of space-time differential operator [2]. The GB-IBEM solution of Pocklington equation in free space for certain values of time

WIT Transactions on Modelling and Simulation, Vol 52,
www.witpress.com, ISSN 1743-355X (on-line)
doi:10.2495/BE110121

domain integration parameters has been presented in [5], while the Hallen equation solution via GB-IBEM have been obtained for thin wire configurations in the presence of a dielectric half-space, e.g. in [6]. The original formulation has been developed for a single wire located horizontally above [2] and a below real ground, respectively [7]. In both cases, the influence of the finitely conducting ground has been taken into account via the corresponding reflection coefficient. Nevertheless, the numerical solution was mostly limited to the cases in which the finite conductivity of the ground could be ignored. This approximation involves cases where the wires are sufficiently far from the two-media interface, or where the ground conductivity is appreciably low or very high, i.e. where the approximation of pure dielectric medium or perfect ground is considered. Through these approximations the time dependent part of the reflection coefficient function vanishes, and the resulting matrix equation simplifies significantly.

However, for the cases where these approximations are not valid, modifications to the original methods are required in order to include the ground conductivity [8]. The related reflection coefficient is space- time dependent, and the resulting convolution integrals have to be included in the matrix system and numerically computed. This leads to a significant increase in the overall computational cost of the method, and requires several modifications.

This paper compares the space-time Pocklington and Hallen equation approaches when the lossy ground effects are taken into account. The related GB-IBEM procedures for the solution of both equations, taking into account a finite value of the ground conductivity, are discussed. Some illustrative computational examples related to transient analysis of overhead wires and grounding electrodes are given in the paper, as well.

2 Hallen integral equation formulation

The straight thin wire of length L and radius a located at height h above a lossy ground is shown in Fig. 1.

The formulation is based on the space-time dependent integral equation of the Hallen type which does not contain a differential operator. The Hallen is therefore particularly useful for a numerical treatment, as this operator is the origin of numerical instabilities [2].

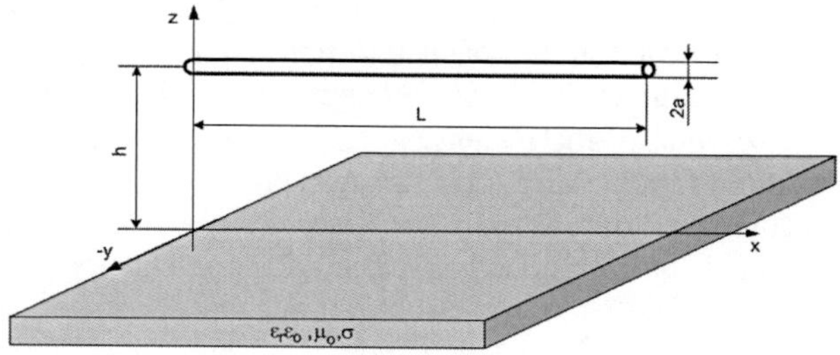

Figure 1: A straight thin wire above a real ground.

The transient current *I(x, t)* induced on the horizontal straight thin wire, depicted in Fig. 1, illuminated by a plane wave electric field can be assessed by solving the space-time Hallen integral equation:

$$\int_0^L \frac{I(x',t-\frac{R}{c})}{4\pi R}dx' - \int_{-\infty}^{t}\int_0^L r(\theta,\tau)\frac{I(x',t-\frac{R^*}{c}-\tau)}{4\pi R^*}dx'd\tau = \frac{1}{2Z_0}\int_0^L E_x^{inc}(x',t-\frac{|x-x'|}{c})dx' + F_0(t-\frac{x}{c}) + F_L(t-\frac{L-x}{c}) \tag{1}$$

where R and R^* are corresponding distances from the observation to the source point located on the real and image wire, respectively, E_x^{exc} is the tangential component of the excitation and F_0, F_L are the unknown functions to account for the reflections at the wire free ends:

$$F_0(t) = \sum_{n=0}^{\infty} K_0\left(t - \frac{2nL}{c}\right) - \sum_{n=0}^{\infty} K_L\left(t - \frac{(2n+1)L}{c}\right)$$
$$F_L(t) = \sum_{n=0}^{\infty} K_L\left(t - \frac{2nL}{c}\right) - \sum_{n=0}^{\infty} K_0\left(t - \frac{(2n+1)L}{c}\right) \tag{2}$$

defined in terms of auxiliary functions K_0, K_L given by:

$$K_0(t) = \int_0^L \frac{I(x',t-\frac{R_0}{c})}{4\pi R_0}dx' - \int_{-\infty}^{t}\int_0^L r(\theta,\tau)\frac{I(x',t-\frac{R_0^*}{c}-\tau)}{4\pi R_0^*}dx'd\tau - \frac{1}{2Z_0}\int_0^L E_x^{inc}(x',t-\frac{x'}{c})dx' \tag{3a}$$

$$K_L(t) = \int_0^L \frac{I(x',t-\frac{R_L}{c})}{4\pi R_L}dx' - \int_{-\infty}^{t}\int_0^L r(\theta,\tau)\frac{I(x',t-\frac{R_L^*}{c}-\tau)}{4\pi R_L^*}dx'd\tau - \frac{1}{2Z_0}\int_0^L E_x^{inc}(x',t-\frac{L-x'}{c})dx' \tag{3b}$$

where subscript 0 and L is related to the distance from the observation point to the source point located on the real and image wire, respectively.

The ground effects are included in the formulation through the space-time dependent reflection coefficient for TM-polarization, [6] which, for convenience, can be written in the form:

$$r(\theta,\tau) = r'(\theta,\tau) + r''(\theta,\tau) \tag{4}$$

where

$$r'(\theta,t) = K\delta(t),\ r''(\theta,t) = \frac{4\beta}{1-\beta^2}\frac{e^{-\alpha t}}{t}\sum_{n=1}^{\infty}(-1)^{n+1}nK^n I_n(\alpha t) \tag{5}$$

$$\tau = \frac{\sigma}{\varepsilon_0\varepsilon_r},\ \beta = \frac{\sqrt{\varepsilon_r - \sin^2\theta}}{\varepsilon_r\cos\theta},\ \gamma = \frac{\tau}{1-\frac{\sin^2\theta}{\varepsilon_r}},\ \theta = arctg\frac{|x-x'|}{2h},\ K = \frac{1-\beta}{1+\beta},\ \alpha = \frac{\tau}{2} \tag{6}$$

Note that I_n is the modified Bessel function of the first order, *n*-th degree.

2.1 BEM procedure for Hallen equation

Applying the weighted residual approach in the spatial domain and GB-IBEM procedure [2], the following local matrix system is obtained:

$$
\begin{aligned}
&[A]\{I\}_i\Big|_{t-\frac{R}{c}} - \left[A^*\right]\{I\}_i\Big|_{t-\frac{R^*}{c}} - \left\{\hat{A}\right\}\Big|_{t-\frac{R^*}{c}} = [B]\{E\}\Big|_{t-\frac{|x-x'|}{c}} \\
&\quad +[C]\left\{\sum_{n=0}^{\infty} I^n\right\}_i\Bigg|_{t-\frac{R_0}{c}-\frac{2nL}{c}-\frac{x}{c}} - \left[C^*\right]\left\{\sum_{n=0}^{\infty} I^n\right\}_i\Bigg|_{t-\frac{R_0^*}{c}-\frac{2nL}{c}-\frac{x}{c}} - \left\{\sum_{n=0}^{\infty} \hat{C}^n\right\}\Bigg|_{t-\frac{R_0^*}{c}-\frac{2nL}{c}-\frac{x}{c}} \\
&\quad -[B]\left\{\sum_{n=0}^{\infty} E^n\right\}\Bigg|_{t-\frac{x'}{c}-\frac{2nL}{c}-\frac{x}{c}} - [D]\left\{\sum_{n=0}^{\infty} I^n\right\}_i\Bigg|_{t-\frac{R_L}{c}-\frac{(2n+1)L}{c}-\frac{x}{c}} \\
&\quad +\left[D^*\right]\left\{\sum_{n=0}^{\infty} I^n\right\}_i\Bigg|_{t-\frac{R_L^*}{c}-\frac{(2n+1)L}{c}-\frac{x}{c}} + [B]\left\{\sum_{n=0}^{\infty} E^n\right\}\Bigg|_{t-\frac{L-x'}{c}-\frac{(2n+1)L}{c}-\frac{x}{c}} \\
&+\left\{\sum_{n=0}^{\infty} \hat{D}^n\right\}\Bigg|_{t-\frac{R_L^*}{c}-\frac{(2n+1)L}{c}-\frac{x}{c}} + [D]\left\{\sum_{n=0}^{\infty} I^n\right\}_i\Bigg|_{t-\frac{R_L}{c}-\frac{2nL}{c}-\frac{L-x}{c}} - \left[D^*\right]\left\{\sum_{n=0}^{\infty} I^n\right\}_i\Bigg|_{t-\frac{R_L^*}{c}-\frac{2nL}{c}-\frac{L-x}{c}} \\
&\quad +\left\{\sum_{n=0}^{\infty} \hat{D}^n\right\}\Bigg|_{t-\frac{R_L^*}{c}-\frac{2nL}{c}-\frac{L-x}{c}} - [B]\left\{\sum_{n=0}^{\infty} E^n\right\}\Bigg|_{t-\frac{L-x'}{c}-\frac{2nL}{c}-\frac{L-x}{c}} - [C]\left\{\sum_{n=0}^{\infty} I^n\right\}_i\Bigg|_{t-\frac{R_0}{c}-\frac{(2n+1)L}{c}-\frac{L-x}{c}} \\
&\quad +\left[C^*\right]\left\{\sum_{n=0}^{\infty} I^n\right\}_i\Bigg|_{t-\frac{R_0^*}{c}-\frac{(2n+1)L}{c}-\frac{L-x}{c}} + [B]\left\{\sum_{n=0}^{\infty} E^n\right\}\Bigg|_{t-\frac{x'}{c}-\frac{(2n+1)L}{c}-\frac{L-x}{c}} - \left\{\sum_{n=0}^{\infty} \hat{C}^n\right\}\Bigg|_{t-\frac{R_0^*}{c}-\frac{(2n+1)L}{c}-\frac{L-x}{c}}
\end{aligned}
\tag{7}
$$

The space-dependent local matrices representing the interaction between i-th source and j-th observation element are defined, as follows:

$$
\begin{aligned}
&[A] = \int_{\Delta l_j}\int_{\Delta l_i} \{f\}_j \{f\}_i^T \frac{1}{4\pi R}\,dx'dx; \quad [B] = \frac{1}{2Z_0}\int_{\Delta l_j}\int_{\Delta l_i} \{f\}_j \{f\}_i^T\,dx'dx \\
&[C] = \int_{\Delta l_j}\int_{\Delta l_i} \{f\}_j \{f\}_i^T \frac{1}{4\pi R_0}\,dx'dx; \quad [D] = \int_{\Delta l_j}\int_{\Delta l_i} \{f\}_j \{f\}_i^T \frac{1}{4\pi R_L}\,dx'dx \\
&\left[A^*\right] = \int_{\Delta l_j}\int_{\Delta l_i} \{f\}_j \{f\}_i^T \frac{r(\theta)}{4\pi R^*}\,dx'dx; \quad \left[C^*\right] = \int_{\Delta l_j}\int_{\Delta l_i} \{f\}_j \{f\}_i^T \frac{r(\theta)}{4\pi R_0^*}\,dx'dx \\
&\left[D^*\right] = \int_{\Delta l_j}\int_{\Delta l_i} \{f\}_j \{f\}_i^T \frac{r(\theta)}{4\pi R_L^*}\,dx'dx;
\end{aligned}
\tag{8}
$$

where $\{f\}$ stands for the shape functions, while additional time dependent vectors are given by:

$$\left\{\hat{A}\right\} = \int_0^{t-\frac{R^*}{c}} \int_{\Delta l_j} \int_{\Delta l_i} \{f\}_j \{f\}_i^T H_1 dx' dx \{I(\tau)\}_i d\tau$$
$$\left\{\hat{C}^n\right\} = \int_0^{t-\frac{R_0^*}{c}-\frac{2nL}{c}-\frac{x}{c}} \int_{\Delta l_j} \int_{\Delta l_i} \{f\}_j \{f\}_i^T H_2 dx' dx \{I(\tau)\}_i d\tau \tag{9}$$
$$\left\{\hat{D}^n\right\} = \int_0^{t-\frac{R_L^*}{c}-\frac{(2n+1)L}{c}-\frac{x}{c}} \int_{\Delta l_j} \int_{\Delta l_i} \{f\}_j \{f\}_i^T H_3 dx' dx \{I(\tau)\}_i d\tau$$

where

$$H_1 = \frac{r''(\theta, t-\frac{R^*}{c}-\tau)}{4\pi R^*}$$
$$H_2 = \frac{r''(\theta, t-\frac{R_0^*}{c}-\frac{2nL}{c}-\frac{x}{c}-\tau)}{4\pi R_0^*} \tag{10}$$
$$H_3 = \frac{r''(\theta, t-\frac{R_L^*}{c}-\frac{(2n+1)L}{c}-\frac{x}{c}-\tau)}{4\pi R_0^*}$$

Assembling the local matrices and vectors into the global ones yields the global matrix system which can be written in the form:

$$\left[A\right]\{I\}\Big|_{t-\frac{R}{c}} = \{g\}\Big|_{\substack{previous\ time\\ instants}} + \{\hat{g}\}\Big|_{\substack{previous\ time\\ instants}} \tag{11}$$

where

$$\begin{aligned}
\{g\} = &\left[A^*\right]\{I\}\Big|_{t-\frac{R^*}{c}} + [B]\{E\}\Big|_{t-\frac{|x-x'|}{c}} \\
&+[C]\left\{\sum_{n=0}^{\infty} I^n\right\}\Big|_{t-\frac{R_0}{c}-\frac{2nL}{c}-\frac{x}{c}} - \left[C^*\right]\left\{\sum_{n=0}^{\infty} I^n\right\}\Big|_{t-\frac{R_0^*}{c}-\frac{2nL}{c}-\frac{x}{c}} - [B]\left\{\sum_{n=0}^{\infty} E^n\right\}\Big|_{t-\frac{x'}{c}-\frac{2nL}{c}-\frac{x}{c}} \\
&-[D]\left\{\sum_{n=0}^{\infty} I^n\right\}\Big|_{t-\frac{R_L}{c}-\frac{(2n+1)L}{c}-\frac{x}{c}} + \left[D^*\right]\left\{\sum_{n=0}^{\infty} I^n\right\}\Big|_{t-\frac{R_L^*}{c}-\frac{(2n+1)L}{c}-\frac{x}{c}} + [B]\left\{\sum_{n=0}^{\infty} E^n\right\}\Big|_{t-\frac{L-x'}{c}-\frac{(2n+1)L}{c}-\frac{x}{c}} \\
&+[D]\left\{\sum_{n=0}^{\infty} I^n\right\}\Big|_{t-\frac{R_L}{c}-\frac{2nL}{c}-\frac{L-x}{c}} - \left[D^*\right]\left\{\sum_{n=0}^{\infty} I^n\right\}\Big|_{t-\frac{R_L^*}{c}-\frac{2nL}{c}-\frac{L-x}{c}} - [B]\left\{\sum_{n=0}^{\infty} E^n\right\}\Big|_{t-\frac{L-x'}{c}-\frac{2nL}{c}-\frac{L-x}{c}} \\
&-[C]\left\{\sum_{n=0}^{\infty} I^n\right\}\Big|_{t-\frac{R_0}{c}-\frac{(2n+1)L}{c}-\frac{L-x}{c}} + \left[C^*\right]\left\{\sum_{n=0}^{\infty} I^n\right\}\Big|_{t-\frac{R_0^*}{c}-\frac{(2n+1)L}{c}-\frac{L-x}{c}} + [B]\left\{\sum_{n=0}^{\infty} E^n\right\}\Big|_{t-\frac{x'}{c}-\frac{(2n+1)L}{c}-\frac{L-x}{c}}
\end{aligned} \tag{12}$$

and

$$\begin{aligned}
\{\hat{g}\} = &\left\{\hat{A}\right\}\Big|_{t-\frac{R^*}{c}} - \left\{\sum_{n=0}^{\infty} \hat{C}^n\right\}\Big|_{t-\frac{R_0^*}{c}-\frac{2nL}{c}-\frac{x}{c}} + \left\{\sum_{n=0}^{\infty} \hat{D}^n\right\}\Big|_{t-\frac{R_L^*}{c}-\frac{(2n+1)L}{c}-\frac{x}{c}} \\
&+\left\{\sum_{n=0}^{\infty} \hat{D}^n\right\}\Big|_{t-\frac{R_L^*}{c}-\frac{2nL}{c}-\frac{L-x}{c}} - \left\{\sum_{n=0}^{\infty} \hat{C}^n\right\}\Big|_{t-\frac{R_0^*}{c}-\frac{(2n+1)L}{c}-\frac{L-x}{c}}
\end{aligned} \tag{13}$$

www.witpress.com, ISSN 1743-355X (on-line)

Applying the weighted residual approach in the time domain, and using the Dirac impulses as weight functions provides the time sampling, and the following recurrent formula is obtained:

$$I_j\Big|_{t_k} = \frac{\sum_{i=1}^{N^s} a_{ji} I_j\Big|_{t_k - \frac{R}{c}} - g_j\Big|_{\substack{previous\ time\\ instants}} - \hat{g}_j\Big|_{\substack{previous\ time\\ instants}}}{a_{jj}} \tag{14}$$

where $I_j\Big|_{t_k}$ is current for the j-th space node at k-th time instant, N is total number of space segments, while the overbar indicates the absence of diagonal members.

It is worth noting that the numerical calculation of convolution integrals is rather tedious task leading to tremendously large computational time of the overall method. The main advantage of the method, on the other hand, is its unconditional stability.

2.2 Numerical results for an overhead wire

Computational example is related to a transient scattering from a straight thin wire of length is L=1m, radius a=2mm, located at height h=0.25m above ground with permittivity ε_r=10, while the conductivity is varied. The wire is illuminated by the tangential electromagnetic pulse (EMP) plane wave:

$$E_x^{inc}(t) = E_0(e^{-at} - e^{-bt}), \quad t \geq 0 \tag{15}$$

with: E_0=1 V/m, $a=4\cdot10^7$ s^{-1}, $b= 6\cdot10^8$ s^{-1}.

Fig. 2 shows the transient current induced at the wire center for different ground conductivities.

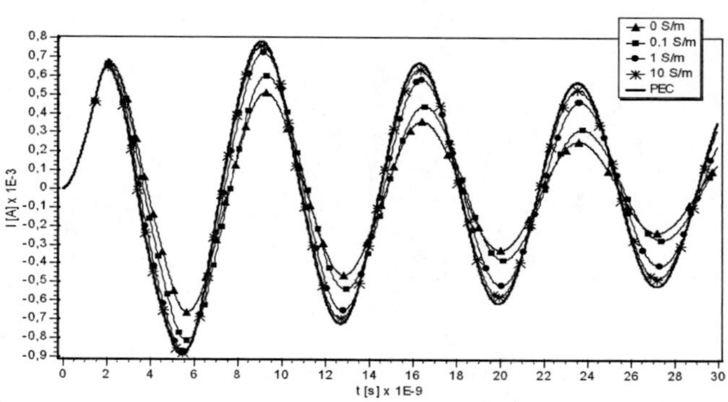

Figure 2: Transient current at the wire center, L=1m, a =2mm, h=0.25, ε_r=10.

The influence of the ground conductivity to the transient response is particularly visible from around 0.1 S/m to 1 S/m.

3 Pocklington integral equation formulation

The geometry of interest, shown in Fig. 3, is the horizontal grounding electrode of length L and radius a, buried in a lossy medium at depth d and excited at one end by an equivalent current source.

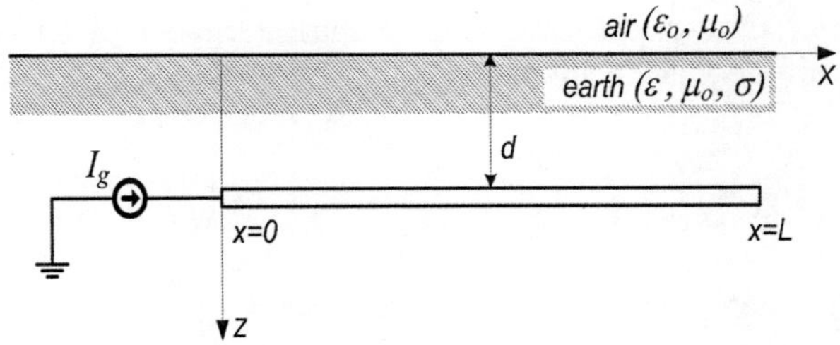

Figure 3: Horizontal grounding wire energized by a current generator Ig.

Since the electric field excitation along the electrode does not exist i.e.:

$$E_x^{exc} = 0 \tag{16}$$

assuming the certain set of approximation [7] the transient current induced along the electrode is governed by the following form of the Pocklington equation:

$$\left[-v^2\frac{\partial^2}{\partial x^2}+\frac{\partial^2}{\partial t^2}+\frac{\sigma}{\varepsilon}\frac{\partial}{\partial t}\right]\cdot \left[\frac{\mu}{4\pi}\int_0^L I(x',t-R/v)\frac{e^{-\frac{t}{\tau_g}\frac{R}{v}}}{R}dx' - \int_{-\infty}^{t}\int_0^L \Gamma_{ref}(\theta,\tau)\frac{I(x',t-R^*/v-\tau)}{4\pi R^*}\frac{e^{-\frac{t}{\tau_g}\frac{R^*}{v}}}{R^*}dx'd\tau\right]=0 \tag{17}$$

where the reflection coefficient is given by [7]:

$$\Gamma_{ref}(t) = -\left[\frac{\tau_1}{\tau_2}\delta(t)+\frac{1}{\tau_2}\left(1-\frac{\tau_1}{\tau_2}\right)e^{-t/\tau_2}\right] \tag{18}$$

where τ_1 and τ_2 are the time constants characterized by a lossy medium [7]:

$$\tau_1 = \frac{\varepsilon_r - 1}{\sigma}\varepsilon_0,\quad \tau_2 = \frac{\varepsilon_r + 1}{\sigma}\varepsilon_0 \tag{19}$$

Note that the current source is included into the integral equation scheme trough the boundary condition:

$$I(0) = I_g \tag{20}$$

which is inserted subsequently in the global matrix system [2].

3.1 BEM procedure for Pocklington equation

It should be emphasized that, for the sake of simplicity, this paper considers only the case of an infinite lossy medium. Applying the weighted residual approach and performing space-discretization i.e. in matrix form one obtains the following time domain differential equation:

$$[M]\frac{\partial^2}{\partial t^2}\{I(t')\} + [C]\frac{\partial}{\partial t}\{I(t')\} + [K]\{I(t')\} = 0 \tag{21}$$

where the space dependent matrices are given by:

$$M_{ji} = \frac{1}{v^2}\int_{\Delta l_j}\int_{\Delta l_i}\{f\}_j\{f\}_i^T \frac{e^{-\frac{T}{\tau}}}{R}dx'dx \tag{22}$$

$$C_{ji} = \frac{\sigma}{\varepsilon}\int_{\Delta l_j}\int_{\Delta l_i}\{f\}_j\{f\}_i^T \frac{e^{-\frac{T}{\tau}}}{R}dx'dx \tag{23}$$

$$K_{ji} = \frac{\mu}{4\pi}\int_{\Delta l_j}\int_{\Delta l_i}\{D\}_j\{D\}_i^T \frac{e^{-\frac{T}{\tau}}}{R}dx'dx \tag{24}$$

where {D} stands for the shape functions derivatives and: $\tau = \frac{2\varepsilon}{\sigma}$ and $T = \frac{R}{v}$. Finally, differential equation (21) is solved performing the marching-on-in-time procedure presented in [5]:

$$\sum_{i=1}^{n}\left[M_{ji} + \beta\Delta t^2 K_{ji}\right]I_i^k = -\sum_{i=1}^{n}\left[-2M_{ji} + \left(\frac{1}{2} - 2\beta + \gamma\right)\Delta t^2 K_{ji}\right]I_i^{k-1} - \sum_{n=1}^{n}\left[M_{ji} + \left(\frac{1}{2} + \beta - \gamma\right)\Delta t^2 K_{ji}\right]I_i^{k-2} \tag{25}$$

where Δt stands for the time increment and the stability of the procedure is achieved by choosing [5] γ=1/2 and β=1/4.

3.2 Numerical results for grounding electrode

Computational example is related to the grounding electrode of length L=10m, radius a=5mm, immersed in the lossy ground with ε_r=10, and σ=0.001S/m. The electrode is excited with the double exponential current pulse:

$$i(t) = I_0 \cdot (e^{-at} - e^{-bt}), \quad t \geq 0$$

defined with I_0=1.1043A, a=0.07924·$10^7$$s^{-1}$, b=4.0011·$10^7$$s^{-1}$.

The transient current induced at the centre of the electrode obtained via the presented direct time domain approach and the indirect frequency domain approach GB-IBEM with Fast Fourier Transform (FFT) is shown in Fig. 4.

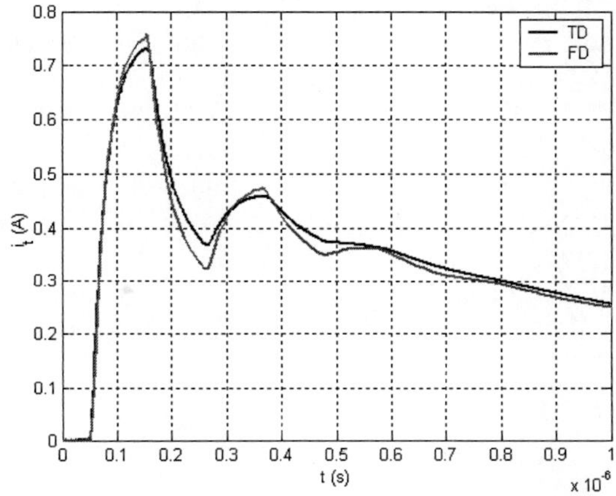

Figure 4: Transient current induced at the centre of the wire.

An acceptable agreement between the results computed via different approaches can be observed.

4 Concluding remarks

The paper deals with the transient analysis of thin wire structures in the presence of a lossy half-space, based on the time domain Hallen and Pocklington integral equation, respectively. The finite conductivity of the ground is taken into account via the corresponding reflection coefficients. The both types of integral equations are handled via the certain scheme of Galerkin-Bubnov Indirect Boundary Element Method (GB-IBEM) The strength and weaknesses of both approaches are emphasized and some illustrative examples related to overhead wires and grounding electrodes are presented within this work.

References

[1] F. M. Tesche, M. Ianoz, and T. Karlsson, EMC Analysis methods and computational models. New York: Wiley Interscience, 1997.
[2] D. Poljak, EMC Advanced Modelling in Computational electromagnetic Compatibility, John Wiley and Sons, New York 2007.

[3] E.K. Miller, J.A. Landt: "Direct time-domain techniques for transient radiation and scattering from wires", IEEE Trans.AP, Vol. 68, No.11, pp. 1396-1424, Nov. 1980.
[4] D. Poljak, C.Y. Tham, A. McCowen: "Transient Response of Nonlinearly Loaded Wires in a Two Media Configuration", IEEE Trans.EMC, Vol. 46, No.1, pp. 121-125, 2004.
[5] V. Doric, D. Poljak, V. Roje, Direct Time Domain Analysis of a Lightning Rod Based on the Antenna Theory, EMC Symposium, 2009.
[6] D. Poljak, Transient Response of Straight Thin Wires Located at Different Heights Above a Ground Plane Using Antenna Theory and Transmission Line Approach, IEEE Trans.EMC, Vol.52, No.1, pp. 108-1116, 2010.
[7] D. Poljak, EABE 2008, Time domain modeling of a thin wire in a two-media configuration featuring a simplified reflection/transmission coefficient approach, Engineering Analysis with Boundary Elements, EABE 33, pp. 283–293., 2009.
[8] S. Antonijevic, D. Poljak, On Time Domain Numerical Modeling of a Thin Wire Above a Lossy Ground, SoftCOM 2010, Split, Brac Island, Croatia, Sept. 2010.

Stability issues in 3D BEM formulations for transient elastodynamics

C. G. Panagiotopoulos[1] & G. D. Manolis[2]
[1]Department of Continuum Mechanics, University of Seville, Spain
[2]Department of Civil Engineering,
Aristotle University of Thessaloniki, Greece

Abstract

Time-domain boundary element method formulations (TD-BEM) are quite versatile in reproducing the transient response of finite and semi-infinite solid media and offer a number of advantages compared to transformed domain (Fourier or Laplace) approaches. They are known, however, to be prone to numerical instability, especially after a very large number of time steps. This seems to be more of a problem with 1D as compared to 3D formulations, although it is present in the latter ones. A careful investigation of the problem reveals that if the conventional BEM formulation using displacement and traction variables is replaced with one using velocity and traction pairs, the problem is ameliorated to a large extent and much better accuracy results.
Keywords: elastodynamics, stability, time-marching schemes, reciprocal theorems, convolution integrals.

1 Introduction

Although the importance of TD-BEM formulations in studying wave motion problems in elastic continua is indisputable, the conventional formulation in terms of the displacement and traction vectors (DBEM) has a latent problem when a large number of time steps are required. More specifically, unstable behaviour may occur that is known as ‘‘intermittent instability’’ [1]. Various schemes have been proposed in recent years, many of them drawn from finite element methodologies (such as the Wilson-theta integration algorithm) to eliminating or reduce this unstable behaviour of the time convolution integrals associated with the transient BEM. All such methods basically seek to modify

WIT Transactions on Modelling and Simulation, Vol 52,
www.witpress.com, ISSN 1743-355X (on-line)
doi:10.2495/BE110131

the underlying boundary integral equations so as to optimize the stability behaviour. Although stability may be enforced, it is usually accompanied by a reduction in accuracy, which is the consequence of the introduction of artificial damping and of natural period elongation [2]. Moreover, these algorithmic techniques are somewhat cumbersome, operating at different levels of complexity, and difficult to implement in existing BEM computer software codes. An alternative approach, as outlined in a recently by the present authors [3], is to introduce a new formulation based on the velocity reciprocal theorem (VBEM). This formulation has been successfully tested on simple problems, namely the single degree of freedom system and one dimensional wave propagation, with superior performance regarding stability as compared to formulations using the displacement reciprocal theorem. In this work, the fully 3D formulation is implemented following work outlined in ref. [4] and tested against benchmark problems. Again, overall superior performance of the VBEM compared to the DBEM is again established.

2 Velocity reciprocal theorem for 3D elastodynamics

Time domain BEM formulations use Graffi's reciprocal theorem [5], which employs displacement and traction fields and essentially is a work statement defined for a regular region V with boundary S (see Fig. 1) and material properties p, c_1, c_2 (mass density, pressure and shear wave speeds). In this work, we adopt an alternative strategy using a power statement that employs velocities in lieu of displacements. Specifically, consider two distinct elastodynamic states $[u_i, t_i, b_i]$ and $[u_i', t_i', b_i']$ comprising displacements, tractions and body forces in V with appropriate initial conditions. Next, define the velocity as the time derivative of displacement. For time $t \geq 0$ and by assuming (for simplicity) zero initial conditions, the following reciprocal theorem holds true

$$\int_S t_i * v'_i dS + \int_V \rho b_i * v'_i \, dV = \int_S t'_i * v_i dS + \int_V \rho b'_i * v_i dV \tag{1}$$

where operation * denotes Riemann's convolution. Utilizing this theorem with one state identified with the solution sought and the other with the fundamental singular solution pair, a velocity-based integral identity is:

$$c_{ij}(\xi) v_i(\xi,t) = \int_S \left(\dot{G}_{ij} * t_i(x) - F_{ij} * v_i(x) \right) dS(x) + p \int_V \dot{G}_{ij} * b_i(x) dV(x) \tag{2}$$

The fundamental velocity solution $\dot{G}_{ij}(x,t;\, \xi, \tau = 0)$ is computed by noting that

$$\int_{c_1^{-1}}^{c_2^{-1}} \delta(t - \lambda r)\lambda d\lambda = \frac{t}{r^2}\left[H(t - \frac{r}{c_1}) - H(t - \frac{r}{c_2}) \right] \tag{3}$$

and is given as:

$$\dot{G}_{ij} = (4\pi\rho)^{-1}\left\{a_{ij}(r)\dot{\delta}(t-\frac{r}{c_1})+b_{ij}(r)\dot{\delta}(t-\frac{r}{c_2})\right.$$
$$\left.+c_{ij}(r)\frac{t}{r^2}\left[\delta(t-\frac{r}{c_1})-\delta(t-\frac{r}{c_2})\right]+c_{ij}(r)\frac{1}{r^2}\left[H(t-\frac{r}{c_1})-H(t-\frac{r}{c_2})\right]\right\} \quad (4)$$

where spatial coefficients a_{ij}, b_{ij}, c_{ij} are the same as those appearing the conventional displacement BEM formulation [6] (i.e., the DBEM). This new formulation can easily be extended to incompressible materials [7], and becomes independent of the dilatational wave speed.

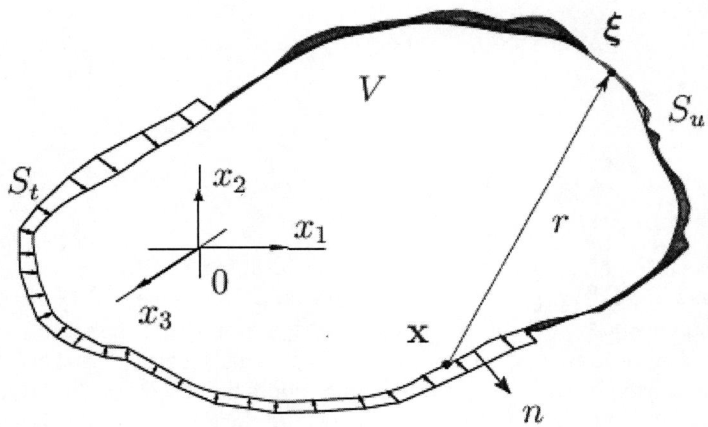

Figure 1: Elastic solid body and BEM notation.

3 BEM formulations for displacements and for velocities

In order to proceed with formulating a BEM algorithm for elastodynamics, we need to establish spatial as well as temporal discretizations of the field variables. First, to integrate the time convolutions analytically, we proceed by dividing the time interval *[0,t]* into *N* equal time steps of length *h* and approximating any of the tractions, displacements (DBEM) or velocities (VBEM) in terms of basis functions. To study accuracy and stability, we consider a linear evolution of any time variable f (t) across a time-step as

$$f(t) = \sum_{n=1}^{N}\phi_n(t)\left(\frac{t_n - t}{h}f_{n-1} + \frac{t - t_{n-1}}{h}f_n\right) \quad (5)$$

where

$$\phi_n(t) = H(t - t_{n-1}) - H(t - t_n) \quad (6)$$

and $t_{n-} = nh$, *H(t)* is the Heaviside function while f_{n-1}, f_n are values at discrete times t_{n-1} ,t_n, respectively.

Regarding spatial discretization, a family of 2D surface isoparametric elements is used, namely the four, eight and nine node quadrilaterals (see Fig. 2). Spatial variation of any field variable f (x) is expressed in terms of the same shape functions M_a used for the geometry, i.e.,

$$f(x)=\sum_{a=1}^{q}M_a(x)f_a \tag{7}$$

In the above, q is the number of element's nodes, while f_a is a nodal value in space. After replacement in the BEM statement, space-time interpolation products arise, which in compact form read as follows:

$$f(x,t)=\sum_{n=1}^{N}\sum_{a=1}^{q}N_{1,n}(t)M_a(x)f_{a,n-1}+N_{2,n}(t)M_a(x)f_{a,n} \tag{8}$$

In the above,

$$N_{1,n(t)}=\phi_n(t)\frac{t_n-t}{h}\ ,\ N_{2,n(t)}=\phi_n(t)\frac{t-t_{n-1}}{h} \tag{9}$$

Analytical spatial integrations of the resulting BEM system matrix coefficients are not in general possible and therefore numerical quadrature is used. Fundamental solution-shape function products are approximated by application of the Gauss-Legendre quadrature formula. There are two basic cases of integrand behaviour, namely non-singular and singular. Regarding the latter, weak singularities in both DBEM and VBEM are done by element subdivisions,

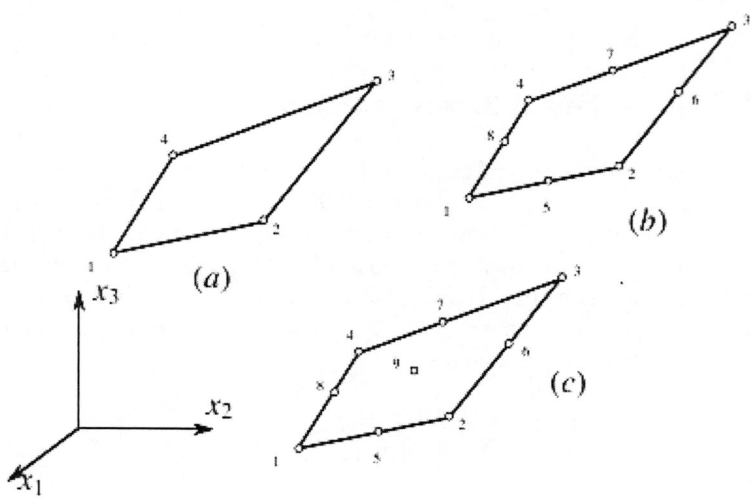

Figure 2: Surface boundary elements: (a) four, (b) eight and (c) nine nodes.

followed by a series of transformations that reduces the order of singularity in the kernels by one and transforms them to regular ones. In the case of strongly singular traction kernels, the diagonal sub-matrices are computed utilizing indirect schemes based on the rigid body motion concept [8].

An important point for accuracy reasons in such formulations is identification of active zones of the integration along an element as it is sweep by the P and S waves. This is accomplished by invoking the causality principle [9] in the numerical integration scheme and employing element subdivision.

4 Stability analysis and BEM performance

We introduce a formal definition of stability as "Stability means that any 'initial' conditions at time t given by errors in the response, which may be due to round-off in the computer, do not grow in the integration". According to this definition [10] we distinguish between accuracy of a numerical method and its stability and focus on the latter. As example, we select a solid cube (see Fig. 3) with side length equal to L_{tot} . We discretize each side into ne × ne square boundary elements of length $L = L_{tot}/ne$. Next, we introduce the dimensionless Courant-Fredricks-Lewy (CFL) parameter as $\beta = cdt/L_W$, where c is either the P or S wave speed and L_W a representative boundary element length, such as the minimum distance between two nodes. We adopt the S wave speed in the CFL measure, since it has the advantage of being applicable to the incompressible material case. Then we assign fixed kinematic (displacements and velocities) boundary conditions on the negative x_1 face of the cube, zero tractions boundary conditions on the remaining faces and introduce unit initial 'error' for all boundary nodal values so as to trace its time evolution.

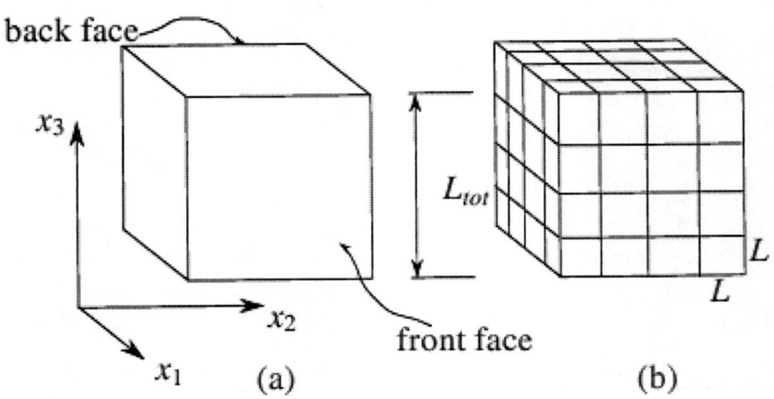

Figure 3: (a) Geometry of solid cube used to study stability (b) BEM discretization.

4.1 Stability aspects of the DBEM formulation

The DBEM stability regions in terms on CFL parameter β_2 can be broken down in two parts: For $\beta_2 \leq 0.6$, the time-stepping algorithm shows instability that becomes more noticeable for diminishing β_2. For values of $\beta_2 > 0.6$, we observe a stable performance since the initial 'error' is damped as time evolves. Meanwhile, if the 'error' evolution is tracked over for a long time, *sudden instability* occurs as shown in Fig. 4. This has been reported elsewhere [11], and if we further increase β_2, we find a small stability region between $1.1 \leq \beta_2 \leq 1.2$, since for a large number of time steps the initial 'error' decreases. For values

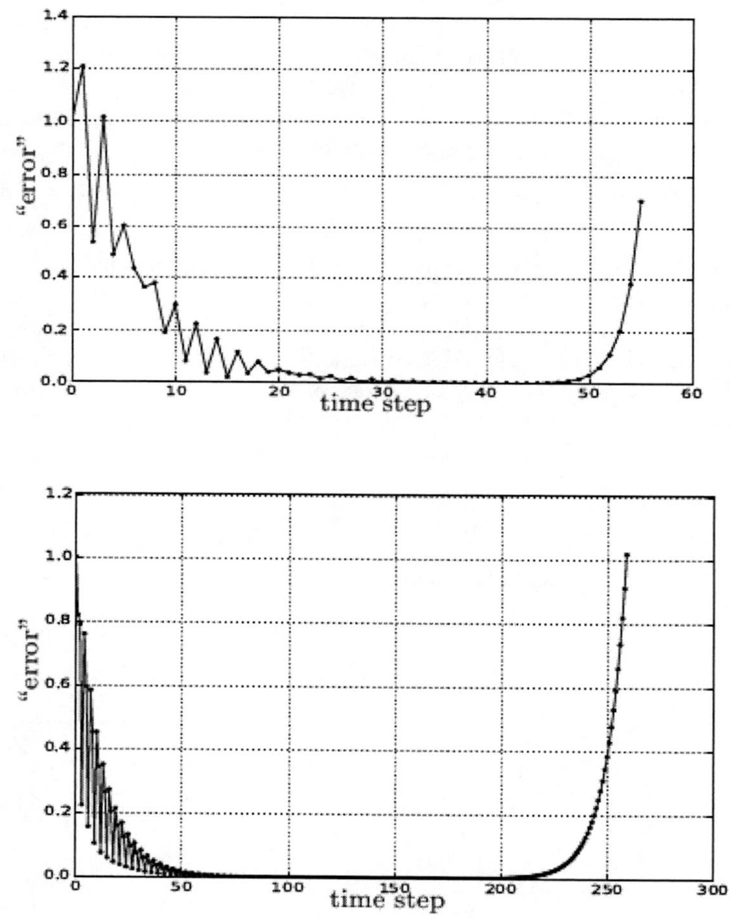

Figure 4: Time evolution of DBEM displacement 'error' (a) $\beta_2 = 0.8$ (b) $\beta_2 = 0.9$.

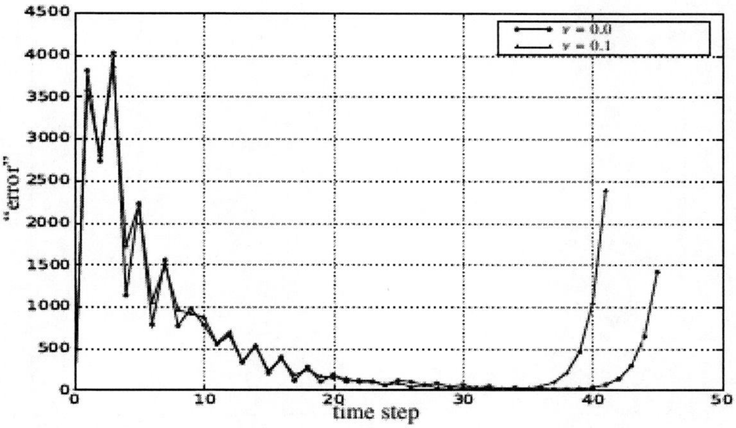

Figure 5: Time evolution of BDEM traction 'error' $\beta_2 = 0.75$ (a) $\nu = 0$ (b) $\nu = 0.1$.

greater than $\beta_2 > 1.2$, instability occurs once again. This kind of behavior, which is referred to in Ref. [12] as *intermittent instability*, is unacceptable as one cannot provide coherent guidelines on an appropriate choice of the time step.

Also, for Poisson's ratio values greater than zero, the DBEM present pronounced instability problems. This is observed in Fig. 5, where the 'error' evolution for tractions it is shown at a fixed β_2 value, but for two distinct values of Poisson's ratio ν. For non-zero ν, instability is manifested even for a small number of time steps. Furthermore, for large values of ν it was impossible to define any stability region regarding the β_2 parameter, in terms of long time evolution, since *sudden instability* occurred.

4.2 Stability aspects of the VBEM formulation

The VBEM stability performance is shown here to be smoother and more predictable. Most important, there exists a lower limit for β_2 beyond which the time-stepping algorithm is found to be stable. Specifically, for Poisson's ratio range $0 \leq \nu \leq 0.5$, the stability region was identified as $0.7 \leq \beta_2 \leq 0.9$. These types of simulations are shown in Fig. 6. Corresponding values for the β_1 parameter may easily be retrieved, and we note that for large values of ν, correspondingly very large values of β_1 result which produce substantial numerical damping. This unwelcome effect, however, may be reduced using with finer boundary element meshes.

In sum, the superior VBEM performance is probably due to the structure of the velocity fundamental solution, and specifically to the presence of the Dirac delta function that has a more 'local' space-time effect. In contrast, the displacement fundamental solution of the BDEM involves use of the Heaviside function.

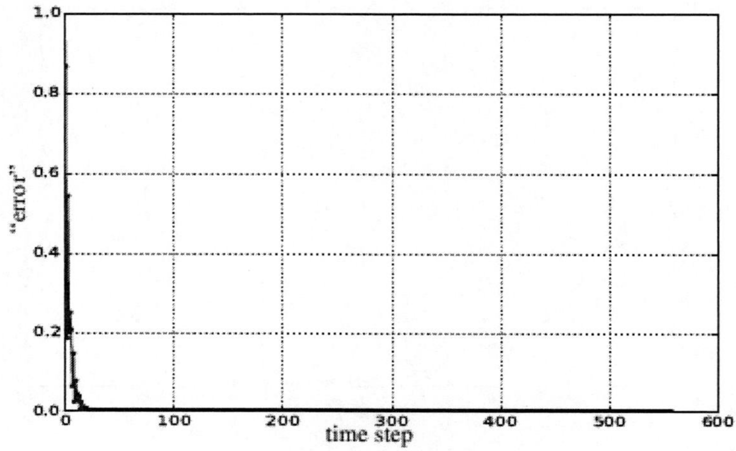

(a)

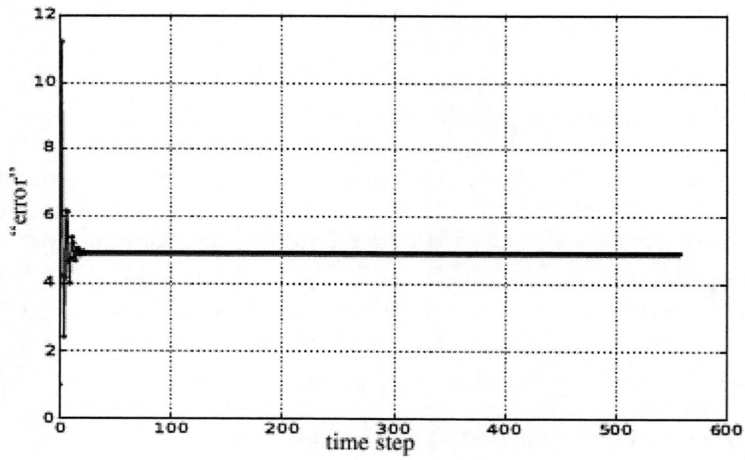

(b)

Figure 6: Evolution of VBEM 'error' for $\nu = 0.0$ and $\beta_2 = 0.8$ (a) velocities (b) tractions.

5 Conclusions

A TD-BEM formulation for 3D elastodynamic problems has been proposed and tested. This new formulation is established in terms of velocities and tractions instead of the conventional one, and is based on a power-type reciprocal theorem. This formulation is found to have a better performance regarding stability and its incorporation into existing BEM codes is simple. It only requires the replacement of the displacement fundamental solution by its respective velocity solution as the kernel function. Furthermore, temporal integration of this new kernel is similar to that of the traction kernel, while spatial integrations are identical to those of the displacement kernel. This way, we avoid here "averaging" methods for decreasing instability problems that have hitherto been used, especially in conjunction with 1D formulations. It is also interesting to note that the 2D elastodynamic case can also be formulated in terms of velocity-based quantities and better stability performance is to be expected in this case.

References

[1] Pierce, A. & Siebrits, E., Stability analysis of model problems for elastodynamic boundary element discretizations. *Numerical Methods for Partial Differential Equations*, **12**, pp. 585-613, 1996.

[2] Soares, D. & Mansur, W.J., An efficient stabilized boundary elements formulation for 2D time-domain acoustics and elastodynamics. *Computational Mechanics*, **40**, pp. 355-365, 2007.

[3] Panagiotopoulos, C.G. & Manolis, G.D., Velocity-based reciprocal theorems in elastodynamics and BIEM implementation issues," *Archives of Applied Mechanics,* 80(12), pp. 1429-1447, 2010.

[4] Panagiotopoulos, C.G. & Manolis, G.D., Three-dimensional BEM for Transient Elastodynamics based on Velocity Reciprocal Theorem, *Engineering Analysis with Boundary Elements*, accepted, to appear in 2011.

[5] Graffi, D., Sul teorema di reciprocita nella dinamica dei corpi elastici. *Memoria della Reale Accademia delle Scienze dell'Istituto di Bologna*, **10**, pp. 103-111, 1946.

[6] Dominguez, J., *Boundary Elements in Dynamics*, Computational Mechanics Publications & Elsevier Applied Science, London, 1993.

[7] Polyzos, D., Tsinopoulos, S.V. & Beskos, D.E., Static and dynamic boundary element analysis in incompressible linear elasticity. *European Journal of Mechanics, A/Solids*, **17**, pp. 515-536, 1998.

[8] Banerjee, P.K, Ahmad, S. & Manolis, G.D, Transient elastodynamic analysis of 3-D problems by Boundary Element Method, *Earthquake Engineering and Structural Dynamics*, **14**, pp. 933-949, 1986.

[9] Frangi, A., Causal shape functions in the time domain boundary element method, *Computational Mechanics* **25**, pp. 533-541, 2000.

[10] Frangi, A. & Novati, G., On the numerical stability of time-domain elastodynamic analyses by BEM, *Computer Methods in Applied mechanics and Engineering*, **173**, pp. 403-417, 1999.

[11] Schanz, M., Ruberg, T. & Kielhorn, L., Time Domain BEM: Numerical aspects of collocation and Galerkin formulations. *Recent Advances in Boundary Element Methods, A Volume to Honor Professor Dimitri Beskos,* editors: G.D. Manolis & D. Polyzos, Springer-Verlag, Dordrecht, pp. 415-432, 2009.

[12] Pierce, A. & Siebrits, E., Stability analysis and design of time-stepping schemes for general elastodynamic boundary element models, *International Journal for Numerical Methods in Engineering*, **40**, pp. 319-342, 1997.

A multi-domain boundary element analysis technique based on a row elimination-back-substitution method for solving large-scale engineering problems

X.-W. Gao & J.-X. Hu
State Key Laboratory of Structural Analysis for Industrial Equipment, Dalian University of Technology, P.R. China

Abstract

This paper presents a novel analysis technique using the multi-domain boundary element method (MDBEM) to solve large-scale engineering problems. Firstly, boundary integral equations for solving general heat conduction and mechanics problems are presented, which are established for a single medium and are formulated in terms of physical quantities at internal, boundary and interface points. Then a sparse system of equations formulated in terms of only interface nodal quantities is assembled based on the three-step variable condensing technique. Finally, a robust linear equation solution method is presented for solving the sparse system based on a row elimination-back-substitution method (REBSM). Combining REBSM and MDBEM makes the boundary element method more efficient for solving large practical engineering problems. A numerical example is given to demonstrate the efficiency of the proposed method.
Keywords: multi-domain boundary element method, Gaussian elimination method, row elimination-back-substitution method, sparse system of equations.

1 Introduction

The boundary element method (BEM) is another extensively used numerical tool in solving engineering problems after the development of the finite element method (FEM). Apart from usual advantages mentioned in references (e.g., [1]), a few important advantages of BEM over FEM can be figured out as: 1) only

WIT Transactions on Modelling and Simulation, Vol 52,
www.witpress.com, ISSN 1743-355X (on-line)
doi:10.2495/BE110141

boundary of the problem needs to be discretized into elements and therefore less labor is required for preparing input data and is easy for modeling complicated problems; 2) it is efficient in solving thin-walled problems after the nearly singular integrals are accurately evaluated [2, 3]; and 3) the gradient of the basic physical quantity has the same accuracy as the physical quantity itself, since its computational formulation can be analytically derived from the basic integral equations. However, BEM has an inherent disadvantage that the formed coefficient matrices are fully-populated and non-symmetric, which limit the scale and speed of solving engineering problems.

In order to solve large-scale problems, researchers developed the multi-domain boundary element method (MDBEM) [1, 4, 5]. In MDBEM, the computational domain of interesting is divided into a number of sub-domains; the BEM algebraic equations are established for each sub-domain; and the global system of equations is formed by assembling results of all sub-domains in terms of the equilibrium and consistence conditions over common interface nodes. The coefficient matrix of the global system of equations based on MDBEM is sparse, and therefore the well-developed solvers for sparse systems can be employed to solve it. The use of MDBEM not only can improve the efficiency both in problem scale and computational speed, but also can solve fracture problems by dividing sub-domains along crack surfaces [6] and multi-media problems by dividing sub-domains along interfaces [7].

In MDBEM, the assembling skill of the system of equations directly affects the computational efficiency. So far, a number of assembling techniques have been proposed [1, 4, 5]. The simple one is to put all unknowns at outer boundary nodes, and displacements and tractions at interface nodes as the unknowns of the system [1]. Such assembling technique is easy for coding, but it makes the size of the system of equations huge, limiting the capability of solving large problems. The efficient assembling technique is the variable condensing methods [4, 5], in which some variables are eliminated first and only a part of variables are served as the final system unknowns. Among the variable condensing methods, the three-step variable condensing technique [5] is very efficient, in which unknowns at internal and outer boundary nodes as well as tractions (or fluxes) at interfaces are eliminated in turn and only displacements (or potentials) at interface nodes are assembled as unknowns of the system. This technique can result in a smallest system of equations and the formed coefficient matrix has a higher sparsity, suitable for solving large-scale problems.

Although the system coefficient matrix of MDBEM is sparse, it is not symmetric. Therefore, the existing powerful equation solvers developed for FEM can not be borrowed to solve the MDBEM systems. New powerful solvers for sparse non-symmetric systems need to be developed. Usually, there are two types of numerical solution methods for linear systems of equations: direct and iterative methods. In direct methods, such as Gaussian elimination, Gauss-Jordan elimination, and LU-factorization methods, the answer can be obtained in a predictable number of operations [8]. In iterative methods, such as the Jacobi method, Conjugate gradients, and GMRES, many steps are necessary in attempting to converge to the desired answer [9]. To keep computational efficiency, the existing direct methods need to store whole coefficient matrix in core and therefore they are not suitable for solving large problems. As iterative methods are operated based on matrix-vector products, large systems can be solved. However, in iterative methods, since the approximation solutions are modified at each iterative step to approach the true answer, a convergent solution is not guaranteed for all systems of equations.

To overcome the deficiency of the existing direct methods in the requirement of large storage, a novel direct method is presented in the paper based on a row elimination-back-substitution method (REBSM). In this method, both elimination and back-substitution procedures are completed in the same row under consideration, and therefore, no later back-substitution procedure is required. Compared to the existing direct methods, such as the Gaussian elimination, the presented REBSM requires less data storage, so it can be used to solve larger system of equations. Also, since REBSM can be applied to systems of non-symmetric matrices, it is adopted in this paper to solve the MDBEM system of equations.

2 Basic boundary integral equations in heat conduction and solid mechanics

In this paper, the heat conduction and solid mechanics problems are served as the research backgrounds. However, the results can be extended to other problems.

2.1 General boundary-domain integral equations for heat conduction

The control equation for general heat conduction problems can be expressed as

$$\frac{\partial}{\partial x_i}\left(k_{\mathrm{ij}}\frac{\partial u}{\partial x_j}\right)+Q=0 \tag{1}$$

where u is the temperature, k_{ij} and Q are heat conductivities and source, respectively. k_{ij} may be the function of coordinates x_i and temperature u in non-homogeneous and non-linear problems. The general boundary-domain integral equation for eqn (1) can be derived from the source isolation method [10] as

$$ku+\int_{\Gamma} q^{*}ud\Gamma=-\int_{\Gamma} Gqd\Gamma+\int_{\Omega} Vud\Omega+\int_{\Omega} GQd\Omega \tag{2}$$

where Γ represents the boundary of the computational domain Ω, G is the Green's function, and q is the heat flux.

$$G=\begin{cases}\dfrac{1}{2\pi}\ln\left(\dfrac{1}{r}\right) & \text{for } 2D\ (\beta=2)\\[2ex] \dfrac{1}{4\pi r} & \text{for } 3D\ (\beta=3)\end{cases} \tag{3}$$

$$q = -k_{ij} \frac{\partial u}{\partial x_j} n_i$$
$$q^* = \frac{\partial G}{\partial x_i} k_{ij} n_j \tag{4a}$$

$$k = \sum_{i=1}^{\beta} k_{ii} / \beta \tag{4b}$$

$$V = \frac{\partial}{\partial x_i}\left(k_{ij} \frac{\partial G}{\partial x_j} \right) = \frac{\partial k_{ij}}{\partial x_i} \frac{\partial G}{\partial x_j} + k_{ij} \frac{\partial^2 G}{\partial x_i \partial x_j} \tag{4c}$$

From eqn (4b) it can be seen that the coefficient k in eqn (2) is the average value of the diagonal terms of k_{ij}. For isotropic problems with constant material parameters, k is reduced to the usual conductivity, while $V=0$.

2.2 General boundary-domain integral equations for mechanics

The equilibrium equation for solid mechanics can be expressed as

$$\sigma_{ij,j} + b_i = 0 \tag{5}$$

where b_i is the body force. The relationship between the stress σ_{ij}, strain ε_{ij} and displacement u_i is [11]

$$\sigma_{ij} = D_{ijkl}\varepsilon_{kl} = D_{ijkl}u_{k,l} \tag{6}$$

in which D_{ijkl} is the stress-strain constitutive tensor which is symmetric about subscripts, i.e., $D_{ijkl} = D_{jilk} = D_{klij}$. For non-homogeneous materials or nonlinear problems, D_{ijkl} may be the functions of coordinates or stresses. From the source isolation method [10], the boundary-domain integral equations for eqn (6) can be derived as:

$$\mu_{ij}u_j + \int_\Gamma T_{ij} u_j d\Gamma = \int_\Gamma U_{ij} t_j d\Gamma + \int_\Omega V_{ij} u_j d\Omega + \int_\Omega U_{ij} b_j d\Omega \tag{7}$$

where U_{ij} is the Kelvin displacement fundamental solutions, t_i is the traction:

$$U_{ij} = \begin{cases} \dfrac{1}{8\alpha\pi(1-\nu)}\{-(3-4\nu)\ln(r)\delta_{ij} + r_{,i} r_{,j}\} & for\ 2D\,(\alpha=1) \\ \dfrac{1}{8\alpha\pi(1-\nu)r}\{(3-4\nu)\delta_{ij} + r_{,i} r_{,j}\} & for\ 3D\,(\alpha=2) \end{cases} \tag{8}$$

$$T_{ij} = U_{ik,l} D_{kljs} n_s \tag{9a}$$

$$\mathrm{V}_{\mathrm{ij}} = (U_{ik,l} D_{kljs})_{,s} = U_{ik,l} D_{kljs,s} + U_{ik,ls} D_{kljs} \tag{9b}$$

$$\mu_{ij} = \frac{1}{2\beta\chi}\left[2(\chi - 1)D_{isjs} - D_{ijss}\right] \tag{10}$$

in which n_s is the outward normal, $\chi = (\beta + 2)(1 - \nu)$ with ν being the Poison's ratio. From eqn (10) it can be seen that the coefficient μ_{ij} in eqn (7) is symmetric, that is $\mu_{ij} = \mu_{ji}$. For isotropic elasticity problems with constant material parameters, $\mu_{ij} = \mu\delta_{\mathrm{ij}}$ with μ being the shear modulus, and $\mathrm{V}_{\mathrm{ij}} = 0$.

It is noted that, in coding, all domain integrals appearing in eqns (2) and (7) are transformed into boundary integrals using the radial integration method (RIM) [12], resulting in a cell-less BEM analysis scheme.

3 MDBEM based on three-step variable condensation method

The boundary-domain integral equations presented above are derived for single domain problems. However, practical engineering problems usually are composite structures consisting of different materials. To solve such problems, the multi-domain boundary element method (MDBEM) usually is employed [1, 4]. For this purpose, the three-step variable condensing MDBEM [5] is adopted in this study. Thus, the domain of problem is divided into a number of sub-domains. For each sub-domain, nodes are classified into three types: *self*-nodes (not shared with other sub-domains), *common* interface nodes, and *internal* nodes. Integral equations (2) and (7) are applied to the three types of nodes, and following algebraic matrix equations can be established for each sub-domain.

$$A_{bs} x_s + H_{bc} u_c + H_{bi} u_i = y_b + G_{bc} t_c \tag{11}$$

$$A_{is} x_s + H_{ic} u_c + H_{ii} u_i = y_i + G_{ic} t_c \tag{12}$$

in which subscripts s, c and i represent quantities corresponding to the *self*, *common* and *internal* nodes, respectively; $b = s + c$ denotes self plus common nodes related to current sub-domain; x_s is the unknown vector consisting of all unknown displacements and unknown tractions at the self nodes of the sub-domain, and y_b and y_i are known vectors formed by multiplying specified displacements and tractions with corresponding elements of related coefficient matrices.

First step: Eliminating internal displacements u_i from eqns (11) and (12), it follows

$$\overline{A}_{bs} x_s + \overline{H}_{bc} u_c = \overline{y}_b + \overline{G}_{bc} t_c \tag{13}$$

where

$$\begin{aligned} \overline{A}_{bs} &= A_{bs} - H_{bi}(H_{ii})^{-1}A_{is} \\ \overline{H}_{bc} &= H_{bc} - H_{bi}(H_{ii})^{-1}H_{ic} \\ \overline{G}_{bc} &= G_{bc} - H_{bi}(H_{ii})^{-1}G_{ic} \\ \overline{y}_b &= y_b - H_{bi}(H_{ii})^{-1}y_i \end{aligned} \tag{14}$$

Second step: Noticing that $b = s + c$ and eliminating unknowns x_s in eqn (13) yield

$$\hat{H}_{cc}u_c = \hat{y}_c + \hat{G}_{cc}t_c \tag{15}$$

where

$$\begin{aligned} \hat{H}_{cc} &= \overline{H}_{cc} - \overline{A}_{cs}(\overline{A}_{ss})^{-1}\overline{H}_{sc} \\ \hat{G}_{cc} &= \overline{G}_{cc} - \overline{A}_{cs}(\overline{A}_{ss})^{-1}\overline{G}_{sc} \\ \hat{y}_c &= \overline{y}_c - \overline{A}_{cs}(\overline{A}_{ss})^{-1}\overline{y}_s \end{aligned} \tag{16}$$

Third step: Eliminating common nodal tractions and forming the system of equations.

Equation (15) holds true for every sub-domain. For assembling the global system, it is written as the following form for the n-th sub-domain:

$$t_c^{(n)} = (\hat{G}_{cc}^{(n)})^{-1}(\hat{H}_{cc}^{(n)}u_c^{(n)} - \hat{y}_c^{(n)}) \tag{17}$$

The common nodal displacement vector $u_c^{(n)}$ for the n-th sub-domain can be expressed in terms of the globally numbered common displacement vector X by using a transfer matrix $Q^{(n)}$ as

$$u_c^{(n)} = Q^{(n)}X \tag{18}$$

The transfer matrix $Q^{(n)}$ consists of 0 and 1, determined by the consistent condition of displacement at the common nodes. Considering contributions of all sub-domains at common nodes, the equilibrium condition of the traction states that

$$\sum_n t_c^{(n)} = 0 \tag{19}$$

Substituting eqn (18) into (17), and the result into (19), the following system of equations can be obtained

$$AX = B \tag{20}$$

where

$$A = \sum_n (\hat{G}_{cc}^{(n)})^{-1}\hat{H}_{cc}^{(n)}Q^{(n)} \tag{21}$$

$$B = \sum_n (\hat{G}_{cc}^{(n)})^{-1}\hat{y}_c^{(n)} \tag{22}$$

Solving eqn (20) for all interface nodal displacements and substituting them back to previous expressions, one can obtain all unknowns. It is noted that the matrix

A in eqn (20) is a non-symmetric sparse matrix and, therefore, an equation solver for such systems needs to be developed. The row elimination-back-substitution method(REBSM) described in the following section is an effective one for solving such type of problems.

4 Row elimination-back-substitution method (REBSM) for solving non-symmetric sparse linear systems of equations

The system of equation (20) can be expressed as the following form:

$$\sum_{j=1}^{n} a_{ij} x_j = b_i \tag{23}$$

where n is the order of the equation set.

The key idea of REBSM is to find solutions of the system by completing both elimination and back-substitution procedures within each row. The main advantage of this treatment over the Gaussian elimination method is the less storage requirement of intermediate data. It is assumed that, after the treatment of the first k-1 rows, the following expressions have been obtained

$$x_i = b_i^{(k-1)} - \sum_{j=k}^{n} a_{ij}^{(k-1)} x_j \quad (i=1,2,\ \dots\ ,k\text{-}1) \tag{24}$$

For the k-th equation, we express it as follows

$$b_k = \sum_{j=1}^{n} a_{kj} x_j = \sum_{j=1}^{k-1} a_{kj} x_j + \sum_{j=k}^{n} a_{kj} x_j \tag{25}$$

Substituting eqn (24) into the first term of the right-hand side of the above equation yields:

$$\sum_{j=k}^{n} a'_{kj} x_j = b'_k \tag{26}$$

where

$$b'_k = b_k - \sum_{l=1}^{k-1} a_{kl} b_l^{(k-1)}, \qquad a'_{kj} = a_{kj} - \sum_{l=1}^{k-1} a_{kl} a_{lj}^{(k-1)} \tag{27}$$

Separating the k-th unknown from eqn (26) gives

$$x_k = b_k^{(k)} - \sum_{j=k+1}^{n} a_{kj}^{(k)} x_j \tag{28}$$

where

$$b_k^{(k)} = \frac{b'_k}{a'_{kk}}, \qquad a_{kj}^{(k)} = \frac{a'_{kj}}{a'_{kk}} \qquad (j = k+1, k+2, \dots, n) \tag{29}$$

Then, substituting eqn (28) back to eqn (24), it follows that

$$x_i = b_i^{(k)} - \sum_{j=k+1}^{n} a_{ij}^{(k)} x_j \quad (i=1,2,\ \dots\ ,k\text{-}1) \tag{30}$$

where

$$b_i^{(k)} = b_i^{(k-1)} - a_{ik}^{(k-1)} b_k^{(k)}, \qquad a_{ij}^{(k)} = a_{ij}^{(k-1)} - a_{ik}^{(k-1)} a_{kj}^{(k)}$$
$$(i = 1, 2, \ldots, k-1; \quad j = k+1, k+2, \cdots, n) \tag{31}$$

Equations (30) and (28) are the new expressions after the treatment of the k-th equation, in which the number of unknowns on the right-hand side is reduced by one compared to the expression of the row k-1 as shown in eqn (24). When the treatment of the last row of the equation (23) is finished, the unknowns of each right-hand side disappear and the remaining terms become the solutions of the system of equations.

From the derivation procedure of above formulations it can be seen that the features of REBSM can be classified as follows:

(1) Elimination and back-substitution are performed in the same row of the system of equations, easy for use in engineering numerical methods such as in MDBEM.
(2) Data storage requirement for intermediate process is different in different row treatments. For a system with full-populated matrix A, the maximum storage occurs in the middle part of A and the required storage size is a quarter of A, being the half of what Gaussian elimination method requires.
(3) From eqn (31) it can be seen that only non-zero elements need to be stored for sparse systems, and no symmetrical and definite properties on A are required. Therefore, REBSM is suitable for use in MDBEM.
(4) If the coefficient matrix A is not dominated by the diagonal elements, the value of a'_{kk} in eqn (29) may be zero or very small. In this case, pivoting is necessary to ensure an accurate result. This is easy to fulfill. What only need to do is that the maximum element among a'_{kj} $(j = k+1, k+2, \ldots, n)$ determined by eqn (27) is picked up and all related elements in this column are exchanged with those in the k-th column.

5 Numerical example

Based on the method presented in this paper, a code named BERIM has been written and a corrugated sandwich structure subjected to distributed load (Fig.1) has been analyzed. The upper and lower cover plates of the structure are made of nickel alloy with the Poison's ratio ν=0.25 and Yong's modulus E=50GMPa; the corrugated brackets are made of titanium with material properties of ν=0.25 and E= 250GMPa. The length, width and thickness of the plates are 4m, 2m and 0.05m, respectively, and two plates are spaced by 1m; the thickness and span of brackets are 0.04m and 0.8m, respectively. In computation, the lower cover is fixed and upper cover is subjected to a distributed pressure load of 0.5MPa. The structure is divided into 22 sub-domains, as shown as Fig.1. The surfaces of the structure is discretized into 7808 eight-noded quadratic boundary elements (Fig.2) with 21236 boundary nodes, among which 1782 are common nodes with

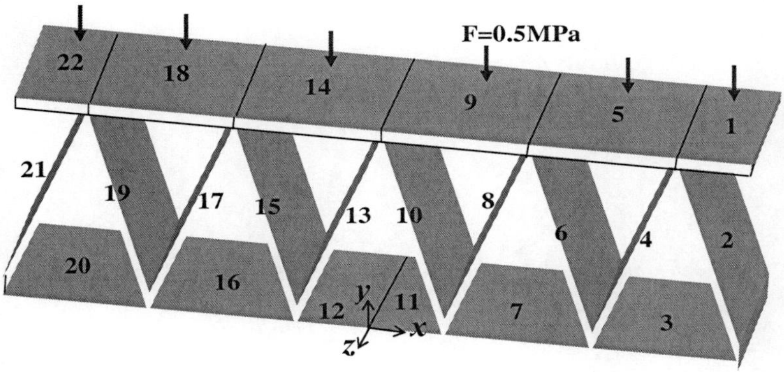

Figure 1: Corrugated sandwich structure under distributed loading.

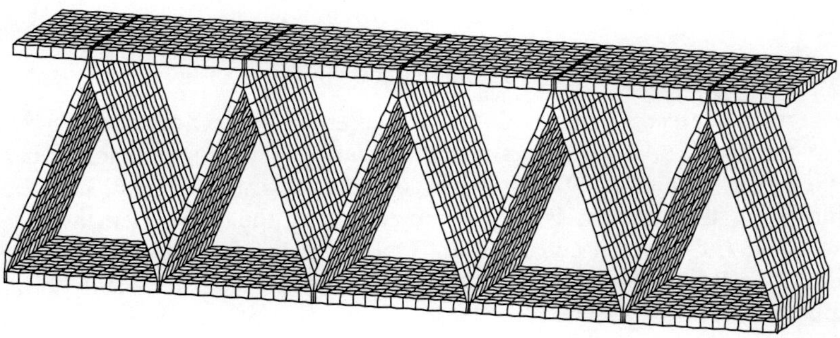

Figure 2: BEM mesh of the corrugated sandwich structure.

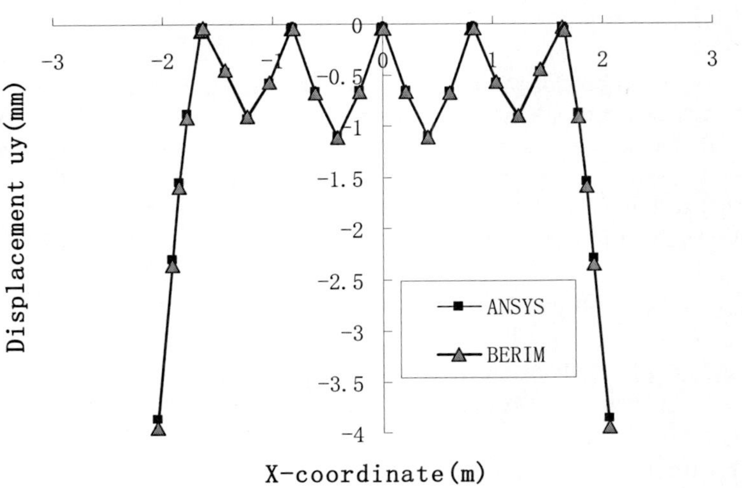

Figure 3: Vertical displacement along mid-line of upper plate surface.

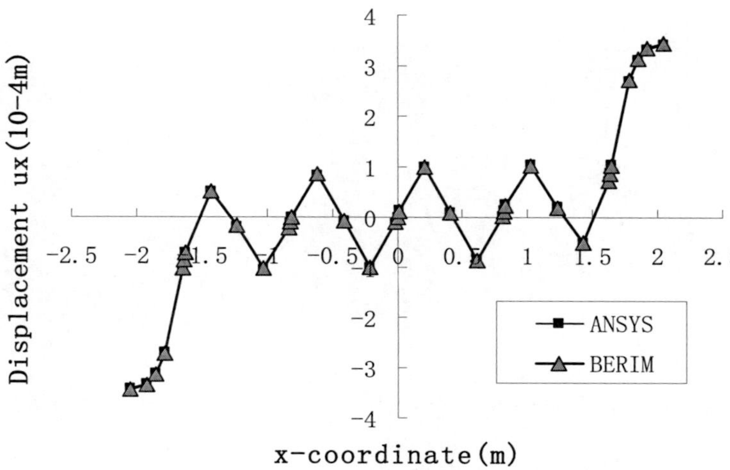

Figure 4: Horizontal displacement along mid-line of upper plate surface.

the degrees of freedom being 5346. The maximum band width of the system of equations is 1743 which shows a higher sparsity. The computational time for solving such system is 107s on a PC computer with the CPU of 3.0GHz. For comparison, the problem is also computed using the finite element software ANSYS with the model consisting of 3360 solid 186 brick elements and 3360 nodes. Figs.3 and 4 show the curves of computed vertical and horizontal displacements along x-direction over the middle line of the upper surface. It can be seen that the current results (BERIM) are in good agreement with those from ANSYS.

6 Conclusions

A multi-domain boundary element method has been presented for solving large-scale engineering problems. The system of equations assembled using the three-step variable condensing technique has the features of the smallest order and higher sparsity; the row elimination-back-substitution method (REBSM) is an efficient technique for solving non-symmetric and indefinite sparse system of equations, requiring less computer storage and suitable for using in MDBEM.

Acknowledgement

The authors gratefully acknowledge the National Natural Science Foundation of China for financial support to this work under Grant NSFC No. 10872050.

References

[1] Brebbia, C.A. & Dominguez, J. *Boundary Elements: an Introductory Course*, McGraw-Hill Book Co., London, 1992.

[2] Luo J. F., Liu Y.J., Berger E. Analysis of two-dimensional thin-structures (from micro- to nano-scales) using the boundary element method. Computational Mechanics, 22, pp.404-412,1998.

[3] J. Wang and X.W. Gao, Structural multi-scale boundary element method based on element subdivision technique, *Chinese Journal of Computational Mechanics*, 27(2),pp.258-263, 2010.

[4] Kane J.H., Kashava Kumar B.L., Saigal S. An arbitrary condensing, noncondensing solution strategy for large scale, multi-zone boundary element analysis. Comput Meth Appl Mech Eng,79, pp.219–44, 1990.

[5] X.W. Gao, L. Guo, Ch. Zhang, Three-step multi-domain BEM solver for nonhomogeneous material problems, Engineering Analysis with Boundary Elements, 31, pp.965–973,2007.

[6] Ch. Zhang, M. Cui, J. Wang, X.W. Gao, J. Sladek, V. Sladek. 3D crack analysis in functionally graded materials. *Engineering Fracture Mechanics,* 78,pp.585–604,2011.

[7] X.W. Gao, K. Yang. Thermal stress analysis of functionally graded material structures using boundary element method, *Chinese Journal of Theoretical and Applied Mechanics*,43(1), pp.136-143, 2011.

[8] Strang, Gilbert. *Introduction to Linear Algebra* (3rd ed.). Wellesley, Massachusetts: Wellesley-Cambridge Press, pp.74–76, 2003.

[9] Saad Y. *Iterative Methods for Sparse Linear Systems*. SIAM, Second edition, 2003.

[10] X.W. Gao. Source point isolation boundary element method for solving general anisotropic potential and elastic problems with varying material properties, *Engineering Analysis with Boundary Elements*, 34, pp.1049–1057,2010.

[11] X.W. Gao and T.G. Davies, *Boundary Element Programming in Mechanics*, Cambridge University Press, 2002.

[12] X.W. Gao. A boundary element method without internal cells for two-dimensional and three-dimensional elastoplastic problems. *ASME Journal of Applied Mechanics,* 69, pp.154-160, 2002.

A boundary element formulation for axi-symmetric problems in poro-elasticity

M. H. Ozyazicioglu[1] & M. Yener Ozkan[2]
[1]*Ataturk University Turkey*
[2]*Middle East Technical University, Turkey*

Abstract

A formulation is proposed for the boundary element analysis of poro-elastic media with axi-symmetric geometry. The boundary integral equation is reduced to a set of line integral equations in the generating plane for each of the Fourier coefficients, through complex Fourier series expansion of boundary quantities in circumferential direction. The method is implemented into a computer program, where the fundamental solutions are integrated by Gaussian Quadrature along the generator, while Fast Fourier Transform algorithm is employed for integrations in circumferential direction. The strongly singular integrands in boundary element equations are regularized by a special technique. The Fourier transform solution is then inverted in to Rθz space via inverse FFT. The success of the method is assessed by problems with analytical solutions. A good fit is observed in each case, which indicates effectiveness and reliability of the present method.

Keywords: poro-elasticity, boundary element method, axi-symmetric, fast Fourier transform, wave propagation.

1 Introduction

Axi-symmetric boundary element formulations for elasto-dynamics [1, 2] and acoustics [3, 4] are available in the literature. However, these formulations are either fully axi-symmetric (both geometry and boundary conditions are axi-symmetric) or they expand the boundary quantities into symmetric and anti-symmetric modes, the final response is obtained by combining solutions for each of these modes. Accurate evaluation of either elliptic integrals or integrations in

WIT Transactions on Modelling and Simulation, Vol 52,
www.witpress.com, ISSN 1743-355X (on-line)
doi:10.2495/BE110151

circumferential direction is required in the first and second approaches respectively, leading to extended run times.

Following the second approach described above, an axi-symmetric formulation for poro-elasticity is given by Dargush and Chopra [5]. Their formulation is fully axi-symmetric.

An alternative method based on complex Fourier series expansion of boundary quantities is developed by Özkan and Mengi [6]. In this method, the integrations in circumferential direction are accomplished by the Fast Fourier Transform (FFT) algorithm, which brings in considerable savings in computations.

The formulation of Tsepoura and Polyzos [7] for gradient elasticity is similar to [6]; however, they used the so called non-periodic FFT for integrations in θ-direction when the source point lies in the integration element.

In this study, we extend the method in [6] to axi-symmetric poro-elastodynamics with general boundary conditions.

The formulation has two major advantages: First, the use of FFT algorithm for integrations over θ-direction increases computational performance considerably compared to Gaussian Quadrature for the same purpose. Second, using complex Fourier expansion obviates the analysis for symmetric and anti-symmetric modes separately therefore, analysis for a general boundary condition is accomplished in a single run.

The formulation is developed in frequency domain, yet solutions in time domain can be obtained by inverse FFT, after solution vectors have been obtained for a sufficient number of frequencies.

2 Biot theory of poro-elasticity

Biot's poro-elasticity theory [8] is a generalization of elastodynamics to materials with fluid filled pore spaces and it includes Terzaghi's 1-D consolidation theory as a special case. This theory led to reformulation of problems in soil mechanics, geophysics, acoustics and biomechanics [9] predicting behaviour beyond that conceivable by classical elasticity theory. The predictions of the theory have been substantially verified experimentally [10, 11]. The material constants involved are easily discernible, physically meaningful and experimentally measurable. The theory also brings in a stronger definition of effective stress, which is a fundamental concept in soil mechanics.

2.1 Governing equations in frequency domain

Governing equations of poro-elasticity [11] in frequency domain are given as

$$\mu u_{i,jj} + (\lambda + \mu) u_{j,ji} - (\alpha + \beta) p_{,i} + f_i = -\omega^2 (\rho + \beta \rho_f) u_i \tag{1}$$

$$-\beta \frac{1}{\omega^2 \rho_f} p_{,kk} + (\alpha + \beta) u_{k,k} + \frac{1}{Q} p = \frac{1}{i\omega} a \tag{2}$$

where, u_i , p are Fourier transforms of displacement components and pore-pressure, f_i , a are body force and fluid generation per unit volume in frequency domain respectively. The Fourier and inverse Fourier transform of a time dependant function $F(t)$ is defined by the following pair

$$f(\omega)=\int_{-\infty}^{\infty} F(t)e^{-i\omega t}dt \quad \leftrightarrow \quad \mathrm{F(t)}=\frac{1}{2\pi}\int_{-\infty}^{\infty} f(\omega)e^{i\omega t}d\omega \tag{3}$$

Other parameters in (1) and (2) are

ω the circular frequency.

$\left.\begin{matrix}\mu\\ \lambda\\ \alpha\\ Q\end{matrix}\right\}$ material constants in constitutive equations

$\left.\begin{matrix}\rho\\ \rho_f\end{matrix}\right\}$ unit mass for bulk material and interstitial fluid

and β is defined as

$$\beta=\frac{n^2\kappa\rho_f\omega^2}{i\omega n^2-\omega^2\kappa(\rho_a+n\rho_f)}$$

where, porosity and permeability are designated respectively by n and κ. Finally, ρ_a is the added mass density.

2.2 The boundary integral equation (BIE)

The weighted residual statement of (1) and (2) is obtained by taking the inner product of the system with a vector of weighting functions $\begin{bmatrix}u_i^* & p^*\end{bmatrix}^T$

$$\int_\Omega \begin{bmatrix}u_i^*\\ p^*\end{bmatrix}^T \left([\mathbf{B}]\begin{bmatrix}u_i\\ p\end{bmatrix} + \{F\}\right)d\Omega=0 \tag{4}$$

where, $\{\mathrm{F}\}=\begin{bmatrix}\mathrm{f_i}\\ \mathrm{a}\end{bmatrix}$, and $[\mathrm{B}]$ represent the partial differential operator pertaining to equations (1) and (2). Using Gauss Integral Theorem, in the absence of body forces and fluid source (4) can be converted to the following BIE

$$\underline{cu}(A)=\int_\Gamma \underline{G'}(A,P)\underline{t}(P)d\Gamma-\int_\Gamma \underline{H'}(A,P)\underline{u}(P)d\Gamma \tag{5}$$

where, an underline designates a matrix; $\underline{G'}$ and $\underline{H'}$ matrices (4x4) contain first and second fundamental solutions of poro-elastodynamics, respectively. These solutions are two point (source at A, response at P) functions and associated with an infinite medium with either a point force in only one coordinate direction in turn or a unit fluid injection rate at a point "A". The point "P" is called the integration point. $\underline{t}$ and $\underline{u}$ are (4x1) column matrices representing generalized traction and displacement vectors at the boundary points, $\underline{c}$ is a (4x4) matrix, which is $c_{ij} = 0.5\delta_{ij}$ on a smooth boundary.

3 Boundary element formulation for axi-symmetric bodies

Consider a poro-elastic axi-symmetric body of boundary S, referred to a cylindrical coordinate system R-θ-z as shown in fig. 1, where z is the axis of revolution. It will be assumed that the *boundary conditions* are *not axi-symmetric*. The method is based on complex Fourier series expansion of the boundary quantities (displacements, pore-pressure, tractions and normal component of fluid flux vector) in circumferencal direction.

There are two main advantages of this method [6] over others available in the literature [1–5]:

i) the evaluation of integrals in θ direction is accomplished by FFT algorithm, which reduces the computational load,
ii) the need for differentiating symmetric and anti-symmetric modes in the analysis is eliminated, which facilitates computer programming.

The transformation from cylindrical to cartesian coordinates is given by:

$$\underline{t} = \underline{Q}\,\underline{t}_c \quad \text{and} \quad \underline{u} = \underline{Q}\,\underline{u}_c \tag{6}$$

where, where, $\underline{u}_c$ and $\underline{t}_c$ are the generalized displacement and traction vectors cylindrical coordinate frame. The transformation (rotation) matrix $\underline{Q}$ is

$$\underline{Q} = \begin{bmatrix} \cos\alpha & -\sin\alpha & 0 & 0 \\ \sin\alpha & \cos\alpha & 0 & 0 \\ 0 & 0 & 1 & 0 \\ 0 & 0 & 0 & 1 \end{bmatrix} \tag{7}$$

Substituting (6) in (5) and multiplying from left by $Q^T(A)$, one gets

$$\underline{c}_c\,\underline{u}_c(A) = \int_C \int_0^{2\pi} \underline{G}_c(A,P)\,\underline{t}_c(P) R\,d\theta\,ds - \int_C \int_0^{2\pi} \underline{H}_c(A,P)\,\underline{u}_c(P) R\,d\theta\,ds \tag{8}$$

Since, $d\Gamma = R\,d\theta\,ds$. (8) is now the BIE in cylindrical coordinates where $\underline{G}_c(A,P)$ and $\underline{H}_c(A,P)$ represent the fundamental solution matrices

$$\begin{aligned} \underline{G}_c(A,P) &= \underline{Q}^T(A)\,\underline{G}'(A,P)\,\underline{Q}(P) \\ \underline{H}_c(A,P) &= \underline{Q}^T(A)\,\underline{H}'(A,P)\,\underline{Q}(P) \end{aligned} \tag{9}$$

and $\underline{c}_c = \underline{c}$ is the free term coefficient.

It can be shown that the fundamental solutions $\underline{G}_c(A,P)$ and $\underline{H}_c(A,P)$ are functions of the form

$$\underline{G}_c(r, \theta - \theta', z - z'),\ \underline{H}_c(r, \theta - \theta', z - z') \tag{10}$$

where, primed variables represent source point coordinates.

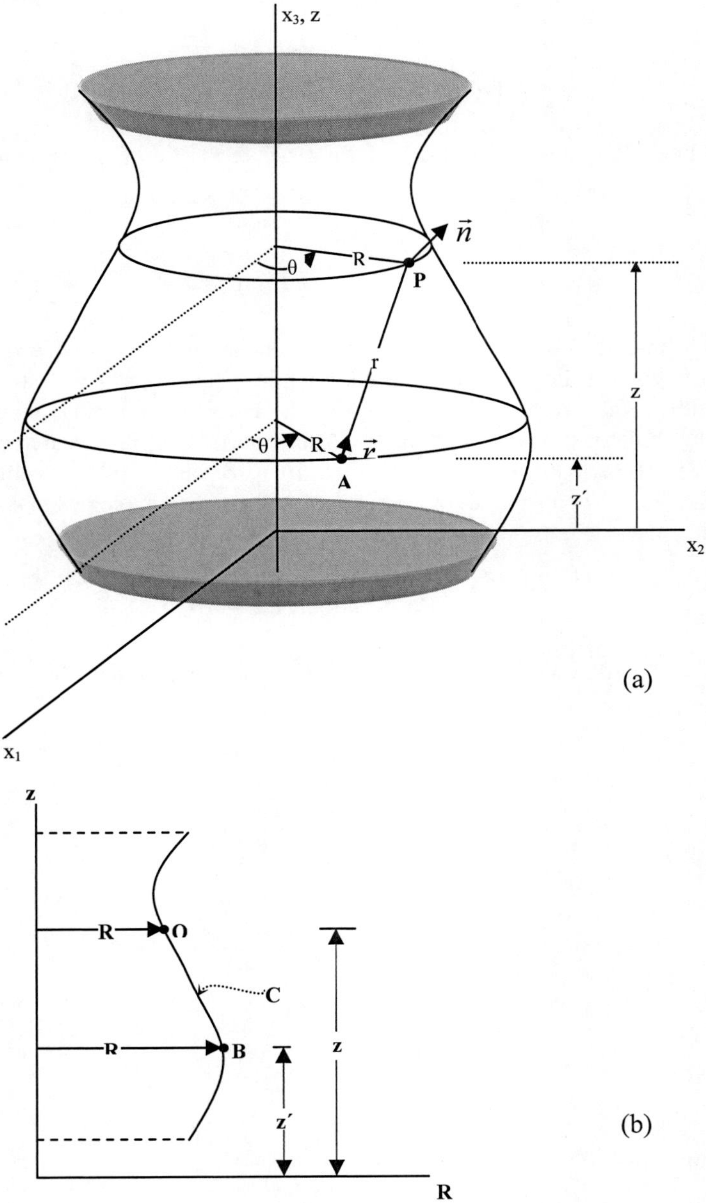

Figure 1: An axi-symmetric body referred to Rθz coordinate system: a) three dimensional body, b) x-section on R-z plane.

3.1 Expansion of field variables in complex Fourier series

It follows from axi-symmetric geometry that the boundary quantities are ***"2π" periodic*** in angular direction, thus can be expand into complex Fourier series:

$$\underline{u}_c(R,\theta,z) = \sum_{k=-\infty}^{\infty} \underline{u}_c^k(R,z)e^{ik\theta}$$
$$\underline{t}_c(R,\theta,z) = \sum_{k=-\infty}^{\infty} \underline{t}_c^k(R,z)e^{ik\theta} \tag{11}$$

when (11) is substituted, (5) is reduced, for a k-th term of the Fourier series,

$$\underline{c}_c\underline{\tilde{u}}_c^k(R',z') = \int_C \underline{G}_c^k(R',z';\mathrm{R},\mathrm{z})\underline{\tilde{t}}_c^k(R,z)Rds -$$
$$\int_C \underline{H}_c^k(R',z';\mathrm{R},\mathrm{z})\underline{\tilde{u}}_c^k(R,z)Rds \tag{12}$$

where the integrals are along the generating curve, thus, *the dimensionality is reduced by one*, at the expense of now k boundary equations. The equations are exact with infinite terms, but a good approximation is generally obtained by truncating the series to only a few terms. After Fourier coefficients are computed from (12) for each k, boundary quantities in Rθz space can be evaluated by inverse FFT. When spatial discretization with constant elements along the generating curve is introduced (12) becomes

$$\underline{c}_c\underline{\tilde{u}}_{c\,i}^k = \sum_j \underline{\tilde{G}}_{ij}^k\underline{\tilde{t}}_{c\,j}^k - \sum_j \underline{\tilde{H}}_{ij}^k\underline{\tilde{u}}_{c\,j}^k \tag{13}$$

where, $\underline{\tilde{u}}_{c\,j}^k$ and $\underline{\tilde{t}}_{c\,j}^k$ are the values of $\underline{\tilde{u}}_c^k$ and $\underline{\tilde{t}}_c^k$ over the element "j", and

$$\underline{\tilde{G}}_{ij}^k = \int_0^{2\pi} \underline{\mathrm{G}}_{ij}(\theta)e^{ik\theta}d\theta$$
$$\underline{\tilde{H}}_{ij}^k = \int_0^{2\pi} \underline{\mathrm{H}}_{ij}(\theta)e^{ik\theta}d\theta \tag{14}$$

where

$$\underline{\mathrm{G}}_{ij}(\theta) = \int_{\mathrm{C}_j} \underline{\mathrm{G}}_c\left(\underbrace{\mathrm{R}',\mathrm{z}'}_{\mathrm{Q}_i};\underbrace{\mathrm{R},\mathrm{z}}_{\mathrm{Q}};\theta-\theta'\right)\Bigg|_{\theta'=0}\mathrm{R\,ds}$$

$$\underline{\mathrm{H}}_{ij}(\theta) = \int_{\mathrm{C}_j} \underline{\mathrm{H}}_c\left(\underbrace{\mathrm{R}',\mathrm{z}'}_{\mathrm{Q}_i};\underbrace{\mathrm{R},\mathrm{z}}_{\mathrm{Q}};\theta-\theta'\right)\Bigg|_{\theta'=0}\mathrm{R\,ds}$$

In system form, (13) may be written as

$$\underline{\tilde{H}}^k\underline{\tilde{u}}^k = \underline{\tilde{G}}^k\underline{\tilde{t}}^k \tag{15}$$

where,

$$\underline{\widetilde{G}}^{k} = \left\lfloor \underline{\widetilde{G}}_{ij}^{k} \right\rfloor$$
$$\underline{\widetilde{H}}^{k} = \left[\underline{\widetilde{H}}_{ij}^{k} + \frac{1}{2} \underline{I} \delta_{ij} \right] \tag{16}$$

Fast and accurate evaluation of the integrals in (14) can be accomplished by FFT. This is the third stage we use FFT algorithm for computations. It should be noted that different number of Fourier terms may be selected for Complex Fourier expansion in (11) and in the integral evaluations in (14).

The computational procedure can be summarised as follows:

- Choose $N'=2^{Mp}$, the number of terms in Complex Fourier Series, and compute Fourier coefficients of boundary excitations by FFT (4th FFT).
- Choose $N=2^{M}$, the number of division along θ for integrals in (14)
- Discretize the generator and let the number of boundary elements be m.
- Compute $\underline{\widetilde{G}}_{ij}^{s}$ and $\underline{\widetilde{H}}_{ij}^{s}$ (s=0…N-1), and form the system matrices $\underline{\widetilde{H}}^{k}$ and $\underline{\widetilde{G}}^{k}$ for $k = 0 \ldots N' - 1$. A frequency shift is necessary when assembling $\underline{\widetilde{G}}_{ij}^{s}$ and $\underline{\widetilde{H}}_{ij}^{s}$ into $\underline{\widetilde{G}}^{k}$ and $\underline{\widetilde{H}}^{k}$, as "s" and "k" run through different ranges. The singular integrals when source point is on the integration element is circumvented by introducing a panel element around the generator, details can be found in [11].
- Solution of the complex algebraic system of equations (15) together with the specified boundary conditions yields the Fourier coefficients $\underline{\widetilde{u}}^{k}$ and $\underline{\widetilde{t}}^{k}$ at frequency points $k = 0, 1, 2, \ldots, N' - 1$.
- By an inverse FFT (5th) evaluate the boundary quantities in (R, θ, z) space.
- The BIE (5) with $\underline{c} = \underline{I}$ is used to compute solution at an internal point if required.

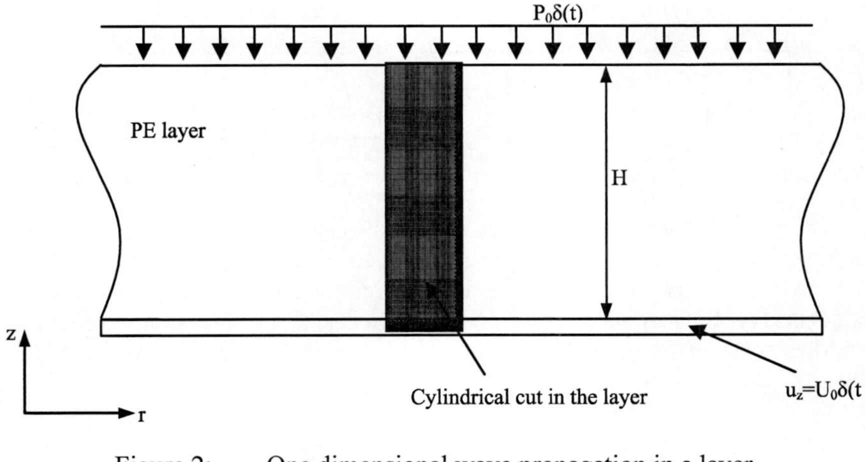

Figure 2: One dimensional wave propagation in a layer.

4 Assessment of the formulation

A computer program called AxiPoro is implemented using above algorithm with standard C language. And the results are assessed against three problems which have exact analytical solutions.

4.1 One dimensional wave propagation in a PE layer

This problem, in the context of poro-elasticity is first mentioned in [12], an analytical solution is also provided in the same reference.

In solving this problem by BEM, we model the PE "layer" by a PE "column" of unit diameter as shown in fig. 3. Although, the PE column is not an exact model for the layer, still is a close approximation. The material properties are given in Table 1 below.

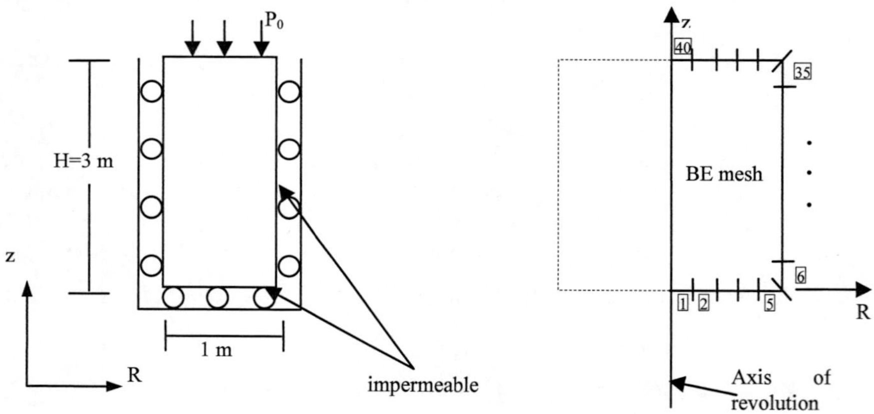

Figure 3: Equivalent column model of PE layer for BE analysis.

Table 1: Material data for Berea sand stone.

N	α	Q (Pa)	μ(Pa)	ν	κ (m^4/N/s)	ρ (kg/m^3)	ρ_f (kg/m^3)	ρ_a (kg/m^3)
0.19	0.778	$1.353*10^{10}$	$6*10^9$	0.2	$1.9*10^{-10}$	2458	1000	125.4

The poro-elastic column problem is solved by program AxiPoro. The column is modelled by 40 axi-symmetric boundary elements. N=128 (2^7) and N=32 (2^5) sub-divisions for circumferential integrations were used for convergence checking. A slight hysteretic damping of 0.3% is also introduced. The results for top displacement are plotted in fig. 4 together with the analytical solution in [12].

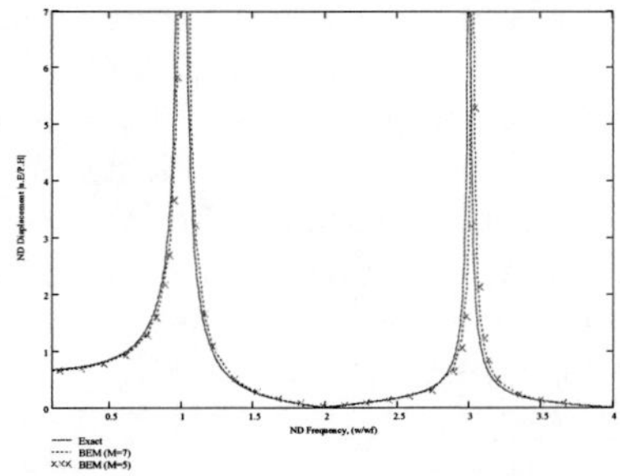

(a) Top displacement amplitude.

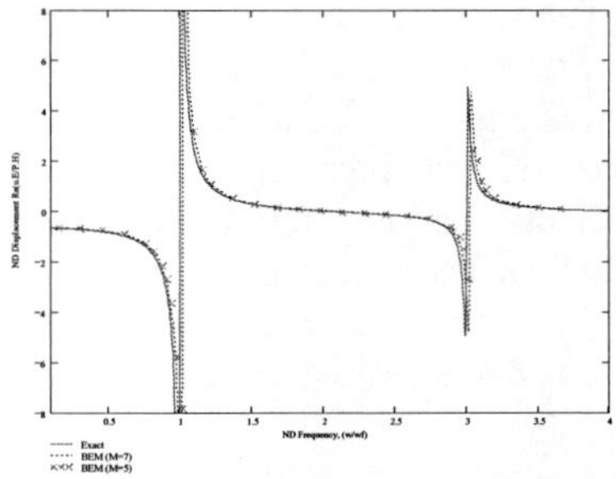

(b) Real part of top displacement.

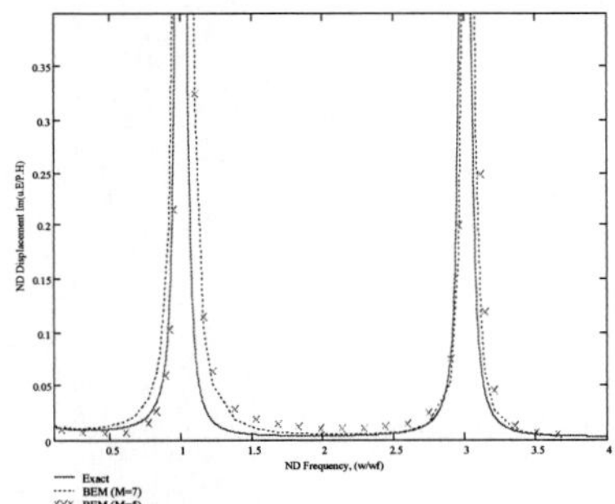

(c) Imaginary part of top displacement, $E=2\lambda+\mu$.

Figure 4: PE column (traction B.C. at top): BEM vs. analytical solution.

4.2 Sudden pressurization of a circular cavity

This problem was investigated by Senjuntichai and Rajapakse [13].

However, the governing equations used in [13] did not include the "continuity equation", therefore, to comply with our BEM formulation, we re-work the solution for sudden pressurization (Dirac loading in time) with permeable wall condition; the analytical solution can be found in [11].

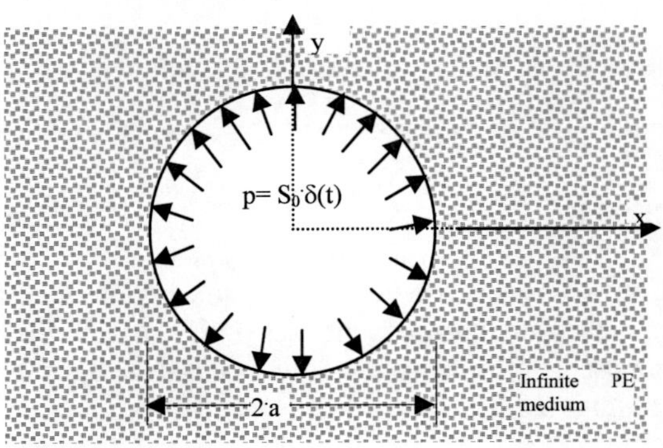

Figure 5: Circular cavity (infinite cylinder) in a poro-elastic full space suddenly pressurized.

www.witpress.com, ISSN 1743-355X (on-line)

The poro-elastic circular cavity problem is solved by program AxiPoro, with the material data of Berea sandstone (Table 1). A finite cylindrical cavity of 10 metres height is modelled by 25 axi-symmetric boundary elements. The angular divisions for circumferential integrations are $N = 2^7 = 128$. A hysteretic damping of 5% is used to eliminate edge reflections as much as possible. The results for radial surface displacements (compliance) at mid-height of the cavity are plotted in fig. 6 together with the analytical solution. Slight noise is observed in the figures due to waves generated at the ends of the cavity.

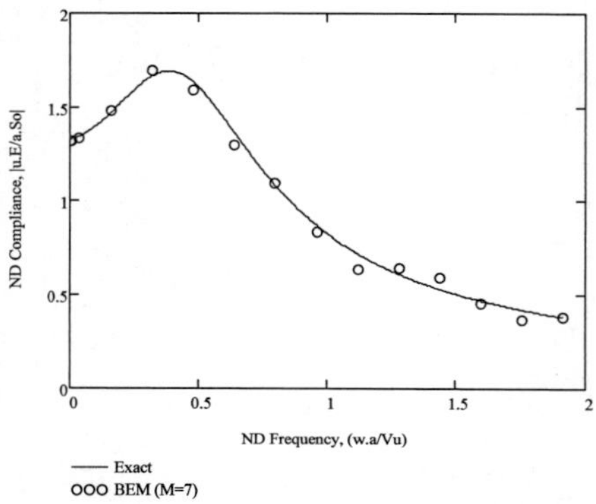

a) Absolute value of ND compliance.

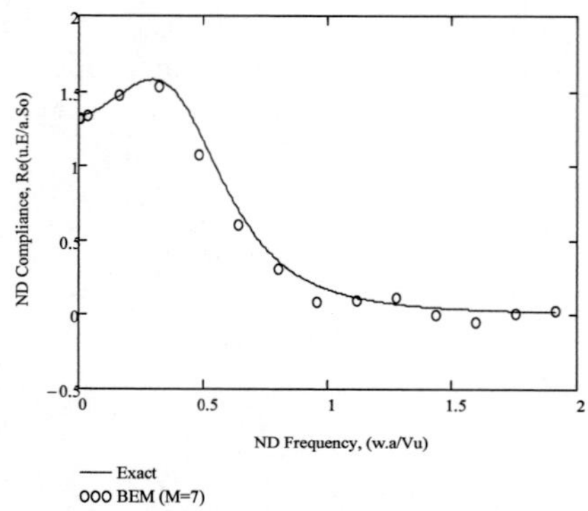

b) Real part of ND compliance.

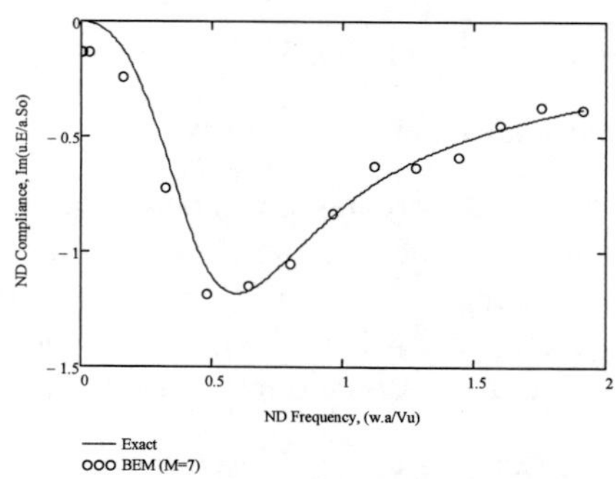

c) Imaginary part of ND compliance.

Figure 6: Circular cavity: BEM vs. analytic solution, E=2λ+μ.

5 Conclusions

In this study, an axi-symmetric BE formulation is presented for dynamic poro-elasticity. The method makes frequent uses of FFT as an effective computational tool. The proposed formulation has several advantages over others in the literature; to summarize

- expansion of boundary variables in complex Fourier series, obviates the need for differentiating symmetric and anti-symmetric modes in the analysis leading to easier coding in the case of arbitrary boundary conditions.

- use of FFT algorithm increases the computational performance and accuracy remarkably.
- convergence of the method for a given BE mesh is controlled by the number of subdivisions in circumferential direction, good accuracy is obtained for $N = 2^8$.

If response in time domain is required, the analysis can be repeated for sufficient number of frequencies ω, and time history of boundary quantities can be evaluated by inverse FFT, again.

The computer implementation of the method is assessed against analytical solutions in the literature, a good fit is observed.

References

[1] Brebbia, C. A. and Dominguez, J., Boundary Elements An Introductory Course, Second Edition, Computational Mechanics Publications, UK, 1992.

[2] Becker, A. A., The Boundary Element Method in Engineering, McGraw Hill, London, 1992.

[3] Juhl, P., An axisymmetric integral equation formulation for free space non-axisymmetric radiation and scattering of a known incident wave, J. Sound and Vibration, V. 163, pp. 397-406, 1993.

[4] Pozrikidis, C., A Practical Guide to Boundary Element Methods with the Software Library BEMLIB, CRC Press, May 15 2002, New York, 2002.

[5] Dargush, G. F. and Chopra, M. B., Dynamic analysis of axisymmetric foundations on poroelastic media, J. Eng. Mech., ASCE, V.122, No.7, pp. 623-632, 1996.

[6] Özkan, G., Mengi, Y., On the use of FFT algorithm for the circumferential coordinate in boundary element formulation of axisymmetric problems, 40(13), pp.2385-2412, 1997.

[7] Tsepoura, K. G. and Polyzos, D., Static and harmonic BEM solutions of gradient elasticity problems with axisymmetry, Computational Mechanics, V.32, pp.89-103, 2003.

[8] Biot, M. A., The theory of propagation of elastic waves in a fluid saturated porous solid, I. Low frequency range, J Acoust. Soc. Am., V.28, pp. 168-178, 1956.

[9] Cowin, S. C., Bone poroelasticity, J. Biomech., V.32, pp.217-238, 1999.

[10] Plona, T. J., Observation of a second bulk comp. wave in a porous medium at ultrasonic frequencies, Appl. Phys. Lett., V.36, pp. 259-261, 1980.

[11] Ozyazicioglu, M., A boundary element formulation for axi-symmetric poroelasticity, Ph. D. Thesis, Middle East Technical University, Dept. of Civ. Eng., Ankara, Turkey, 2006.

[12] Cheng, A. H-D.; Badmus, T.; Beskos, D. E., Integral equation for dynamic poroelasticity in frequency domain with BEM solution, J. Eng. Mech., ASCE, Vol. 117, No. 5, pp. 1136-1157, 1991.

[13] Senjuntichai, T. and Rajapakse, R. K. N. D., Transient response of a circular cavity in a poroelastic medium, Int. J. Num. Analyt. Meth. Geomech., V.17, pp.357-383, 1993.

Section 4
Advanced formulations

The expedite boundary element method

N. A. Dumont & C. A. Aguilar
Department of Civil Engineering,
Pontifical Catholic University of Rio de Janeiro, Brazil

Abstract

The present developments combine the variationally-based, hybrid boundary element method with a consistent formulation of the conventional, collocation boundary element method in order to establish a computationally less intensive procedure, although not necessarily less accurate, for large-scale, two-dimensional and three-dimensional problems of potential and elasticity, including time-dependent phenomena. Both the double-layer and the single-layer potential matrices, **H** and **G**, whose evaluation usually requires dealing with singular and improper integrals, are obtained in an expedite way that circumvents almost any numerical integration – except for a few regular integrals in the case of **H**. A few numerical examples are shown to assess the applicability of the method, its computational effort and some convergence issues.
Keywords: boundary elements, meshless methods, hybrid boundary elements.

1 Introduction

The *collocation boundary element method* (CBEM), whenever applicable, is a simple, powerful numerical analysis tool [1]. The present contribution is an attempt to show that the CBEM can be still more efficient and powerful – and still easier to implement computationally. (A not lesser contribution is the demonstration that simplicity can be achieved without resorting to exotic concepts such as node displacements from corner points or *regularizations*.)

Some precursory works have already been published on the subject [9] or are being prepared [11]. However, this is the first attempt to summarize the basic concepts that lead to the *expedite* boundary element method (EBEM) and to show its main features and possibilities of application in an outline that is meant to be itself expedite.

WIT Transactions on Modelling and Simulation, Vol 52,
www.witpress.com, ISSN 1743-355X (on-line)
doi:10.2495/BE110161

2 Problem formulation

An elastic body is submitted to body forces b_i in the domain Ω and traction forces $\bar{t}_i$ on part Γ_σ of the boundary. Displacements $\overline{u}_i$ are known on the complementary part Γ_u of Γ. One is looking for an adequate approximation of the stress field that satisfies equilibrium in the domain,

$$\sigma_{ji,j} + b_i = 0 \quad \text{in} \quad \Omega \tag{1}$$

also satisfying the boundary equilibrium and compatibility equations,

$$\sigma_{ji} n_j = \bar{t}_i \quad \text{along} \quad \Gamma_\sigma, \qquad u_i = \overline{u}_i \quad \text{on} \quad \Gamma_u \tag{2}$$

where n_j is the outward unit normal to Γ. Indices i, j, (also k, l) may assume values 1, 2 or 3, as they refer to the coordinate directions x, y or z, respectively, for a general 3D analysis. Summation is indicated by repeated indices. Particularization to 2D analysis as well as to potential problems is straightforward.

2.1 Stress and displacement assumptions

Three independent fields are used in the following developments. The displacement field is explicitly approximated along the boundary by u_i^d, where $(\,)^d$ means *displacement assumption*, in terms of polynomial functions u_{im} with compact support and nodal displacement parameters $\mathbf{d} = [d_m] \in \mathbb{R}^{n^d}$, for n^d displacement degrees of freedom of the discretized model. An independent stress field σ_{ij}^s, where $(\,)^s$ stands for *stress assumption*, is given in the domain in terms of some particular solution σ_{ij}^p plus a series of fundamental solutions $\sigma_{ij\,m}^*$ with global support, multiplied by force parameters $\mathbf{p}^* = [p_m^*] \in \mathbb{R}^{n^*}$ applied at the same boundary nodal points m to which the nodal displacements d_m are attached $(n^* = n^d)$. Displacements u_i^s are obtained from σ_{ij}^s. Then,

$$u_i^d = u_{im}\, d_m \quad \text{on} \quad \Gamma \quad \text{such that} \quad u_i^d = \bar{u}_i \quad \text{on} \quad \Gamma_u \quad \text{and} \tag{3}$$

$$\sigma_{ij}^s = \sigma_{ijm}^*\, p_m^* + \sigma_{ij}^p \quad \text{such that} \quad \sigma_{jim,j}^* = 0 \quad \text{and} \quad \sigma_{ji,j}^p = b_i \quad \text{in} \quad \Omega \tag{4}$$

$$\Rightarrow \quad u_i^s = u_{im}^*\, p_m^* + u_i^p + u_{is}^r\, C_{sm}\, p_m^* \quad \text{in} \quad \Omega \tag{5}$$

where u_{im}^* are displacement fundamental solutions corresponding to σ_{ijm}^*. Rigid body motion is included in terms of functions u_{is}^r multiplied by in principle arbitrary constants $C_{sm} \in R^{n^r \times n^*}$, where n^r is the number of rigid body displacements (r.b.d.) of the discretized problem, as dealt with formally in Definition 1, introduced in Section 4 [7, 10]. The fundamental solutions σ_{ijm}^* are used as weight functions in the CBEM. In the variational BEMs and in the EBEM, in particular, they represent domain interpolation functions.

The third independent field is used to approximate traction forces along the boundary by t_i^t, where $(\,)^t$ means *traction assumption*, as required in the conventional boundary element method, given as

$$t_i^t = u_{i\ell} t_\ell \quad \text{in the CBEM,} \tag{6}$$

where $u_{i\ell}$ are polynomial interpolation functions with compact support and $\mathbf{t} = [t_\ell] \in \mathbb{R}^{n^t}$ are traction-force parameters. The index i refers to the coordinate directions whereas the index ℓ refers to any of the n^t traction-force degrees of freedom of the problem (thus denoting both location and orientation), for nodes adequately distributed along boundary segments of Γ. The interpolation functions $u_{i\ell}$ have the same properties of u_{in}, as presented in eqn (3). Equation (6) holds as $\bar{t}_i = u_{i\ell}\bar{t}_\ell$ along Γ_σ, in particular, according to eqn (2). An improved version of eqn (6) is proposed for problems with curved boundaries,

$$t_i^t = u_{i\ell} \frac{|J|_{(at\,\ell)}}{|J|} t_\ell \equiv t_{i\ell} t_\ell \quad \text{in the MBEM,} \tag{7}$$

which leads to the *modified* boundary element method (MBEM) [8]. In this equation, $|J|_{(at\,\ell)}$ is the value of the Jacobian of the global (x, y, z) to natural (ξ, η) coordinate transformation at the nodal point ℓ and the term $|J|_{(at\,\ell)}/|J|$ features a term in the denominator that cancels the Jacobian term of the infinitesimal boundary segment $\mathrm{d}\Gamma = |J|\mathrm{d}\xi\mathrm{d}\eta$ in the numerator of two integral expressions introduced in eqns (10) and (14). This not only improves the capacity of t_i^t to represent the traction forces along curved boundary segments but also simplifies the numerical integration of the related terms. In the subsequent developments, one refers to the approximation of the traction forces on Γ generically as given in eqn (7), $t_i^t = t_{i\ell}t_\ell$, explaining in the text, whenever explicitly required, whether eqn (6) or (7) is meant.

The numbers of degrees of freedom for traction forces n^t and displacements n^d are not necessarily the same, since one may need more than one traction-force parameter to represent tractions that are not single valued at the boundary surface, generally at nodes where adjacent boundary segments present different outward normals [8]. Then, it results that $n^t \geq n^d$, as t_ℓ in eqns (6) and (7) are *traction-force* attributes on boundary segments, whereas u_{in} in eqn (3) are *displacement* attributes at nodal points. The fact that $n^t \geq n^d$ leads to some rectangular matrices – the same eqns (14) and (10) of the CBEM, which have been just referred to, plus a third one, introduced in eqn (21).

2.2 Boundary approximation of the particular solution

Although neither conceptually nor formally necessary, the following approximation may render all subsequent equations simpler and more elegant [7]. Given a sufficiently refined boundary mesh, the displacements u_i^p and the traction forces t_i^p related to an arbitrary particular solution of the non-homogeneous governing eqn (1), whenever available, can be approximated accurately enough by nodal displacement parameters $\mathbf{d}^p = [d_n^p] \in \mathbb{R}^{n^d}$ and traction force parameters $\mathbf{t}^p = [t_\ell^p] \in \mathbb{R}^{n^t}$, respectively, in terms of the interpolation functions of eqns (3) and (6) or (7):

$$u_i^p \approx u_{in} d_n^p \,, \quad t_i^p \approx t_{i\ell} t_\ell^p \quad \text{on} \quad \Gamma \tag{8}$$

One assumes with the above equations that a particular solution for the domain forces b_i in eqn (1) is known in terms of displacements u_i^p and stresses σ_{ij}^p. The means to obtain such particular solutions other than in close form are not discussed herein (see, for instance, Partridge et al [14]).

3 Conventional and modified boundary element methods

The matrix equation of the CBEM [1] may be expressed as [8]

$$\mathbf{H}\,(\mathbf{d} - \mathbf{d}^p) = \mathbf{G}\,(\mathbf{t} - \mathbf{t}^p) \tag{9}$$

where $\mathbf{H} = [H_{mn}] \in \mathbb{R}^{n^d \times n^d}$ is a kinematic transformation matrix [5, 7, 10] and $\mathbf{G} = [G_{m\ell}] \in \mathbb{R}^{n^d \times n^t}$ is a flexibility-like matrix (that is in general rectangular, as proposed). The formal definition of these matrices is

$$H_{mn} = \int_\Gamma \sigma_{jim}^* \eta_j u_{in} \mathrm{d}\Gamma \,, \quad G_{m\ell} = \int_\Gamma t_{i\ell} u_{im}^* \mathrm{d}\Gamma \tag{10}$$

The double-layer and single-layer potential matrices H_{mn} and $G_{m\ell}$ comprise in their definition singular and improper integrals, respectively, when source (m) and field (either n or ℓ) indexes refer to the same nodal points. The singular integrals can be always evaluated mathematically in correspondence to simple mechanical meanings [1]. A conceptual assessment of eqn (10) is given in Reference [8].

4 Some virtual-work statements

Several virtual-work statements are outlined in the following. They are actually theorems that must be proven from some mechanical axioms. Some of them have already been dealt with in References [6–10]. In this paper, one attempts to keep full consistency of the equations while being not too formal. This compromise is hopefully achieved with the following definition.

Definition 1 *Let n^r be the number of independent r.b.d. of an elasticity problem, in general. Then, $n^r = 3$ or 6 for 2D or 3D problems (and $n^r = 1$ for potential problems) Problems involving symmetry present different values of n^r. One may eventually have $n^r = 0$, as for an infinite domain. The r.b.d. $W \in R^{n_d}$ are spanned by the columns of a matrix $\mathbf{W} \in R^{n_d \times n_r}$, which is orthogonal, for convenience.*

4.1 Displacement virtual-work statement

Part of the Hellinger-Reissner potential [5, 10] leads to the equilibrium equation

$$H_{mn}\, p_m^* = p_n - p_n^p \quad \text{or} \quad \mathbf{H}^{\mathrm{T}}\,\mathbf{p}^* = \mathbf{p} - \mathbf{p}^p \tag{11}$$

in which $\mathbf{H} = [H_{nm}] \in \mathbb{R}^{n^d \times n^*}$ is the same double layer potential matrix of the collocation boundary element method [1], already introduced in eqn (10).

Moreover, $\mathbf{p} = [p_n] \in \mathbb{R}^{n^d}$ and $\mathbf{p}^p = [p_n^p] \in \mathbb{R}^{n^d}$, defined as

$$p_n = \int_\Gamma \sigma_{ji} n_j u_{in}\, \mathrm{d}\Gamma\,, \quad p_n^p = \int_\Gamma \sigma_{ji}^p n_j u_{in}\, \mathrm{d}\Gamma \tag{12}$$

are vectors of equivalent nodal forces corresponding respectively to applied boundary tractions, as given in eqn (2) and to the particular solution of eqn (4).

4.2 Virtual-work relations between the approximate fields given by d and t

It may be convenient to express the boundary traction approximations of eqns (6) or (7) in terms of equivalent nodal forces, obtained from the virtual work statement:

$$\begin{aligned} &\delta d_m p_m(\mathbf{t}) = \delta d_m \textstyle\int_\Gamma u_{im} t_{i\ell} \mathrm{d}\Gamma t_\ell \\ &\Rightarrow p_m(\mathbf{t}) = L_{\ell m} t_\ell \quad \text{or} \quad \mathbf{p}(\mathbf{t}) = \mathbf{L}^{\mathrm{T}}\mathbf{t} \end{aligned} \tag{13}$$

where the interpolation functions of eqns (3) and (7) (as for the MBEM) were used, thus defining

$$\mathbf{L} = [L_{\ell m}] \in \mathbb{R}^{n^t \times n^d} = \int_\Gamma t_{i\ell} u_{im} \mathrm{d}\Gamma \tag{14}$$

As given in eqn (13), $\mathbf{L}^{\mathrm{T}}$ performs an equilibrium transformation of traction-force parameters $\mathbf{t}$ to equivalent nodal forces $\mathbf{p}(\mathbf{t})$. The argument $(\mathbf{t})$ shows that $\mathbf{p}(\mathbf{t})$ is a function of $\mathbf{t}$, thus one of three possible approximations. Observe that, according to Definition 1, $\mathbf{W}^{\mathrm{T}}(\mathbf{p}(\mathbf{t}) - \mathbf{p}^p) = \mathbf{W}^{\mathrm{T}}\mathbf{L}^{\mathrm{T}}(\mathbf{t} - \mathbf{t}^p) = \mathbf{0}$ for a problem consistently formulated.

One may express the contragradient statement

$$\mathbf{p}(\mathbf{t}) = \mathbf{L}^{\mathrm{T}}\mathbf{t} \quad \Rightarrow \quad \mathbf{d}^t(\mathbf{d}) = \mathbf{L}\mathbf{d} \tag{15}$$

where $\mathbf{d}^t(\mathbf{d})$ are equivalent nodal displacements defined such that $\delta\mathbf{t}^{\mathrm{T}}\mathbf{d}^t(\mathbf{d})$ has the meaning of virtual work. This contragradient statement is part integrand of the hybrid *displacement* BEM, which may be derived from the Hu potential [4,7,13].

4.3 Virtual-work relations between the approximate fields given by d and p*

One obtains from eqn (11) the contragradient relation

$$\mathbf{p}(\mathbf{p}^*) = \mathbf{H}^{\mathrm{T}}\mathbf{p}^* \quad \Rightarrow \quad \mathbf{d}^*(\mathbf{d}) = \mathbf{H}\mathbf{d} \tag{16}$$

where $\mathbf{d}^*(\mathbf{d})$ are equivalent nodal displacements defined such that $\delta\mathbf{p}^{*\mathrm{T}}\mathbf{d}^*(\mathbf{d})$ has the meaning of virtual work.

4.4 Subspaces of admissible forces for the field approximations

The matrix $\mathbf{W}$ of nodal r.b.d. was introduced in Definition 1, which is also the subspace of forces $\mathbf{p}$ that are not in balance. As remarked after eqn (14), the

columns of $\mathbf{W}^{\mathrm{T}}\mathbf{L}^{\mathrm{T}}$ span the subspace of forces $\mathbf{t}$ that are not in balance [6]. For a finite domain, the columns of $\mathbf{W}$ are the null space of $\mathbf{H}$. Then, one obtains for consistency of eqn (11) that balanced forces $\mathbf{p}^*$ must be orthogonal to the null space $\mathbf{V}$ of $\mathbf{H}^{\mathrm{T}}$ [5]. These conclusions are formalized in the following theorem.
Theorem 1 *The columns of the matrices* $\mathbf{W}$, $\mathbf{LW}$ *and* $\mathbf{V}$ *span the subspaces of r.b.d. of the approximating fields represented by the parameters* $\mathbf{d}$, $\mathbf{d}^t$ *and* $\mathbf{d}^*$, *respectively. Each one of the vectors* $\mathbf{p}$, $\mathbf{t}$ *and* $\mathbf{p}^*$ *represent nodal forces that are in equilibrium if and only if* $\mathbf{W}^{\mathrm{T}}\mathbf{p} = 0$, $\mathbf{W}^{\mathrm{T}}\mathbf{L}^{\mathrm{T}}\mathbf{t} = \mathbf{0}$ *and* $\mathbf{V}^{\mathrm{T}}\mathbf{p}^* = 0$, *respectively.*

4.5 An approximation of the double-layer potential matrix H

Equation (5) may be applied to the boundary nodes [7, 10], thus asserting that u_i^d – from eqn (3) – and u_i^s should coincide along Γ:

$$\mathbf{U}^*\mathbf{p}^* + \mathbf{WCp}^* = (\mathbf{d} - \mathbf{d}^p) \tag{17}$$

where $\mathbf{WCp}^*$ accounts for an amount of r.b.d. that cannot be transformed between the approximating fields whose parameters are $\mathbf{p}^*$ and $\mathbf{d}$. The above equation is a very simple statement, except that there is an embedded amount of r.b.d. and – most important – that the terms of $\mathbf{U}^* = [U^*_{mn}] \in \mathbb{R}^{n^d \times n^d}$ for m and n referring to the same node cannot be directly evaluated.

One may assert on the basis of Theorem 1 that, if the set of parameters $\mathbf{p}^*$ in eqn (17) corresponds to forces in balance, then $\mathbf{V}^{\mathrm{T}}\mathbf{p}^* = \mathbf{0} \Rightarrow \mathbf{WCp}^* = \mathbf{0}$ [10] and the following contragradient statement holds:

$$\mathbf{U}^*\mathbf{p}^* = \mathbf{d}(\mathbf{p}^*) \Rightarrow \mathbf{U}^{*\mathrm{T}}\mathbf{p} = \mathbf{d}^*(\mathbf{p}), \quad \text{provided that } \mathbf{V}^{\mathrm{T}}\mathbf{p}^* = \mathbf{0},\ \mathbf{W}^{\mathrm{T}}\mathbf{p} = \mathbf{0} \tag{18}$$

Then, if one uses eqn (13) to define a set of equivalent nodal forces $\mathbf{p}(\mathbf{t})$ and eqn (16) to define a set of equivalent nodal displacements $\mathbf{d}^*(\mathbf{d})$, the right-hand side of the above equation becomes

$$\mathbf{U}^{*\mathrm{T}}\mathbf{L}^{\mathrm{T}}\mathbf{t} = \mathbf{Hd} \tag{19}$$

By comparing this equation with eqn (9), one concludes that

$$\mathbf{U}^{*\mathrm{T}}\mathbf{L}^{\mathrm{T}} \approx \mathbf{G} \tag{20}$$

which can be formally obtained in the frame of an energy theorem [10, 11].

4.6 An approximation of the single-layer potential matrix G

Equation (17) was obtained by simply asserting that eqn (3) should hold for nodal points along Γ (it actually has a variational basis [5, 7, 10]). A similar assertion may be made for traction forces along Γ,

$$\mathbf{T}^*\mathbf{p}^* = \mathbf{t}(\mathbf{p}^*) \tag{21}$$

with the introduction of the matrix $\mathbf{T}^* = [T^*_{\ell m}] \in \mathbb{R}^{n^t \times n^*}$ of traction forces, obtained by measuring the effect $\sigma^*_{jim} n_j$ at a boundary node and direction characterized by ℓ caused by a unit force p^*_m, according to eqn (4). Application of a contragradient statement leads to [11]

$$\mathbf{T}^*\mathbf{p}^* = \mathbf{t}(\mathbf{p}^*) \quad \Rightarrow \quad \mathbf{T}^{*\mathrm{T}}\mathbf{d}^t = \mathbf{d}^*(\mathbf{d}^t), \quad \text{r.b.d. excluded} \tag{22}$$

The expression on the right-hand side may be written in an amenable format if one resorts to the expressions of $\mathbf{d}^t$ and $\mathbf{d}^*$ in eqns (15) and (16):

$$\mathbf{T}^{*\mathrm{T}}\mathbf{L}\mathbf{d} = \mathbf{H}\mathbf{d}, \quad \text{r.b.d. excluded} \tag{23}$$

which involves only nodal displacements $\mathbf{d}$. Since $\mathbf{H}\mathbf{W} = \mathbf{0}$ for a finite Ω, r.b.d. are automatically excluded in the term on the right-hand side.

Then, one may conclude that

$$\mathbf{T}^{*\mathrm{T}}\mathbf{L} \approx \mathbf{H} \tag{24}$$

provided that the terms of $\mathbf{T}^* = T^*_{\ell m}$, for m and ℓ referring to the same nodal point, are somehow evaluated and that, at least, one makes sure that

$$\mathbf{T}^{*\mathrm{T}}\mathbf{L}\mathbf{W} = 0 \quad \text{for a finite } \Omega \tag{25}$$

5 The expedite boundary element method

Equations (20) and (24) are together the expression of the *expedite* boundary element method – EBEM,

$$\mathbf{T}^{*\mathrm{T}}\mathbf{L}\left(\mathbf{d} - \mathbf{d}^p\right) = \mathbf{U}^{*\mathrm{T}}\mathbf{L}^{\mathrm{T}}\left(\mathbf{t} - \mathbf{t}^p\right) \tag{26}$$

as a reasonable approximation of eqn (9) for the CBEM, here repeated for clarity,

$$\mathbf{H}\left(\mathbf{d} - \mathbf{d}^p\right) = \mathbf{G}\left(\mathbf{t} - \mathbf{t}^p\right)$$

provided that the puzzle of obtaining the still undefined coefficients of $\mathbf{U}^*$ and $\mathbf{T}^*$ is solved. If one prefers to work in terms of equivalent nodal forces, as in the finite element method, eqn (26) can be alternatively written, according to eq. (13), as

$$\mathbf{T}^{*\mathrm{T}}\mathbf{L}\left(\mathbf{d} - \mathbf{d}^p\right) = \mathbf{U}^{*\mathrm{T}}\left(\mathbf{p} - \mathbf{p}^p\right) \tag{27}$$

which is an additional, operational advantage of the proposed EBEM.

5.1 Evaluation of the *undefined* coefficients of $\mathbf{T}^*$

The matrix $\mathbf{T}^*$ of traction forces introduced in eqn (21) is rectangular. However, one is actually interested in obtaining the undefined values of the square matrix given as the product $\mathbf{T}^{*\mathrm{T}}\mathbf{L}$ of either eqn (26) or (27).

The matrix $\mathbf{L}$, as defined in eqn (14), has the same numbers of rows and columns as $\mathbf{T}^*$, but is banded, with non-zero coefficient $L_{\ell m}$ only if the nodal

www.witpress.com, ISSN 1743-355X (on-line)

displacement δd_m and the *traction-force* attribute t_ℓ refer to the same boundary segment (element).

Since $\mathbf{L}$ is banded, the undefined coefficients of $\mathbf{T}^*$ affect the coefficients (m,n) of the product $\mathbf{T}^{*\mathrm{T}}\mathbf{L}$ that refer to the same boundary element. One concludes that the coefficients (m,n) of this matrix product, when referring to different boundary elements, can be obtained directly. On the other hand, the number of undefined coefficients of the product $\mathbf{T}^{*\mathrm{T}}\mathbf{L}$ is larger than the number of undefined coefficients of $\mathbf{T}^*$. Since there is a strong singularity affecting the coefficients of $\mathbf{T}^{*\mathrm{T}}\mathbf{L}$, this must be taken into account, as proposed in the following algorithm [11].

Algorithm for the evaluation of the coefficients of $\mathbf{T}^{*\mathrm{T}}\mathbf{L}$

1. If the indices (m,n) of $\mathbf{T}^{*\mathrm{T}}\mathbf{L}$ refer to a boundary segment that is not adjacent to a singularity, then just evaluate the coefficient as the indicated product.
2. If the indices (m,n) refer to a node that is adjacent to a singularity, then replace the coefficient with the corresponding value of $\mathbf{H}$, eqn (10), which requires the evaluation of a regular integral ($u_{in} = 0$ at the singularity point).
3. If the indices (m,n) refer to a node directly affected by a singularity, evaluate the coefficient by forcing the matrix to be orthogonal to rigid-body displacements. (For an unbounded domain, use the complementary, bounded domain. In case of symmetries, when the number of r.b.d. is not sufficient, additionally apply the problem to a simple analytical solution.)

5.2 Evaluation of the *undefined* coefficients of $\mathbf{U}^*$

Once the undefined coefficients of the product $\mathbf{T}^{*\mathrm{T}}\mathbf{L}$ are evaluated, the best and only way of obtaining the undefined coefficients of $\mathbf{U}^*$ is by applying either eqn (26) or (27) to a sufficient number of simple solutions. Let either $(\mathbf{D}^a, \mathbf{T}^a)$ or $(\mathbf{D}^a, \mathbf{P}^a)$ represent a set with a sufficiently large number of simple analytical solutions of the homogeneous differential equations (1) of the problem that is being modeled, given in terms of displacements and of either surface tractions parameters or equivalent nodal forces. The coefficients about the main diagonal of $\mathbf{U}^*$ are obtained in such a way that either

$$\left|\mathbf{T}^{*\mathrm{T}}\mathbf{L}\mathbf{D}^a - \mathbf{U}^{*\mathrm{T}}\mathbf{L}^\mathrm{T}\mathbf{T}^a\right| = \min \quad \text{or} \quad \left|\mathbf{T}^{*\mathrm{T}}\mathbf{L}\mathbf{D}^a - \mathbf{U}^{*\mathrm{T}}\mathbf{P}^a\right| = \min \qquad (28)$$

with actually more equations than unknowns, in order to have no direction preferences in the evaluation of the coefficients of $\mathbf{U}^*$, which leads to a solution in terms of least squares. For potential problems, the number of constant fluxes is either two or three, for 2D or 3D problems, and just one unknown per node. For general elasticity problems, there are either three or six constant stress states, for 2D or 3D problems, and either 2×2 or 3×3 unknowns (if symmetry is not enforced). This solution scheme is similar to the one adopted in the HBEM for the evaluation of the undefined coefficients of the flexibility matrix $\mathbf{F}$ [2, 3, 5, 10–13].

www.witpress.com, ISSN 1743-355X (on-line)

6 Some numerical examples

Figure 1 shows on the left an irregularly shaped domain for which some of the equations and concepts outlined in this paper are assessed numerically in the frame of the solution of the 2D Laplace equation. The figure has corner coordinates $(0,0)$, $(10,20)$, $(20,0)$, $(15,35)$, $(0,20)$, $(17,19)$, $(16,22)$, $(21,24)$ and $(22,20)$. The four curved boundary segments have radii of curvature 20, 15, 4 and -4. The numerical model is implemented in Fortran with double precision. The scheme shown in Figure 1 has a total of 124 nodes [8]. The problem is modeled using linear, quadratic and cubic elements with differently refined meshes, for the total numbers of nodes shown in Table 1. A series of patch tests is run for potential fields applied to the models, according to Table 2, where the potential $\ln r/2\pi$ refers to either of the indicated source points $A = (-5,2)$, $B = (10,2)$ or $C = (19.5, 20.5)$.

In order to estimate the highest numerical accuracy to be expected in the tests, convergence is first assessed for eqn (9) in the frame of either the CBEM, thus using eqn (6), or of the MBEM, which uses eqn (7) for the interpolation of traction

Table 1: Total numbers of nodal points for the numerical model of Figure 1.

Element type	Total number of nodal points							
Linear	31	62	124	248	496	992	1984	–
Quadratic	–	62	124	248	496	992	1984	3968
Cubic	–	93	186	372	774	1488	2976	

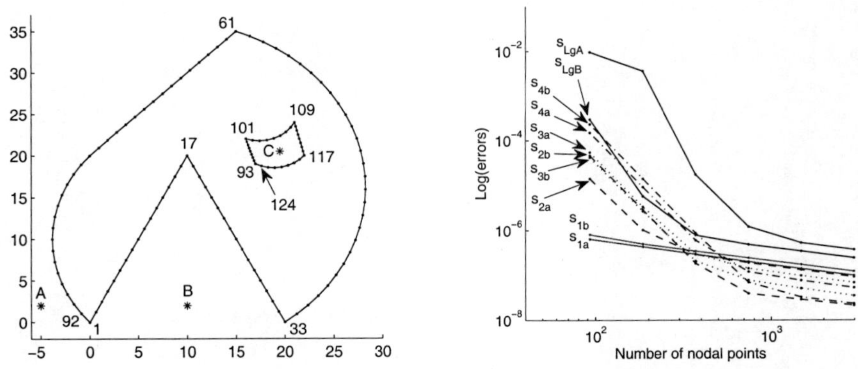

Figure 1: Discretization scheme according to Table 1 and convergence assessments of eqn (9) for the potential fields of Table 2 to establish a benchmark.

Table 2: Potential fields applied to test the numerical model of Figure 1.

S_{1a}	S_{1b}	S_{2a}	S_{2b}	S_{3a}	S_{3b}	S_{4a}	S_{4b}	S_{Lg}
x	y	xy	x^2-y^2	x^3-3xy^2	y^3-3x^2y	$x^4+y^4-6x^2y^2$	x^3y-xy^3	$\ln r/2\pi$

gradients. The results are expressed in terms of the error norm

$$\epsilon\,(\mathbf{Hd}-\mathbf{Gt}) = \sqrt{\sum_m \left(\sum_n H_{mn}d_n - \sum_\ell G_{m\ell}t_\ell\right)^2 \Big/ \sum_m \left(\sum_n H_{mn}d_n\right)^2} \tag{29}$$

for sets of boundary solutions $(\mathbf{d}, \mathbf{t})$ corresponding to the applied potential fields of Table 2. Convergence results for the MBEM with cubic elements are shown on the right of Figure 1. The results for the CBEM are almost indistinguishable from these ones, except when linear potential fields S_{1a} or S_{1b} are applied, as in such a case the error norm of eqn (29) is equal to zero (within numerical integration errors) in the MBEM for curved boundaries, which does not happen with the CBEM. Figure 1 shows the expected convergence pattern of a consistently formulated numerical method up to an error norm $\epsilon \approx 10^{-6}$, when numerical integration errors tend to prevail and accuracy hardly improves with increasing mesh refinement.

Figure 2 shows the convergence patterns for the best and worst numerical results obtained for the sets of potential fields of Table 2. The results on the left correspond to the linear potential field S_{1a}. The CBEM and the MBEM coincide for linear elements (Con_l and Mod_l), as there are only piecewise straight boundary segments. However, the higher accuracy of the MBEM is flagrant for quadratic and cubic elements (Mod_q and Mod_c compared to Con_q and Con_c), with exact values given the threshold of numerical integration errors. The corresponding results with

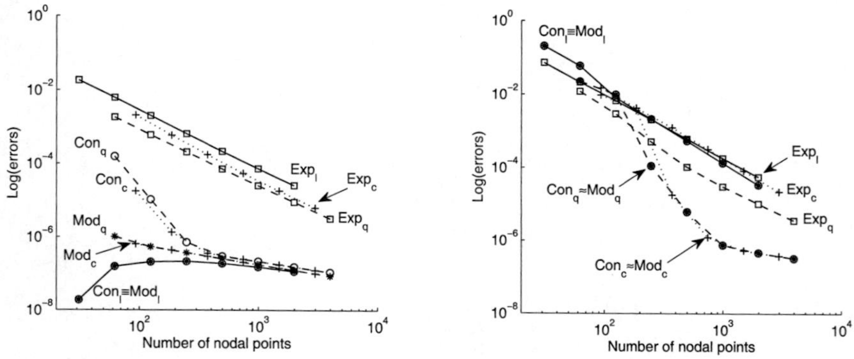

Figure 2: Error norm of eqn (29) for the linear potential field S_{1a} (on the left) and for a logarithmic field with source at point C in Figure 1.

the EBEM for linear, quadratic and cubic elements (Exp_l, Exp_q and Exp_c are also shown. These graphics actually show the error residuals of the least-square procedure used to evaluate the diagonal coefficients of $\mathbf{U}^*$ in eqn (28). Accuracy improves monotonically, although not in the same rate of the CBEM.

The graphics on the right of Figure 2 correspond to the same type of analysis on the left, but for a logarithmic source centered at point C of Figure 1, with error results of eqn (29) that are almost indistinguishable from the ones for a source at point A. For high gradient fields, the results with the CBEM and the MBEM are almost the same [8]. The results with the EBEM are initially comparable to the ones of the CBEM. (In the example shown, the results with the EBEM for a coarse mesh are actually better, but no general conclusions can be drawn.) However, the convergence rate is smaller for the EBEM than in the case of the CBEM. The best results with the EBEM, in this and in other examples for 2D potential problems, are obtained in the implementation with quadratic elements. The results of the EBEM are consistently more accurate than in the implementations of the CBEM using linear elements, a pattern that is also observed in other numerical examples.

Conclusions

An *expedite* formulation of the boundary element method is proposed. No integrations are required, except for a few regular ones for a narrow band of coefficients above and below the main diagonal of the matrix that approximates the double-layer potential matrix $\mathbf{H}$. The improved treatment of boundary traction forces leads to simplifications of the conventional BEM itself and to the construction of the auxiliary kinematic/equilibrium transformation matrix $\mathbf{L}$ in a way that also circumvents integration. Although not shown, one may combine the formulation with an efficient technique (GMRES) for the iterative solution of very large equation systems, which also enables the evaluation of results at internal points with no further post-processing. As proposed, the EBEM promises to be superior to the fast multi-pole methods in concept, implementation and computational efficiency. Application of the formulation to time-dependent problems in the frequency domain is straightforward. An extended version of the present manuscript is being prepared, in which numerical examples of three-dimensional problems are also shown.

Acknowledgements

This project was supported by the Brazilian agencies CAPES, CNPq and FAPERJ.

References

[1] C. A. Brebbia, J. F. C. Telles, and L. C. Wrobel. *Boundary Element Techniques*. Springer-Verlag, Berlin, 1984.

[2] R. A. P. Chaves. *The Simplified Hybrid Boundary Element Method Applied to Time-Dependend Problems (in Portuguese)*. PhD thesis, Pontifical Catholic University of Rio de Janeiro, 2003.

[3] M. F. F. de Oliveira and N. A. Dumont. Conceptual completion of the simplified hybrid boundary element method. In E. J. Sapountzakis and M. H. Aliabadi, editors, *BETeq 2009 - International Conference on Boundary Element Techniques*, pages 49–54, Athens, Greece, July 2009.

[4] T. B. B. DeFigueiredo. *A New Boundary Element Formulation in Engineering*. Lecture Notes in Engineering. Springer-Verlag, Berlin, 1991.

[5] N. A. Dumont. The hybrid boundary element method: an alliance between mechanical consistency and simplicity. *Applied Mechanics Reviews*, 42(11):S54–S63, 1989.

[6] N. A. Dumont. An assessment of the spectral properties of the matrix G used in the boundary element methods. *Computational Mechanics*, 22:32–41, 1998.

[7] N. A. Dumont. Variationally-based, hybrid boundary element methods. *Computer Assisted Mechanics and Engineering Sciences (CAMES)*, 10:407–430, 2003.

[8] N. A. Dumont. The boundary element method revisited. In C. A. Brebbia, editor, *Boundary Elements and Other Mesh Reduction Methods XXXII*, pages 227–238. WITPress, Southampton, U.K., 2010.

[9] N. A. Dumont. From the collocation boundary element method to a meshless formulation. In M. A. Storti E. N. Dvorkin, M. B. Goldschmit, editor, *Mecánica Computacional, MECOM 2010 – IX Argentinean congress on Computational Mechanics and II South American congress on Computational Mechanics, XXXI CILAMCE – XXXI Iberian Latin-American Congress on Computational Methods in Engineering*, pages 4635–4659 (on CD), Buenos Aires, Argentina, 2010.

[10] N. A. Dumont. The hybrid boundary element method – fundamentals (*to be submitted*). *Engineering Analysis with Boundary Elements*, 2011.

[11] N. A. Dumont. Toward a meshless formulation of the simplified hybrid boundary element method (to be submitted). 2011.

[12] N. A. Dumont and R. A. P. Chaves. General time-dependent analysis with the frequency-domain hybrid boundary element method. *Computer Assisted Mechanics and Engineering Sciences*, (10):431–452, 2003.

[13] M. F. F. Oliveira. Conventional, hybrid and simplified boundary element methods (in Portuguese). Master's thesis, Pontifical Catholic University of Rio de Janeiro, 2004.

[14] P. W. Partridge, C. A. Brebbia, and L. C. Wrobel. *The Dual Reciprocity Boundary Element Method*. Computational Mechanics Publications, Southampton, 1992.

A fundamental solution based FE model for thermal analysis of nanocomposites

H. Wang[1,2] & Q. H. Qin[2]
[1]*Institute of Scientific and Engineering Computation, Henan University of Technology, China*
[2]*School of Engineering, Australian National University, Australia*

Abstract

This paper presents a fundamental solution (FS) based finite element (FE) formulation for analyzing the axisymmetric thermal behavior of composites enhanced with carbon nanofibers (CNFs) or carbon nanotubes (CNTs), which are modeled by a cylindrical representative volume element (RVE). The proposed approach utilizes the axisymmetric FS to construct an intra-element approximate field within the element and describes the element boundary field using conventional shape functions. A new hybrid variational functional is developed to establish a linkage between the independent intra-element field and the element boundary fields and generate the final force-displacement equations. Several numerical examples are considered to assess the efficiency and accuracy of the proposed model. The results show that the radius of the nanofiller and the thickness of the interface have little effect on thermal conductivity of the composites, whereas the length of the nanofiller and the material parameters of the interface play an important role in the effective thermal conductivity of the composites.
Keywords: nanocomposites, fundamental solution, hybrid FEM, thermal conductivity, cylindrical representative volume element.

1 Introduction

Over the past decades, nanomaterials have been used increasingly as ideal additives to polymers, due to their excellent thermal, mechanical, and electrical properties over conventional fillers like carbon fiber and glass fiber [1]. Here we

www.witpress.com, ISSN 1743-355X (on-line)
doi:10.2495/BE110171

direct our attention to the thermal analysis of CNF/CNT reinforced composites to simulate the heat dissipation in them, rather than their mechanical simulation.

Currently, the effective thermal properties of nanocomposites have been studied by utilizing experimental techniques [2], molecular scale simulation (i.e. a molecular dynamics (MD) approach [3]), and continuum mechanical models [4]. Among studies based on continuum models, very little numerical work has been carried out to predict the thermal property of nanocomposites. Nishimura and Liu [5] analyzed the thermal behavior of nanocomposites embedded with rigid CNT inclusions by the boundary integral equation (BIE) method with the help of fast multipole expansion. Zhang et al. [6] developed a hybrid boundary node method (BNM) for a multi-domain model, in which both polymer matrix and CNT were modeled separately and the necessary continuity condition on the interface of matrix and CNT was required, and a simplified single-domain model with the assumption that the surface temperature of the CNT filler keeps constant [7]. Singh et al. [8] studied the thermal behavior of nanocomposites by the element-free Galerkin method (EFGM). Furthermore, interphases formed due to chemical reactions between the CNT/CNF and matrix, or the use of protective coatings on the filler during manufacturing, can significantly affect the overall thermal properties of composites. Singh et al. [9] applied the EFGM to analyze interface effects by adjusting interface thickness and interface material parameters.

In the present work, a new hybrid FE formulation called HFS-FEM is developed to analyze the size effect of CNT/CNF and the interface effect on the overall thermal conductivity of nanocomposites. As an extension of our previous works [10–12], in the current computational model, a nanoscale axisymmetric cylindrical representative volume element (RVE) containing a single centered effective solid nanofiber is considered. The FS of axisymmetric problems are employed to construct the intra-element temperature field and an independent frame field along the element boundary is defined using conventional shape functions. Then, a new hybrid variational functional is presented to link two assumed fields and to generate the final solving system including boundary integrals only. During the solution process, no any hypersingular integral is involved, though the FS is used, and the element shape can theoretically be arbitrary.

The paper begins with a basic description of axisymmetric heat conduction problems in a nanoscale cylindrical RVE in Section 2. Then, the derivation of the proposed approach is described in Section 3 to provide full insight into the hybrid FE model. The effects of the dimensions of the nanofiller, including its length and diameter, as well as the interface between the nanofiller and matrix are analyzed numerically in Section 4 and some concluding remarks are presented in Section 5.

2 Mathematical model for axisymmetric problems

2.1 General mathematical model of axisymmetric heat conduction

Without loss of generality, consider heat conduction in an orthotropic anisotropic homogeneous medium. In a cylindrical coordinate system (r,θ,z) the governing equation can be expressed as

$$\frac{\partial}{\partial r}\left(k_r\frac{\partial u}{\partial r}\right)+k_r\frac{1}{r}\frac{\partial u}{\partial r}+\frac{1}{r^2}\frac{\partial}{\partial\theta}\left(k_\theta\frac{\partial u}{\partial\theta}\right)+\frac{\partial}{\partial z}\left(k_z\frac{\partial u}{\partial z}\right)=0,\qquad\forall\mathbf{x}\in\Omega.\tag{1}$$

and the heat flux components related to the temperature gradient are modeled by Fourier's law given in the following formulation

$$q_r=-k_r\frac{\partial T}{\partial r},\ q_\theta=-\frac{k_\theta}{r}\frac{\partial T}{\partial\theta},\ q_z=-k_z\frac{\partial T}{\partial z}.\tag{2}$$

where k_r, k_θ and k_z denote the thermal conductivity along the r, θ and z directions, respectively; u is the sought temperature field in terms of spatial cylindrical coordinate $\mathbf{x}=(r,\theta,z)$.

When there is an axisymmetric geometry and boundary conditions about a reference axis, say z axis, that is, all quantities such as temperature and heat flux components are independent of the circumferential direction (θ direction), the governing equation (1) for three-dimensional cases reduces to

$$\frac{\partial}{\partial r}\left(k_r\frac{\partial u}{\partial r}\right)+k_r\frac{1}{r}\frac{\partial u}{\partial r}+\frac{\partial}{\partial z}\left(k_z\frac{\partial u}{\partial z}\right)=0,\qquad\forall\mathbf{x}\in\Omega.\tag{3}$$

To keep the system solvable, temperature and heat flux boundary conditions should be added

$$u=\overline{u},\ \text{on }\Gamma_u;\quad q=\overline{q},\ \text{on }\Gamma_q.\tag{4}$$

where $q=-k_iu_{,i}n_i$ is the boundary normal heat flux, where in the axisymmetric model the subscript i represents the coordinates r (i=1) and z (i=2), respectively. n_i are the direction cosines of the unit outward normal vector $\mathbf{n}$ to the boundary $\Gamma=\Gamma_u\cup\Gamma_q$ of the domain of interest, and $\overline{u}$ and $\overline{q}$ are specified functions on the corresponding boundaries, respectively. For convenience, the space derivatives are indicated above by a comma, i.e. $u_{,r}=\partial u/\partial r$.

For multi-material problems, the interface continuity conditions for temperature and heat flux should be complemented, i.e. for the subdomains Ω_i and Ω_j with different thermal properties k_i and k_j, respectively. On the interface $\Omega_i\cap\Omega_j$ we have continuity of temperature and reciprocity of heat flux

$$u_i=u_j,\qquad q_i+q_j=0.\tag{5}$$

2.2 Nanoscale cylindrical representative volume element (RVE)

In the present work, a nanoscale cylindrical RVE (see Fig. 1a) is used to evaluate the effective thermal conductivity of the nanocomposites. The nanofiller is placed symmetrically at the center of the cylindrical RVE such that the axis of the RVE of interest coincides with the axis of the nanofiller. Assume that there is perfect bonding on the interface between the nanofiber and the interphase, and the interphase and the surrounding matrix. The top and bottom surfaces of the RVE are maintained at two different constant temperatures u_1 and u_0, respectively, and the outer surface is kept insulated. In Fig. 1, the cylindrical RVE has length L and radius R_0, and the nanofiber has length l and radius r_0.

Because the specified geometry and boundary conditions are axisymmetric, a simplified 2D axisymmetric computational model is chosen as a typical rotating plane shown in Fig. 1b, which will be solved by means of the hybrid FE model developed in the present work.

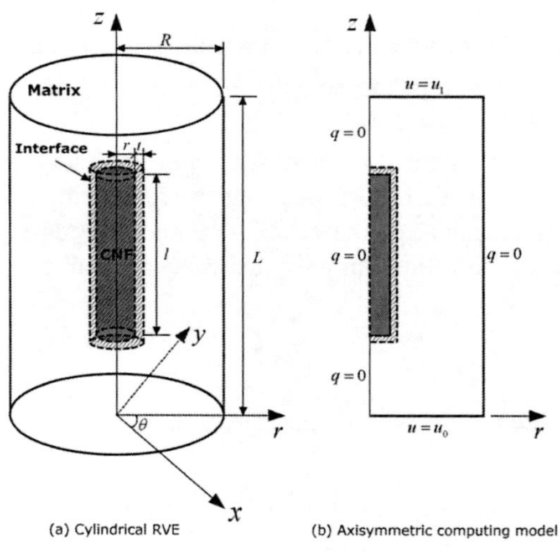

Figure 1: Nanoscale axisymmetric cylindrical RVE containing single CNT.

According to Fourier's law, the thermal conductivity along the z-direction is defined as

$$k_z = -\frac{q_z}{\partial u / \partial z} = -\frac{q_z}{(u_1 - u_0)/l} = -\frac{q_z l}{u_1 - u_0}. \tag{6}$$

which can be used to determine the effective thermal conductivity k_e of the nanocomposite in the direction parallel to the fiber alignment by simply adjusting as

$$k_e = -\frac{q_z^{\text{ave}} l}{u_1 - u_0}. \tag{7}$$

where q_z^{ave} denotes the average values on the data-collection surfaces, i.e. the top and bottom surfaces of the cylindrical RVE. In Eq. (6), the fact that while the temperature varies linearly along the z-direction, then the heat flux component q_z remains constant, has been used.

2.3 FS for axisymmetric model

The FS is critically important for implementation of the proposed algorithm. If the medium is isotropic, that is $k = k_r = k_z$, the FS for axisymmetric problems is usually derived by integrating the FS of three-dimensional steady-state heat conduction along the circumstance direction [13] and setting the source point $\mathbf{x}_s = (r_s, z_s)$ on the Oxz plane. As a result, we obtain

$$u^*(\mathbf{x}, \mathbf{x}_s) = -\int_0^{2\pi} \frac{1}{4\pi\rho k} \mathrm{d}\theta = -\frac{K(m)}{k\pi\sqrt{a+b}}. \tag{8}$$

with

$$m = \frac{2b}{a+b}, \quad a = r^2 + r_s^2 + (z - z_s)^2, \quad b = 2rr_s, \quad K(m) = \int_0^{\pi/2} \frac{\mathrm{d}t}{\sqrt{1 - m\sin^2 t}}.$$

where $\mathbf{x} = (r, z)$ is a field point to be considered, ρ is the distance between the field point and source point, and $k(m)$ represents the complete elliptic integral of the first kind.

3 FS based FE formulation

In this section, a hybrid FE model with FS as an intra-element trial function is developed for the boundary value problem (BVP) defined by Eqs. (3) and (4).

3.1 Non-conforming intra-element field

Motivated by the idea of the method of FS (MFS) [14] to remove the singularity of the FS, for a particular element e occupying sub-domain Ω_e, the intra-element temperature field is extracted from a linear combination of FS, that is,

$$u_e(\mathbf{x}) = \sum_{j=1}^{n_s} G(\mathbf{x}, \mathbf{x}_{sj}) c_{ej} = \mathbf{N}_e(\mathbf{x})\mathbf{c}_e, \qquad \forall \mathbf{x} \in \Omega_e, \mathbf{x}_{sj} \notin \Omega_e. \tag{9}$$

where c_{ej} is undetermined coefficients and n_s is the number of virtual sources outside the element e, which can be generated by the formulation in literatures [10–12].

To guarantee the requirement of $rr_s \geq 0$ in the evaluation of the elliptic integrals of the first and second kinds, the local coordinate system $(\breve{r}, \breve{z})$ must be employed, whose origin can be determined by means of the minimum values of the coordinates of source points.

The corresponding normal heat flux on Γ_e is given by

$$q_e = \begin{bmatrix} n_r & n_z \end{bmatrix} \begin{bmatrix} q_r \\ q_z \end{bmatrix} = \mathbf{Q}_e \mathbf{c}_e \cdot \tag{10}$$

where

$$\mathbf{Q}_e = -k \begin{bmatrix} n_r & n_z \end{bmatrix} \begin{bmatrix} \mathbf{N}_{e,r} \\ \mathbf{N}_{e,z} \end{bmatrix}. \tag{11}$$

3.2 Auxiliary conforming frame field

In order to enforce conformity on the field variable *u*, for instance, $u_e = u_f$ on $\Gamma_e \cap \Gamma_f$ of any two neighboring elements *e* and *f*, an auxiliary inter-element frame field $\tilde{u}$ is used and expressed in terms of the same degrees of freedom (DOF), $\mathbf{d}_e$, as used in conventional FEM

$$\tilde{u}_e(\mathbf{x}) = \tilde{\mathbf{N}}_e(\mathbf{x})\mathbf{d}_e, \qquad \mathbf{x} \in \Gamma_e. \tag{12}$$

where $\tilde{\mathbf{N}}_e$ represents the conventional shape functions used in conventional FEM.

3.3 Modified variational principle and stiffness equation

For the BVP defined in Eqs. (3) and (4), since the stationary conditions of the traditional potential or complementary variational functional cannot guarantee satisfaction of the inter-element continuity condition required in the proposed HFS FE model, a modified potential functional is developed as follows:

$$\frac{\Pi_{me}}{\pi} = -\int_{A_e} \left\{ k_r \left(\frac{\partial u}{\partial r} \right)^2 + k_z \left(\frac{\partial u}{\partial z} \right)^2 - 2Gu \right\} r\mathrm{d}A - \int_{\Gamma_{qe}} \bar{q}\tilde{u} r\mathrm{d}\Gamma + \int_{\Gamma_e} (\tilde{u} - u) qr\mathrm{d}\Gamma. \tag{13}$$

in which the governing equation (3) is assumed to be satisfied, *a priori*, due to the use of FS as intra-element trial functions.

Having independently defined the intra-element field and frame field in a particular element *e*, the next step is to generate the element stiffness equation by substituting those fields into the proposed functional (13).

Due to the constant coefficient π appearing in Eq. (13) having no effect on the final result, it is discarded. As a result, the variational functional Π_{me} without the heat generation *G* can be rewritten as

$$\Pi_{me} = -\frac{1}{2} \int_{A_e} \left\{ k_r \left(\frac{\partial u}{\partial r} \right)^2 + k_z \left(\frac{\partial u}{\partial z} \right)^2 \right\} r\mathrm{d}A - \int_{\Gamma_{qe}} \bar{q}\tilde{u} r\mathrm{d}\Gamma + \int_{\Gamma_e} (\tilde{u} - u) qr\mathrm{d}\Gamma. \tag{14}$$

Applying the Gaussian theorem to the element functional (14), we finally have the following functional defined on the element boundary

$$\Pi_{me} = -\frac{1}{2} \int_{\Gamma_e} qur\mathrm{d}\Gamma - \int_{\Gamma_{qe}} \bar{q}\tilde{u} r\mathrm{d}\Gamma + \int_{\Gamma_e} q\tilde{u} r\mathrm{d}\Gamma. \tag{15}$$

which, by substituting Eqs. (9), (10) and (12) into the functional (15), yields

$$\Pi_e = -\frac{1}{2}\mathbf{c}_e^{\mathrm{T}}\mathbf{H}_e\mathbf{c}_e - \mathbf{d}_e^{\mathrm{T}}\mathbf{g}_e + \mathbf{c}_e^{\mathrm{T}}\mathbf{G}_e\mathbf{d}_e. \tag{16}$$

with

$$\mathbf{H}_e = \int_{\Gamma_e} \mathbf{Q}_e^{\mathrm{T}}\mathbf{N}_e r\mathrm{d}\Gamma, \quad \mathbf{G}_e = \int_{\Gamma_e} \mathbf{Q}_e^{\mathrm{T}}\tilde{\mathbf{N}}_e r\mathrm{d}\Gamma, \quad \mathbf{g}_e = \int_{\Gamma_{eq}} \tilde{\mathbf{N}}_e^{\mathrm{T}}\bar{q}r\mathrm{d}\Gamma. \tag{17}$$

Minimization of the functional Π_e with respect to $\mathbf{c}_e$ and $\mathbf{d}_e$, respectively, gives

$$\frac{\partial\Pi_e}{\partial\mathbf{c}_e^{\mathrm{T}}} = -\mathbf{H}_e\mathbf{c}_e + \mathbf{G}_e\mathbf{d}_e = \mathbf{0}, \quad \frac{\partial\Pi_e}{\partial\mathbf{d}_e^{\mathrm{T}}} = \mathbf{G}_e^{\mathrm{T}}\mathbf{c}_e - \mathbf{g}_e = \mathbf{0}. \tag{18}$$

from which the optional relationship between $\mathbf{c}_e$ and $\mathbf{d}_e$, and the stiffness equation, can be produced

$$\mathbf{K}_e\mathbf{d}_e = \mathbf{g}_e, \qquad \mathbf{c}_e = \mathbf{H}_e^{-1}\mathbf{G}_e\mathbf{d}_e. \tag{19}$$

where $\mathbf{K}_e = \mathbf{G}_e^{\mathrm{T}}\mathbf{H}_e^{-1}\mathbf{G}_e$ stands for the element stiffness matrix.

Assembling the element stiffness matrix $\mathbf{K}_e$ and the right-hand vector $\mathbf{g}_e$ gives the following set of equations

$$\mathbf{Kd} = \mathbf{g}\,. \tag{20}$$

where **K**, **d** and **g** are global quantities.

4 Numerical results and discussion

In this section, the proposed axisymmetric hybrid FE model for the thermal analysis of the cylindrical RVE is first validated by comparing the results obtained to those from ABAQUS. Then, the hybrid model is applied to study the effects of nanofiber and the interphase on the effective thermal conductivity of the resulting composites. The material elastomer S160, having a thermal conductivity 0.56 W/mK, is chosen as matrix, while the nanofiber is assumed to have a thermal conductivity 1000 W/mK. The thermal property of the interphase is assigned over the interval $[0.01 \sim 1] \times 0.56$ W/mK. Additionally, for the sake of convenience, the remaining data used in the analysis is tabulated in Table 1.

In the calculation, a total of 160 eight-node elements with 569 nodes are used to model the computing domain for different cases, i.e. $t_i = 1.25\mathrm{nm}$ and $t_i = 2.5\mathrm{nm}$, illustrated in Fig. 2, in which the red region represents the interface elements, the yellow region denotes the nanofiber elements, and the white region represents the matrix elements. The computing codes are written in MATLAB and used to evaluate the HFS-FEM results of the problem.

Table 1: Geometry and boundary conditions for cylindrical RVE.

Parameters	*Values*
RVE length L	100nm
RVE radius R_0	10nm
Nanofiber length l	25nm, 50nm, 75nm, 100nm
Nanofiber radius r_0	2.5nm, 5nm, 7.5nm
Interface thickness t_i	1.25nm, 2.5nm
Temperature at the upper surface u_1	100K
Temperature at the lower surface u_0	200K

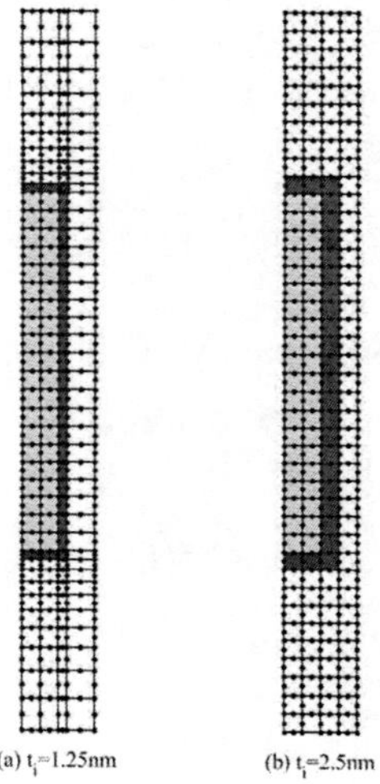

Figure 2: Two mesh divisions used for HFS-FEM.

4.1 Validation of the proposed approach

To validate the proposed algorithm, the temperature distributions in the RVE considered are calculated and compared with those from ABAQUS, in the absence of an interfacial layer. Since the thermal conductivity of the filler is about 1786 times higher than that of the matrix, this difference is sufficient to permit most of the heat flux to pass through the fiber portion of the composites. To examine the thermal effect in the cylindrical system, the length and radius of the fiber are assumed to be 50nm and 5nm, respectively. Fig. 3 plots the temperature distribution along the line $r = 0$, 5, 7.5, 10 nm, and all results are compared to those from ABAQUS with the same meshes. As expected, there is a reasonably good agreement between them. Moreover, the temperature distribution for the case without the filler is also provided in Fig. 3 to demonstrate the effect of the filler. It can be clearly seen that the presence of the filler assists in reducing the overall temperature in the lower region ($0 \le 2z \le L-l$), whereas the overall temperature in the upper region, that is $L+l \le 2z \le 2L$, increases, compared to the corresponding temperature without fiber. Simultaneously, we also find that the temperature in the fiber is almost constant (about 149.8K). Moreover, near the tip of the fiber, the temperature changes at a more rapid rate than the surface temperature of the filler.

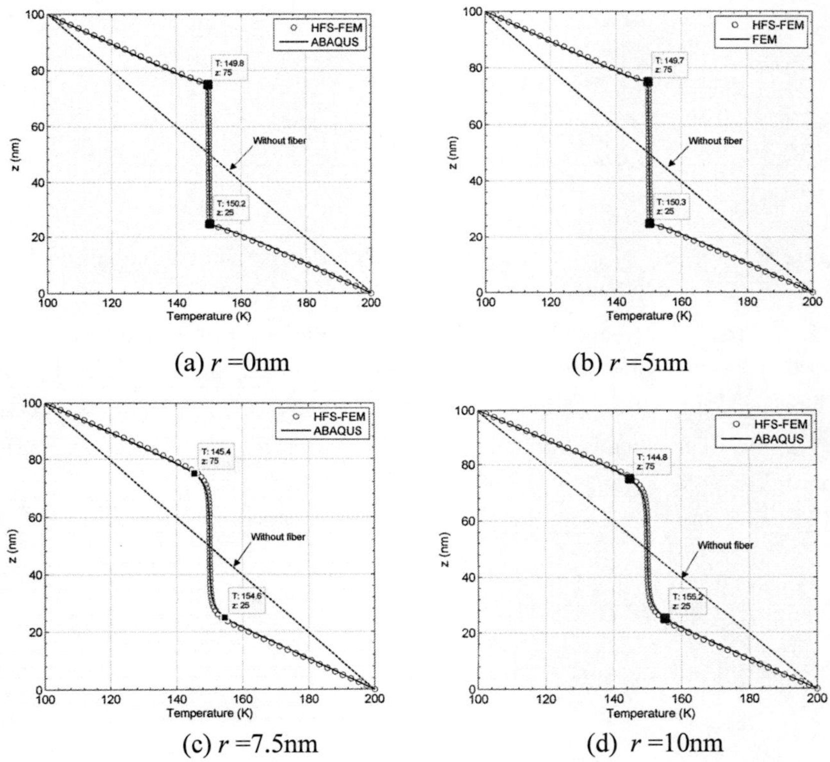

Figure 3: Temperature distribution with $l = 50$ nm and $r_0 = 5$ nm.

4.2 Effect of nanofiber size on thermal properties of the composites

To investigate the size effect of the nanofiber on the overall thermal properties of the nanocomposite in the absence of an interfacial layer, two different cases are considered. In the first case the length of the nanofiber is kept constant and its radius changes; for example, we take $l = 50$ nm and $r_0 = 2.5$nm, 5.0nm and 7.5nm. The results given in Fig. 4 show that an increase in the radius of the fiber induces a larger value of the effective thermal conductivity k_e of the composites. However, this effect is not very significant as the effective thermal property of the nanocomposite changes only slightly even when the radius is increased from 2.5nm to 7.5nm.

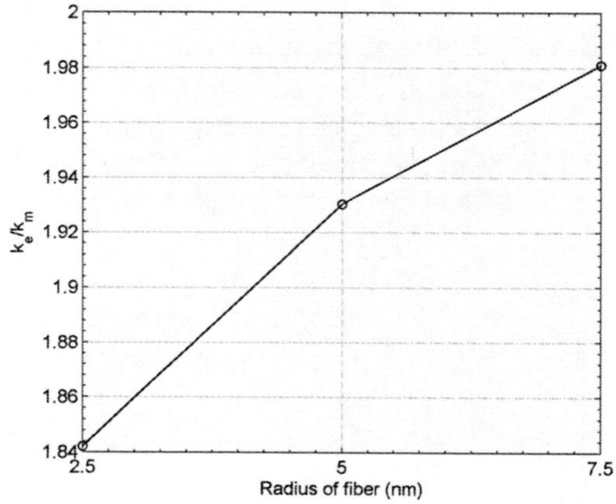

Figure 4: Effect of nanofiber radius on the effective thermal conductivity of nanocomposite with constant length $l = 50$ nm.

In contrast, the length of the fiber has a more significant effect on the effective thermal conductivity k_e. The results in Fig. 5 show that for a given radius $r_0 = 5\text{nm}$, when the length l of the filler increases from 25nm to 100nm, that is, the filler extends through the matrix, the ratio k_e / k_m rapidly increases from 1.3 to 14.5.

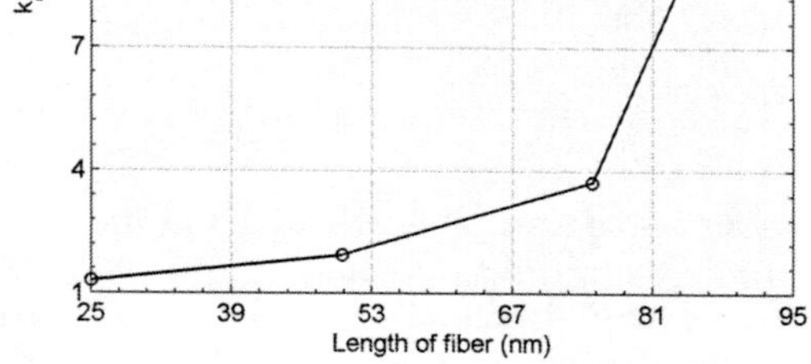

Figure 5: Effect of nanofiber length on the effective thermal conductivity of nanocomposite with constant radius $r_0 = 5$ nm.

4.3 Effect of the interphase on effective thermal conduction

To investigate the interface effect on the thermal conductivity of the nanocomposites, various values of interface conductivity and thickness are considered. In the computation, the length and radius of the fiber remain invariant at 50nm and 5nm, respectively, while the thermal conductivity of the interphase changes in the range [0.01, 1]×0.56 W/mK, making the thickness of the interphase 1.25nm and 2.5nm, respectively. Fig. 6 shows the variation of effective thermal conductivity k_e of the composites by considering the existence of interphase. It can be seen that the interface thickness has an insignificant effect on the effective property of the composite. Furthermore, variation of the interface property has a significant effect on the effective conductivity of the composite, and a decrease in the value of interface conductivity induces a decrease in the overall conductivity of the composite.

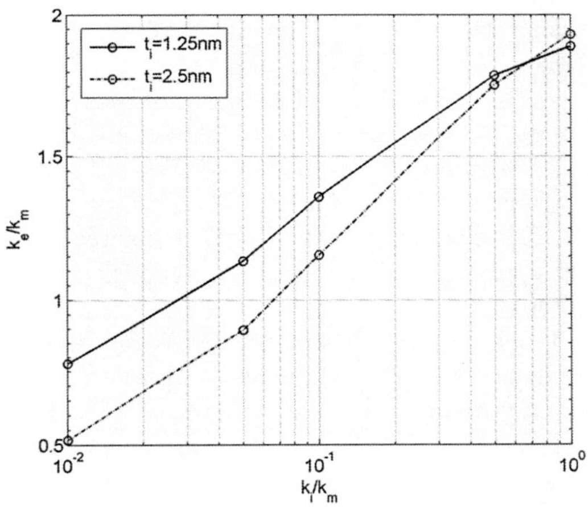

Figure 6: Effect of interface thickness and conductivity on the effective thermal conductivity of the composite.

5 Conclusion

In this paper an axisymmetric model of heat conduction in a cylindrical RVE is developed and used to study the effective thermal properties of nanocomposites. The proposed hybrid FE formulation involves element boundary integrals only, by virtue of use of the FS of the problem as an intra-element trial function. Using the proposed model, the effective thermal conductivity is calculated for various model parameters, including various sizes of nanofiller, and various thickness and material parameters of the interface. The numerical results show that the surface temperature of the nanofiller remains almost constant during heat transfer, and the overall thermal properties of the nanocomposites are affected largely by the length of the nanofiller and the material parameters of the interface, but only slightly affected by the radius of the nanofiller and the interfacial thickness.

References

[1] Moniruzzaman, M., Sahin, A. & Winey, K.I., Improved mechanical strength and electrical conductivity of organogels containing carbon nanotubes. *Carbon*, **47**, pp. 645-650, 2009.

[2] Agarwal, S., Khan, M.M.K. & Gupta, R.K., Thermal conductivity of polymer nanocomposites made with carbon nanofibers. Polymer Engineering and Science, **48**, pp. 2474-2481, 2008.

[3] Clancy, T.C. & Gates, T.S., Modeling of interfacial modification effects on thermal conductivity of carbon nanotube composites. *Polymer*, **47**, pp. 5990-5996, 2006.

[4] Liu, Y.J. & Chen, X.L., Continuum models of carbon nanotube-based composites using the boundary element method. *Electronic Journal of Boundary Elements*, **1**, pp. 316-335, 2003.

[5] Nishimura, N. & Liu, Y.J., Thermal analysis of carbon-nanotube composites using a rigid-line inclusion model by the boundary integral equation method. *Computational Mechanics*, **35**, pp. 1-10, 2004.

[6] Zhang, J.M., Tanaka, M., Matsumoto, T. & Guzik, A., Heat conduction analysis in bodies containing thin-walled structures by means of hybrid BNM with an application to CNT-based composites. *JSME-Series A*, **47**, pp. 181-188, 2004.

[7] Zhang, J.M., Tanaka, M. & Matsumoto, T., A simplified approach for heat conduction analysis of CNT-based nano-composites. *Computer Methods in Applied Mechanics and Engineering*, **193**, pp. 5597-5609, 2004.

[8] Singh, I.V., Tanaka, M. & Endo, M., Thermal analysis of CNT-based nano-composites by element free Galerkin method. *Computational Mechanics*, **39**, pp. 719-728, 2007

[9] Singh, I.V., Tanaka, M. & Endo, M., Effect of interface on the thermal conductivity of carbon nanotube composites. *International Journal of Thermal Sciences*, **46**, pp. 842-847, 2007.

[10] Wang, H. & Qin, Q.H., Hybrid FEM with Fundamental Solutions as trial functions for Heat Conduction Simulation. *Acta Mechanica Solida Sinica*, **22**, pp. 487-498, 2009.

[11] Wang, H. & Qin, Q.H., FE approach with Green's function as internal trial function for simulating bioheat transfer in the human eye. *Archives of Mechanics*, **62**, pp. 493-510, 2010.

[12] Wang, H. & Qin, Q.H., Fundamental-solution-based finite element model for plane orthotropic elastic bodies. *European Journal of Mechanics-A/Solids*, **29**, pp. 801-809, 2010.

[13] Brebbia, C.A. Telles, J.C.F. & Wrobel, L.C., *Boundary element techniques*, Springer-Verlag: Berlin and New York, 1984.

[14] Wang, H. & Qin, Q.H., A meshless method for generalized linear or nonlinear Poisson-type problems. *Engineering Analysis with Boundary Elements*, **30**, pp. 515-521, 2006.

On STEM of tectonic stress fields in tsunami regions

J. Irsa[1,2] & A. N. Galybin[1,3]
[1]*Wessex Institute of Technology, Southampton, UK*
[2]*Florida Institute of Technology, USA*
[3]*Institute of Physics of the Earth, RAS, Moscow, Russia*

Abstract

We apply the STEM (Stress Trajectories Element Method) to investigate changes in the stress states as a result of a great earthquake. Two regions are considered: 1) the region affected by the great Sumatran earthquake (26th of December 2004, 9.1M), taking into account data collected before and after the event separately; 2) the region where the recent Japanese earthquake (11th of March 2011, 9.0M) occurred. The latter is analysed taking into account data collected before the event only while the former case allows one to consider the changes in data due to the significant increase of measurements since the 2004 event. The analysis is based on the stress orientation data from the World Stress Map (WSM) database and the geometrical model of tectonic plate boundaries. The method has proved to be able to recover the stress fields from this type of data with satisfactory accuracy. The modelling utilizes the Trefftz type approach for the complex potentials in plane elastic problems and assumes continuity of the stress vector across the tectonic plate boundaries. As a result of the modelling we obtain stress trajectories, maximum shear stress and mean stress, and the plate driving forces expressed in terms of normal and tangential stresses. The results for the two time subsets revealed changes in the stress patterns induced by the 26/12/2004 event and moreover the stress pattern before the great Sumatran Earthquake was very similar to the one near Japan.
Keywords: world stress map, stress reconstruction, stress changes, Sumatran earthquake, Japan earthquake.

WIT Transactions on Modelling and Simulation, Vol 52,
www.witpress.com, ISSN 1743-355X (on-line)
doi:10.2495/BE110181

1 Introduction

The great Sumatra-Andaman earthquake on the 26.12.2004 with the magnitude of 9.1 is the 3rd greatest earthquake since 1900, and the Japanese earthquake on the 11.03.2011 is the 4th largest [4]. The motivation of this study is to investigate the changes in tectonic stress fields induced by the 26.12.2004 event and also analyse the region where the Japan earthquake occurred, based on the developed numerical method, STEM [3]. The first area for the analysis is the Sunda trench region where data before and after the event are available. The second is the Japan region where data are available only before the date of the event.

The approach uses the data on stress orientations from the World Stress Map database [1] and the tectonic plate boundaries database PB2002 [2]. The numerical algorithm is based on the previous study [3] but it has been significantly modified in order to take the multiple plate margins into account. Using the modified STEM approach we determine not only the stress state inside the tectonic plates but also the boundary forces on the tectonic plates. It is assumed that the tangential and normal stresses are continuous across the plate margins.

The method is applied to model the stress states of the following regions: *Sunda region* (longitude 80E–110E, latitude 20S–10N) and *Japan region* (longitude 125°E–150°E, latitude 27°N–45°N). The results allow us to observe the stress changes after the great Sumatran earthquake and also to compare the two regions based on their stress patterns.

2 Methodology

2.1 Stress orientations data and plate boundaries data

The stress orientations data are taken from the WSM database, release 2008, containing over 21000 data. Most of the data are of quality A-C, which is considered to have accuracy ±25° in the stress orientation [1]. A single entry in the database shows location, azimuth of the major principal stress, quality, depth, stress regime, date/time, method, magnitude of the earthquake (if from earthquake), some other information as focal mechanism, etc. The Sumatra region includes of 854 data, which is further divided into two groups of 269 data for the CASE 1 (before event) and 585 data for the CASE 2 (after event). The Japan region includes 982 data (all before the event).

The digital boundaries of the tectonic plates are taken from the database PB2002 [2], describing 52 plates. The Bird's data and WSM data for the Sumatra region are shown in Figure 1, and for the Japan region in Figure 2.

2.2 Assumptions in modelling

The following assumptions are accepted:

The lithosphere is considered to be elastic and of constant (effective) thickness.

www.witpress.com, ISSN 1743-355X (on-line)

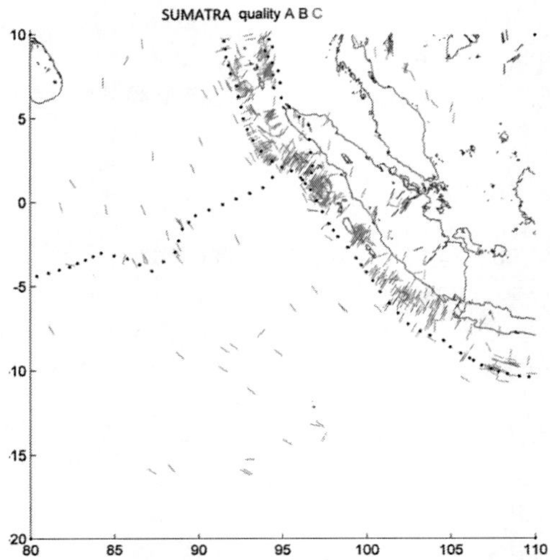

Figure 1: Sumatra region, with WSM data on stress orientations and (short segments) bird's plate boundaries data (dots).

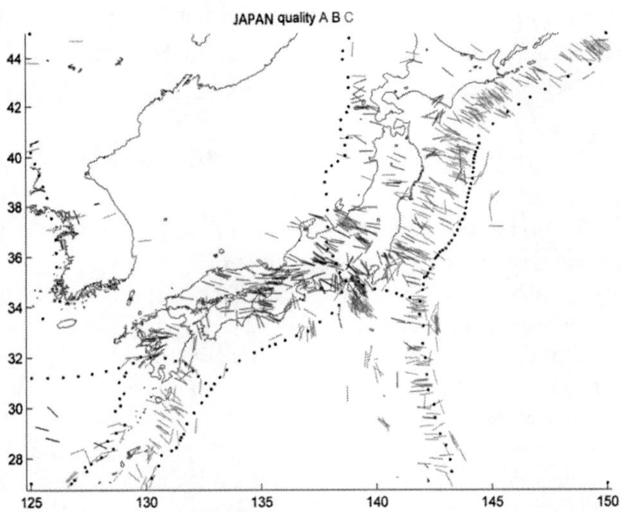

Figure 2: Japanese region with WSM data on stress orientations (short segments) and bird's plate boundaries data (dots).

- The tectonic stresses fields are analysed based on the plane stress assumption that is a consequence of the fact that one of the principal stresses is usually oriented vertically [5–7]. Therefore the 3D stress field in the Earth's crust can be decomposed into 2D plane stress and 1D, which mainly presents the effect of gravity.
- The Earth's curvature is neglected due to the sizes.

- No changes in the stress fields are assumed to be before the earthquake and after it (Sumatra region) and no changes in the stress fields are assumed to be before the Japan earthquake, i.e. the change has occurred right after the earthquake.

2.3 Mathematical model

The 2D stress state in an elastic body in the absence of body forces can be fully described by the complex potentials $\Phi(z), \Psi(z)$ [8], where $z = x + iy$ is complex variable. The mean stress and the deviatoric stress can be determined from the complex potentials as:

$$P(z,\bar{z}) = 2\,\mathrm{Re}[\Phi(z)]$$
$$D(z,\bar{z}) = \bar{z}\Phi'(z) + \Psi(z)\text{,} \tag{1}$$

where $\bar{z} = x - iy$ is the conjugated complex variable.

The stress deviator function can be expressed in the exponential form $D(z,\bar{z}) = |\tau_{\max}(z,\bar{z})|e^{i\alpha(z,\bar{z})}$ where the modulus represents the maximum shear stress $\tau_{\max}(z,\bar{z})$ and the argument is connected with the principal directions θ as $\alpha(z,\bar{z}) = -2\theta(z,\bar{z})$ (compression is assumed to be positive). The complex conjugation stress vector on the contour is found from the stress functions as follows:

$$\sigma_n(\zeta) - i\sigma_t(\zeta) = P(\zeta) + e^{2i\gamma(\zeta)}D(\zeta), \zeta \in \Gamma \tag{2}$$

where γ is the angle between the tangent to the contour and the x-axis, $P(\zeta)$ and $D(\zeta)$ are the boundary values of the stress functions.

The stress components are found from the complex potentials by the Kolosov-Muskhelishvili formulae [8], thus the stress field is fully determined everywhere within the 2D body by the complex potentials.

It is assumed that the entire domain is composed from a number of subdomains. The subdomains are subdivided into triangular elements. The collocation method is used to ensure continuity across the interfaces between the adjacent elements. For this purpose several collocation points are placed on every element interface and the continuity conditions are forced at these points as explained below. Two types of element interfaces are present: *boundary element interfaces BEI* (presenting the plate boundaries) and *domain element interfaces DEI* (interfaces inside the subdomains). The collocation points placed on the DEI assume continuity of the complex potentials, while the collocation points on BEI assume continuity of the stress vector. The previous approach [3] assumed continuity of two potentials $\Psi(z), \Phi'(z)$ across the adjacent elements, the subdomain boundaries have not been taken into account. As a result only the complex-valued stress deviator function (resp. maximum shear stress and stress trajectories) has been determined.

In the present study the continuity equations are imposed on $\Phi(z)$, $\Psi(z)$, $\Phi'(z)$ (across the elements) and the stress vector (across the plate boundaries). The approximations for the complex potentials $\Psi(z)$, $\Phi'(z)$ within the elements are taken in the form of linear polynomials. The approximation for $\Phi(z)$ is obtained by the integration of $\Phi'(z)$. Thus the approximating functions are sought in the form:

$$\Phi'(z) = a + bz,\ \Phi(z) = az + \frac{1}{2}bz^2 + e,\ \Psi(z) = c + dz\ , \tag{3}$$

These functions satisfy *a-priory* the governing equations within the elements. The parameters a,b,c,d are complex while the parameter e is real. This is due to the fact that the constant in the imaginary part of $\Phi(z)$ does not affect the stresses.

The continuity of the complex potentials yields the following set of equations:

$$\begin{aligned} \Phi'(z_c)\big|_{nc} - \Phi'(z_c)\big|_{mc} &= 0 \\ \Phi(z_c)\big|_{nc} - \Phi(z_c)\big|_{mc} &= 0 \\ \Psi(z_c)\big|_{nc} - \Psi(z_c)\big|_{mc} &= 0 \end{aligned}\ , \tag{4}$$

where n_c and m_c are the numbers of adjacent elements and z_c is the position of a particular collocation point on the boundary between these elements lying in the same subdomain, $\gamma_c=\gamma(z_c)$.

Continuity of the stress vector results in the following equations:

$$P(z_s,\bar{z}_s)\big|_{n_s} + e^{2i\gamma_s} D(z_s,\bar{z}_s)\big|_{n_s} - P(z_s,\bar{z}_s)\big|_{m_s} - e^{2i\gamma_s} D(z_s,\bar{z}_s)\big|_{m_s} = 0\ , \tag{5}$$

where n_s and m_s are the numbers of adjacent elements and z_S is the position of a particular collocation point on the boundary between subdomains, $\gamma_s=\gamma(z_s)$.

The data on stress orientations are directly implemented in the system of linear algebraic equations, based on the fact that the maximum shear stress is a real-valued function of the complex variables. Thus, the following equation is valid at every data point:

$$\mathrm{Im}\left[D(z_j,\bar{z}_j)\, e^{2i\theta(z_j)}\right] = 0\ , \tag{6}$$

where N is the number of known data $\theta(z_j)$, $j = 1,\ldots,N$.

Equations (4)–(6) form the system of equations for the determination of unknown coefficients in (3). Due to the homogenous right hand side of this system, at least one extra equation has to be used to provide consistency of the system. This equation expresses normalization of the complex stress deviator by unknown positive constant that has dimension of stresses. Thus, the maximum shear stress depends upon a free multiplicative parameter A (A>0), which cannot be determined in the framework of this modelling. This condition yields that the sum of all $\tau_{\max}(z,\bar{z})$ placed on the known data points is equal to the number of data points:

$$\sum_{j=1}^{N} D(z_j,\bar{z}_j)\, e^{2i\theta(z_j)} = N\,, \tag{7}$$

Thus, the complete system consists of the following equations:

- For every *domain element interface*'s collocation point three equations providing continuity of the complex potentials eq. (4).
- For every *boundary element interface*'s collocation point one equation providing continuity of stress vector, this presents continuity of stress vector along the fault eq. (5).
- *N* equations (6) directly implementing the data on stress orientations into the system.
- One equation averaging the maximum shear stress values at the points of measurement eq. (7).

The total number of equations, *q*, is the sum of all equations above; it will be specified for particular examples considered further on. The total number of real unknowns is 9*n*, where *n* is the number of elements.

The system of equations (4)–(7), can be written in a matrix form as $\mathbf{Mr}=\mathbf{s}$. $\mathbf{M}\in R^{q\times 9n}$, $\mathbf{s}\in R^{q}$ and $q>9n$. The 9*n* vector of unknowns $\mathbf{r}$ is formed from the unknown coefficients a,b,c,d,e for each element. Right-hand side $\mathbf{s}$ is formed as $\mathbf{s}\in(0,...0,N)^{T}$. Detailed description of the matrix formation is found in Irsa and Galybin [2010]. This system is over-determined and further solved with the LSQR iterative method [9].

3 Results of modelling

This section presents the results of stress modelling in two regions shown in Figures 1 and 2. Apart from the stress trajectories, the other characteristics of the full stress tensor cannot be reconstructed uniquely. Thus, due to (7) the maximum shear stress is normalized such that its average over the domain is unity. The mean stress is found by integration of the differential equation of equilibrium, which introduces an additive arbitrary constant B, therefore the stress components include two arbitrary parameters, expressed through the multiplicative and additive arbitrary constants. We compare the results of reconstructions for the Sumatra region, the CASE1 and the CASE2, by plotting the stress trajectory patterns, the normalized maximum shear stresses, the normalized mean stresses, and profiles of the normalized normal σ_n and shear σ_t stresses along all some interfaces.

3.1 Sumatra region (longitude 80E–110E, latitude 20S–10N)

The Sumatra region contains the Sunda trench which is a subduction zone on the junction of the Sunda (SU), Burma (BU), Indian (IN) and Australian (AU) plates. Following the previous analysis [10], we analyze the region 30 by 30 degrees between 80E–110E and 20S–10N. The plate boundaries are represented by 91 points from (PB2002).

There are 854 data of A-C quality in total separated into 269 data for the CASE 1 and 585 data for the CASE 2. One can notice the changes of the stress orientations near the southern interface between AU-SU plates comparing these two subsets. This area includes new stress orientations data in the Sunda plate with the direction sub-parallel to the trench (perpendicular to the previously observed data). Similar changes are observed for the northern part of the BU plate inside of the selected region.

The domain is subdivided into 319 triangular elements (shown in Figure 3); among them 80 in SU, 161 in AU, 15 in BU and 63 in IN. Every interface between the adjacent elements within a tectonic plate contains 2 collocation points. Every interface on the plate boundaries contains 3 collocation points. For the configuration shown in Figure 3, the conditions expressing continuity of the complex potentials across the adjacent elements resulted in 5042 equations of which 216 equations have been imposed by the continuity of the stress vector across the plate margins. The percentage of the data equations out of the total number of equations was 4.8% for the *CASE 1* and 10% for the *CASE 2*.

The data in the *CASE 1* are distributed over the whole region with different density. Most of the data are located at the Sunda trench. The observed orientations are shown together with the obtained stress trajectories in Figure 4a. The reconstructed maps of maximum shear stress and mean stress are shown in Figure 4(b,c) respectively. The new data in the *CASE 2* are located near the plate boundaries and therefore the results of stress reconstructions in the regions remote from the plate boundaries are less accurate and have to be considered as extrapolation. This extrapolation, however, fully satisfy all the governing equations accepted in modelling. The results of the stress state reconstruction in the *CASE 2* are presented in Figure 4(d–f).

Comparison of the two cases, the *CASE 1* (before 26.12.2004) and the *CASE 2* (after 27.12.2004), shows the change in the stress trajectory field on the AU/SU interface. One singular (isotropic) point (where the maximum shear stress vanishes) has been detected in the southern part of the Sunda Trench. The mean stress has significantly increased in this area. There is also significant decrease of the maximum shear stress after the event, particularly inside the Burma plate and alongside the interfaces of AU/SU, IN/BU and BU/SU. It should be noted that hereafter the changes in the stress field are discussed relatively to the average values over the considered region, i.e. for the functions τ_{max} and p and for the normalized boundary stresses.

The absence of new data in some areas in the *CASE 2* does not affect the results of the stress reconstructions alongside the plate boundaries. The distributions of normal and shear stresses along the interfaces between all interfaces are shown in Figure 5 for both, the *CASE 1* and the *CASE 2*. It is evident from the figures that the normal stresses have changed considerably along the AU/SU boundary after the event.

It is also evident from the distributions presented in Figure 5 that the magnitudes of the shear stresses are comparable with the normal stresses. The boundary shear stresses are often neglected in the conventional modelling, which

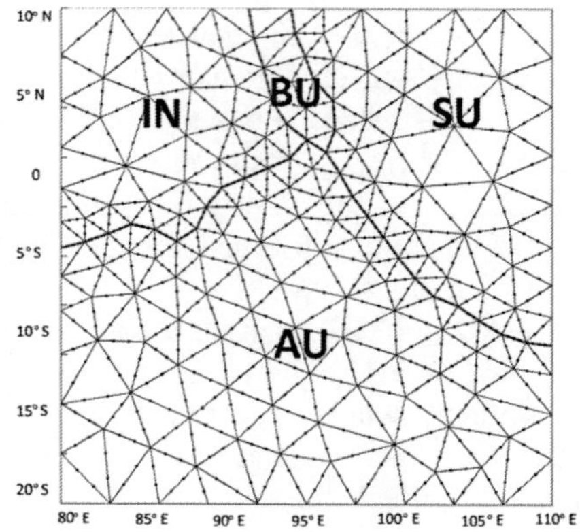

Figure 3: Sumatra region discretization. The boundary element interfaces are marked with the thick (green) line.

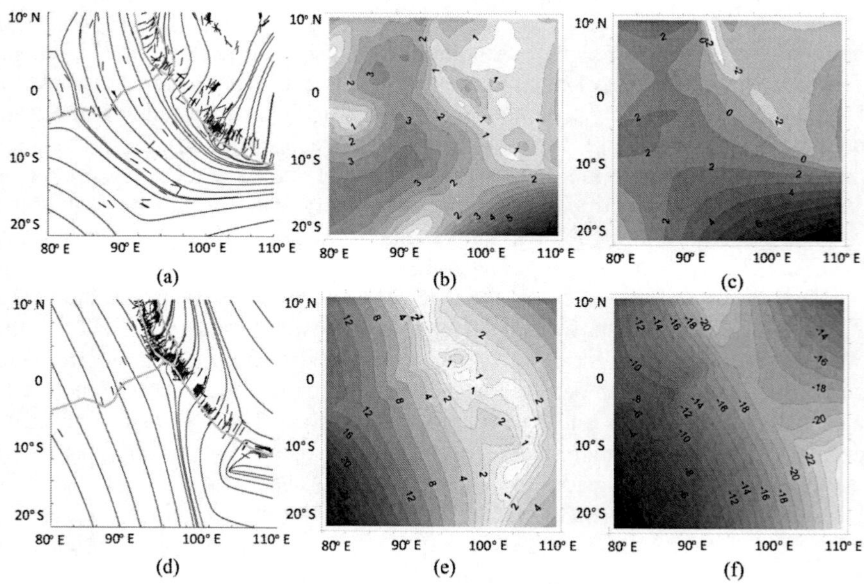

Figure 4: Stress field reconstructions in the Sunda Trench region. The CASE 1 with the data up to 26.12.2004 (a) data and stress orientations, (b) maximum shear stress, (c) mean stress; the CASE 2 with the data after 27.12.2004 (d) data and stress orientations, (e) maximum shear stress, (f) mean stress.

may result in large errors. Some parts of the boundary, e.g. AU/SU and BU/IN have been found to be mainly driven by the normal stress.

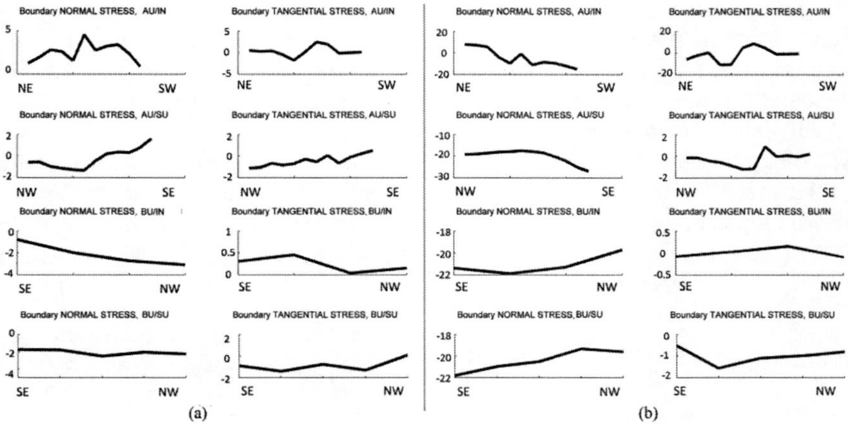

Figure 5: Reconstructed boundary stresses along the margins of the tectonic plates within the Sunda trench region: normal stress (1st and 3rd columns) and tangential stress (2nd and 4th columns), (a) CASE 1, (b) CASE 2.

3.2 Japanese region (longitude 125E–150E, latitude 27N–45N)

The investigated Japanese region consists of the following tectonic plates: Pacific (PA), Okhotsk (OK), Eurasia (EU), Philippine Sea (PS), Okinawa (ON), and Yangtze (YA). The analysed region is 25 by 18 degrees between 125E–150E and 27N–45N. The plate boundaries are represented by 165 points from (PB2002).

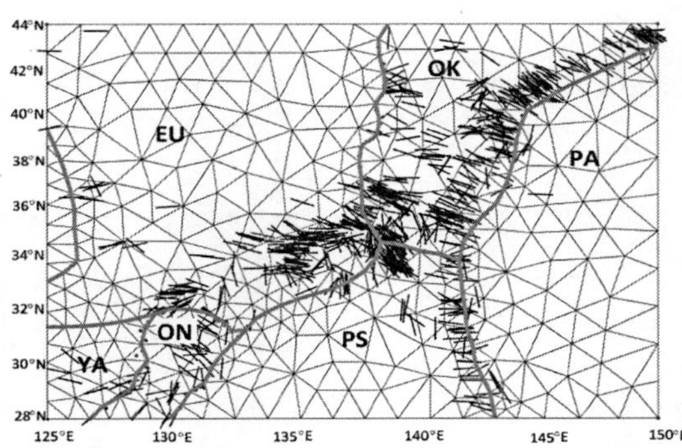

Figure 6: Japan region discretization. The boundary element interfaces are marked with the thick (green) line.

There are 830 data of A-C quality. The domain is subdivided into 722 triangular elements (shown in Figure 6). Every interface between the adjacent elements within a tectonic plate contains 2 collocation points. Every interface on the plate boundaries contains 3 collocation points. For the configuration shown in Figure 6, the conditions expressing continuity of the complex potentials across the adjacent elements resulted in 12839 equations of which 279 equations have been imposed by the continuity of the stress vector across the plate margins. The percentage of the data equations out of the total number of equations was 6.4%.

This region contains much higher quality data compared to the previous ones (see Figures 1 and 2). The data are distributed more homogeneously and thus the results are expected to be of higher accuracy. The observed orientations are shown together with the obtained stress trajectories in Figure 7a. The reconstructed maps of maximum shear stress and mean stress are shown in Figure 7(b,c) respectively.

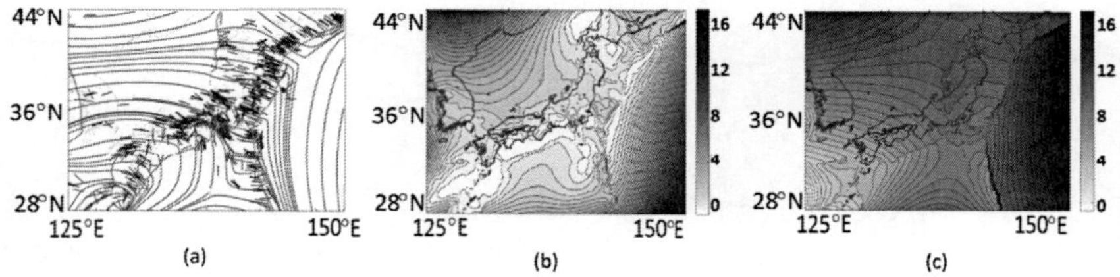

Figure 7: Stress field reconstructions in the Japanese region. (a) data and stress orientations, (b) maximum shear stress, (c) mean stress.

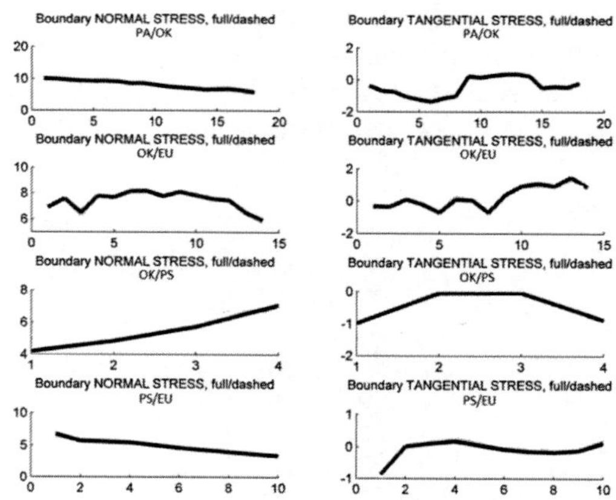

Figure 8: Reconstructed boundary stresses along the margins of the tectonic plates within the Japan region, normal stress and tangential stress.

The resulting boundary conditions are shown in Figure 8. The boundary conditions results show that the PA/OK boundary is mainly driven by the normal stress. This is also evident from the physical movement of the tectonic plates, which indicates correctness of the solution. It is remarkable that the analysis reveals relatively high tangential stresses for the remaining boundaries, which have similar magnitudes as the normal stresses. Therefore the commonly neglected tangential stresses are indeed important in the analyses.

Comparison of the stress patterns of the Sumatran region before the great Sumatran earthquake and the Japan region shows very similar patterns as for stress trajectories, so does for the maximum shear stress and mean stress. The areas where the earthquakes occurred have very low, if not singular, maximum shear stress values. Also the comparison of the normal and tangential stresses on the AU/SU boundary and PA/OK boundary shows similar behaviour, i.e. the normal stresses are much greater as compared to the shear stresses. This result also reflects the directions of the physical movement of the tectonic plates.

4 Discussion and conclusions

Two regions, affected by similar events, have been analysed using the modification of the STEM technique developed earlier [3]. The first one was the region where the great Sumatran earthquake (26.12.2004, 9.1M) happened. The second one was the region of the recent Japan earthquake. Both these events are similar, not only by the magnitudes of the earthquakes and the triggered tsunamis but also by similar geodynamic conditions of tectonic plate interactions in the subduction zones.

The proposed modification of the STEM [3] takes into account both the stress orientations data from the WSM database and the tectonic plate boundaries from Birds PB2002 database. This allowed us to reconstruct the stresses inside the tectonic plates as well as the stresses along the plate margins. Overall, four outputs are obtained: the stress trajectories field (determined uniquely), the map of normalized maximum shear stress, the mean stress and boundary conditions in terms of normal and tangential stress (these fields include two arbitrary parameters).

The results of this study demonstrate that the stress fields have changed after the great Sumatran earthquake. Moreover the results show that both these regions have somewhat similar stress patterns prior to the events and, thus, the stresses in the Japan region might be expected change in a similar way as the ones in the Sumatran region after the earthquake.

Acknowledgement

The second author is grateful to RFBR for partial support of this research.

References

[1] Heidbach, O., M. Tingay, A. Barth, J. Reinecker, D. Kurfess, and B. Müller (2010), Global crustal stress pattern based on the World Stress Map database release 2008, *Tectonophysics*, 482 (1-4), 3-15, doi:10.1016/j.tecto.2009.07.023.

[2] Bird, P. (2003), An updated digital model of plate boundaries, *Geochem. Geophys. Geosyst.*, 4, 1027, 55, doi:10.1029/2001GC000252.

[3] Irsa, J., and A.N. Galybin (2010), Stress trajectories element method for stress determination from discrete data on principal directions, *Eng. Anal. Bound. Elem.*, 34, 423-432, doi:10.1016/j.enganabound.2009.12.004.
[4] U.S. Geological Survey, obtained 20.03.2011, (http://earthquakes.usgs.gov /earthquakes/world/10_largest_world.php)
[5] Zoback, M. L., M. D. Zoback, J. Adams, M. Assumpção, S. Bell, E. A. Bergman, P. Blümling, N. R. Brereton, D. Denham, L. Ding, K. Fuchs, N. Gay, S. Gregersen, H. K. Gupta, A. Givishiani, K. Jacob, R. Klein, P. Knoll, M. Magee, J. L. Mercier, B. C. Müller, C. Paquin, K. Rajendran, O. Stephansson, G. Suarez, M. Sutter, A. Udias, Z. H. Xu, M. Zhizhin (1989), Global patterns of tectonic stress, *Nature*, 341, 291-298, doi:10.1038/341291a0.
[6] Amadei, B., and O. Stephansson (1997), *Rock stress and its measurement*, Chapman & Hall, London.
[7] Fairhurst, C. (2003), Stress estimation in rock: a brief history and review, *Int. J. Rock Mech. Min. Sci.*, 40, 957-973, doi:10.1016/j. ijrmms.2003.07.002.
[8] Mushkelishvili, N. I. (1953), *Some basic problems of the mathematical theory of elasticity*, P. Noordhoff Ltd., Groningen-Holland.
[9] Paige, Ch. C., M. A. Saunders (1982), LSQR: an algorithm for sparse linear equations and sparse least squares, *ACM T. Math. Software*, 8 (1), 43-71.
[10] Mukhamediev, Sh.A., and A.N. Galybin (2006), Where and how did the ruptures of December 26, 2004 and March 28, 2005 earthquakes near Sumatra originate?, *Dokl. Earth Sci.*, 406 (1), 52-55, doi: 10.1134 /S1028334X06010132.

Section 5
Advanced meshless and mesh reduction methods

Time integrations in solution of diffusion problems by local integral equations and moving least squares approximation

V. Sladek[1], J. Sladek[1] & Ch. Zhang[2]
[1]*Institute of Construction and Architecture, Slovak Academy of Sciences, Bratislava, Slovakia*
[2]*Department of Civil Engineering, University of Siegen, Germany*

Abstract

The paper deals with the numerical solution of initial-boundary value problems for diffusion equation with variable coefficients by using a local weak formulation and a meshless approximation of spatial variations of the field variable. The time variation is treated either by the Laplace transform technique or by the linear Lagrange interpolation in the time stepping approach. Advanced formulation for local integral equations is employed. A comparative study of numerical results obtained by the Laplace transform and the time stepping approach is given in a test example for which the exact solution is available and utilized as a benchmark solution.
Keywords: transient heat conduction, weak formulation, Laplace transform, time stepping, accuracy, computational efficiency.

1 Introduction

The diffusion and the transient heat conduction problems in functionally graded materials belong to frequent engineering problems. From the mathematical point of view, the solution of initial-boundary value problems for the diffusion equation with variable coefficients is rather complex task and therefore there is a demand to have sophisticated and efficient numerical techniques. The local weak formulations appear to be appropriate. The local integral equations replacing the governing equations are acceptable from the physical point of view, since such equations express the balance principles. The variation of material coefficients is involved naturally without any complication as compared with the formulations

WIT Transactions on Modelling and Simulation, Vol 52, © 2011 WIT Press
www.witpress.com, ISSN 1743-355X (on-line)
doi:10.2495/BE110191

for homogeneous media. Furthermore, the local formulation enables us to develop truly meshless formulations and so eliminate all inappropriateness of finite size elements employed in standard element based discretization methods. Besides the advantages of the meshless formulations one should name also their main drawback which consists in prolongation of creation of the discretized system matrix. It is given by the fact that the shape functions in meshless approximations are not given in a closed form and a computational procedure is required for evaluation of the shape functions at each point. Nevertheless this disadvantage can be eliminated by decreasing the amount of the integration points which can be done by performing the integrations analytically [1]. The other disadvantage is the failure of accuracy of higher order derivatives of the shape functions at points near the boundary of the analyzed domain. This can be avoided by expressing such derivatives in terms of the first order derivatives [1].

The other task which should be solved in the case of transient problems is the treatment of the time variable. Besides the well known time stepping approaches, the Laplace transform technique is often used alternatively. In this paper, both these techniques are used in combination with the local integral equation formulation and the MLS approximation in the comparative study for accuracy in a test example.

2 Governing equations and local integral equation formulations

The governing equation for diffusion problems is the same as for transient heat conduction with the absent volume density of heat sources $w(\mathbf{x},t)$. Thus, we shall consider initial-boundary value problems for the partial differential equation [2]

$$\left(\lambda(\mathbf{x})u_{,k}(\mathbf{x},t)\right)_{,k} - \rho(\mathbf{x})c(\mathbf{x})\frac{\partial u(\mathbf{x},t)}{\partial t} = 0 \ , \quad \text{in} \quad \Omega\times[0,T] \tag{1}$$

where $u(\mathbf{x},t)$ can be interpreted as the temperature field. In isotropic and continuously non-homogeneous media, the material parameters, such as the mass density $\rho(\mathbf{x})$, the volume density of the specific heat per unit mass $c(\mathbf{x})$, and the thermal conductivity coefficient $\lambda(\mathbf{x})$ are spatially dependent and the governing equation is the partial differential equation (PDE) of parabolic type with variable coefficients. The first term on the left-hand side of Eq.(1) is the divergence of the heat flux vector

$$q_k(\mathbf{x},t) = -\lambda(\mathbf{x})u_{,k}(\mathbf{x},t) \tag{2}$$

and the second term is the rate of the temporal change of the volumetric density of heat.

Three types of physically reasonable boundary conditions are applicable (Dirichlet b.c., Neumann b.c., and Robin b.c.). The boundary conditions are to be supplemented by the initial condition, which in the present parabolic problem is the initial value of the temperature $u(\mathbf{x},0) = v(\mathbf{x})$ in $\Omega \cup \partial\Omega$.

Since the energy balance should be satisfied in an arbitrary finite part of the continuum Ω^c bounded with the boundary $\partial\Omega^c$, we may write the integral relationship

$$\int_{\partial\Omega^c} n_i(\boldsymbol{\eta})\lambda(\boldsymbol{\eta})u_{,i}(\boldsymbol{\eta},t)d\Gamma(\boldsymbol{\eta}) - \int_{\Omega^c} \rho(\mathbf{x})c(\mathbf{x})\frac{\partial}{\partial t}u(\mathbf{x},t)d\Omega(\mathbf{x}) = 0 \qquad (3)$$

It is easy to show that equation (3) is the integral equivalent of the differential governing equation (1).

Sometimes, the Laplace transform (LT) technique is an efficient tool for treatment of the time evolution. Then, the time variable is eliminated temporarily and replaced by the Laplace transform parameter p. The governing equations (1) and/or (3) can be rewritten for the Laplace transform of the temperature as

$$\left(\lambda(\mathbf{x})\bar{u}_{,k}(\mathbf{x},p)\right)_{,k} - p\rho(\mathbf{x})c(\mathbf{x})\bar{u}(\mathbf{x},p) = -\rho(\mathbf{x})c(\mathbf{x})v(\mathbf{x}) \qquad (4)$$

$$\int_{\partial\Omega^c} n_i(\boldsymbol{\eta})\lambda(\boldsymbol{\eta})\bar{u}_{,i}(\boldsymbol{\eta},p)d\Gamma(\boldsymbol{\eta}) - p\int_{\Omega^c} \rho(\mathbf{x})c(\mathbf{x})\bar{u}(\mathbf{x},p)d\Omega(\mathbf{x}) =$$

$$= -\int_{\Omega^c} \rho(\mathbf{x})c(\mathbf{x})v(\mathbf{x})d\Omega(\mathbf{x}) . \qquad (5)$$

The boundary conditions for the Laplace transform of the temperature can be obtained by direct application of the Laplace transformation to the prescribed boundary conditions. In the LT-approach, the numerical inversion of the LT is a key issue, since it is an ill-posed problem. Various Laplace-inversion algorithms are available in literature. Regarding good experience with the Stehfest's algorithm [3], we shall use this algorithm in the present analysis with taking 10 values of the transform parameter for each time instant.

The derived local integral equations (LIE) (3) and/or (5) are the restriction relationships, which should be satisfied together with the prescribed initial and boundary conditions in solving initial-boundary value problems. In order to solve these equations, the spatial variation of field variables is usually approximated in terms of certain shape functions and unknowns related to a finite number of nodal points. Implementation of the spatial approximation in the LIE (3) results in the ordinary differential equations (ODE) for the nodal unknowns which is known as a semi-discrete formulation. On the other hand, having used the spatial approximation in the LIE (5) for certain value of the LT parameter, we obtain a system of algebraic equations for the nodal unknowns of the Laplace transforms of the field variable (temperature).

3 MLS-approximation for spatial variations

Recall that the Moving Least Squares approximation belongs to mesh free approximations since no predefined connectivity among nodal points is required. In this paper, we shall consider the Central Approximation Node (CAN) concept of the MLS-approximation [4].

Let $\mathbf{x}^q$ be the CAN for the approximation at a point $\mathbf{x}$. Then, the amount of nodes involved into the approximation at $\mathbf{x}$ is reduced a-priori from N_t (total number of nodes) to N^q, where N^q is the number of nodes supporting the approximation at the CAN $\mathbf{x}^q$, i.e. the amount of nodes in the set $\mathcal{M}^q = \left\{\forall \mathbf{x}^a;\ w^a(\mathbf{x}^q) > 0\right\}_{a=1}^{N_t}$, where $w^a(\mathbf{x})$ is the weight function associated with the node $\mathbf{x}^a$ at the field point $\mathbf{x}$. In this paper, we employ the Gaussian weights [4]. The MLS-CAN approximation for spatial variation of the field variable $f(\mathbf{x}) \in \{u(\mathbf{x},t), \overline{u}(\mathbf{x},p)\}$ is given by

$$f(\mathbf{x}) \approx \sum_{a=1}^{N^q} \hat{f}^{n(q,a)} \phi^{n(q,a)}(\mathbf{x}), \tag{6}$$

where $n(q,a)$ is the global number of the a-th local node from $\mathcal{M}^q$, $\hat{f}^{n(q,a)} \in \{\hat{u}^{n(q,a)}(t), \hat{\overline{u}}^{n(q,a)}(p)\}$, with $\hat{f}^{n(q,a)}$ being the nodal unknowns, which are different from the nodal values of physical quantities, in general. In this paper, we shall specify the CAN $\mathbf{x}^q$ as the nearest node to the approximation point $\mathbf{x}$. Recall that the shape functions $\phi^m(\mathbf{x})$ are not known in closed form and a computational procedure must run for evaluation at each approximation point $\mathbf{x}$. This is the main handicap of mesh-free approximations as compared with mesh-based approximations utilizing mostly polynomial interpolations.

Besides the approximation of field variables, we need also their gradients which can be approximated as gradients of approximated fields (6)

$$f_{,j}(\mathbf{x}) \approx \sum_{a=1}^{N^q} \hat{f}^{n(q,a)} \phi_{,j}^{n(q,a)}(\mathbf{x}) \tag{7}$$

and similarly, one can approximate also higher-order derivatives. Recall that the evaluation of the derivatives of the shape functions is still more complicated than the evaluation of shape functions and also the accuracy of such approximations is worse.

Substituting these approximations into the governing equation (5) considered at nodal points $\mathbf{x}^c$ ($\Omega^c \ni \mathbf{x}^c$ is a sub-domain around the node $\mathbf{x}^c$), one obtains the system of algebraic equations

$$\sum_g \left(K^{cg} - p_n M^{cg}\right)\hat{\overline{u}}^g(p_n) = -\overline{R}^c(p_n), \quad (c = 1,2,...,N_t), \ (n = 1,2,...,N) \tag{8}$$

where the matrix elements are given as

$$K^{cg} = \int_{\partial\Omega^c} n_i(\boldsymbol{\eta})\lambda(\boldsymbol{\eta})\phi_{,i}^{n(q_\eta,a)}(\boldsymbol{\eta})d\Gamma(\boldsymbol{\eta}), \quad M^{cg} = \int_{\Omega^c} \rho(\mathbf{x})c(\mathbf{x})\phi^{n(q_x,a)}(\mathbf{x})d\Omega(\mathbf{x}),$$

$$\overline{R}^c(p_n) = \int_{\Omega^c} \rho(\mathbf{x})c(\mathbf{x})v(\mathbf{x})d\Omega(\mathbf{x}), \tag{9}$$

with g being global numbers of nodes generated by $n(q_\eta, a)$ and/or $n(q_x, a)$, where q_x is the nearest nodal point to the integration point $\mathbf{x}$.

Similarly, substituting the approximations for the temperature and its gradients into the governing equation (3), we obtain the system of the ODE

$$\sum_g \left(K^{cg}\hat{u}^g(t) - M^{cg}\frac{\partial \hat{u}^g(t)}{\partial t} \right) = 0, \qquad (c = 1,2,...,N_t). \tag{10}$$

The integrations in the integrals defined in Eqn. (9) can be performed analytically by using the Taylor series expansions for the material coefficients and the shape functions [1]. For this purpose the circular shape is chosen for sub-domains and their radius should be sufficiently small to restrict the order for the derivatives of shape functions.

In order to solve the ODE (11), we employ a polynomial interpolation for the time variation of the nodal unknowns.

4 Linear time interpolations (LLI)

Let us split the time interval $[0,T]$ by discrete time instants t_i into a finite number of subintervals $[t_i, t_{i+1}]$ to complete the discretization in the semi-discrete formulation. In the case of Linear Lagrange Interpolation (LLI) the element T_i is defined as the interval $T_i = [t_i, t_{i+1}]$ with the interior points being parametrized as

$$t\big|_{T_i} = \sum_{a=1}^{2} t_{i-1+a}N^a(\tau) = t_i + \frac{\Delta t_i}{2}(1+\tau), \qquad \tau \in [-1, 1] \tag{11}$$

since $N^1(\tau) = (1-\tau)/2$, $N^2(\tau) = (1+\tau)/2$. The time dependence of a physical variable $u(t)$ is approximated on T_i by the interpolation

$$u(t)\big|_{T_i} = \sum_{a=1}^{2} u_{i-1+a}N^a(\tau) = \frac{1}{2}\left(u_{i+1} + u_i\right) + \frac{\tau}{2}\left(u_{i+1} - u_i\right), \qquad u_k = u(t_k). \tag{12}$$

Then, the time derivative $\dot{u}(t) = du(t)/dt$ is approximated by the constant

$$\dot{u}(t)\big|_{T_i} = \frac{1}{J(\tau)}\frac{du}{d\tau}\bigg|_{T_i} = \frac{1}{\Delta t_i}\left(u_{i+1} - u_i\right), \tag{13}$$

since the Jacobian of the transformation (11) is given as $J(\tau) = dt/d\tau\big|_{T_i} = \Delta t_i/2$.

Making use a different parametrization $\theta = (1+\tau)/2$ with $\theta \in [0,1]$, we obtain from (12) and (13)

$$u(t_i + \theta\Delta t_i) = \theta u_{i+1} + (1-\theta)u_i \quad , \; \dot{u}(t_i + \theta\Delta t_i) = \left(u_{i+1} - u_i\right)/\Delta t_i \tag{14}$$

Considering the system of the ODE (11) at $t = t_i + \theta\Delta t_i$, we obtain

$$\sum_g \left(K^{cg} - \frac{1}{\theta\Delta t_i} M^{cg} \right) \hat{u}^g_{i+1} = \sum_g \left((1 - \frac{1}{\theta}) K^{cg} - \frac{1}{\theta\Delta t_i} M^{cg} \right) \hat{u}^g_i \text{ , } (i = 0,1,2,...) \quad (15)$$

which is the well known θ-method used in time stepping approaches for solution of the ODE with $\theta \in (0,1]$.

5 Numerical tests

In our numerical experiments, it is important to have the exact solution which could be used as a benchmark solution. In this paper, we consider a square domain $L \times L$ occupied by medium with exponentially graded heat conduction as well as specific heat while constant mass density: $\rho = \text{const}$., $\lambda(\mathbf{x}) = \lambda_0 \exp(2\delta x_2 L) = c(\mathbf{x})\lambda_0 / c_0$. If constant values of the temperature are prescribed on the bottom u_0 and top u_L of the square, while the lateral sides are thermally insulated for $t \in [0, T]$ and constant initial value of temperature $v(\mathbf{x}) = \text{const} = v$ is assumed, the exact solution is available [5]. In numerical computations, we have used $\lambda_0 = 1 = c_0 = \rho$, $\delta = 1$, $u_0 = 1 = v$, $u_L = 20$. The uniform distribution of nodal points is employed with h being the distance between two neighbour nodes. Furthermore, in the MLS-approximation we have used quadratic polynomials as the basis functions, Gaussian weights with the shape factor $c = h$, radius of the influence domain $R_i = 3.001 \times h$, and the radius of circular sub-domains Ω^c in the local weak formulation is $r^c = 0.3 \times h$. In the analytical integrations of the LIE, we have used the third order of the derivatives of the shape functions as maximal and the Taylor series expansions of the integrands have been stopped on the 8th power of the radius of sub-domain.

It can be seen from Fig. 1 that considerable errors are sharply localized in both the space and time for short time steps. The inaccuracy of the numerical results by the LLI approach is partially decreased and remarkably delocalized with increasing the time step, while in the case of LT approach the delocalization is marginal as compared to the substantial increase of accuracy. For more detailed study of errors near the top side of the analyzed domain, we have used 1296 nodes and the test point is $(L/2,\ 0.9L)$.

Fig. 2 shows the numerical results at the point (x_1, x_2) which is the nearest node to the point $(L/2,\ 0.9L)$. The sequences of time instants at which the numerical results are received are given as follows: $t_i = 4 \times 10^{-4} + (i-1)\Delta t$, $(i = 1,2,....)$ for three different time steps Δt. Thus, the time step is used for specification of time instants in the calculations by both the LT approach and LLI approach, but in the latter one, the time step is also the length of the time interval within which the interpolation is assumed.

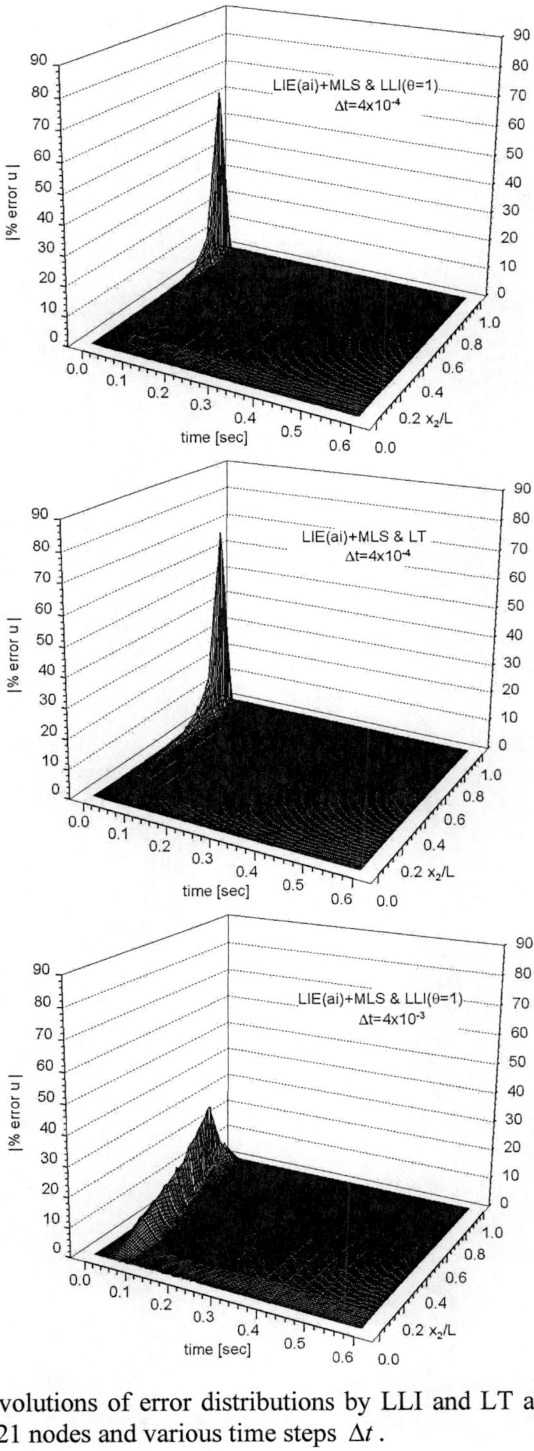

Figure 1: Evolutions of error distributions by LLI and LT approaches using 121 nodes and various time steps Δt .

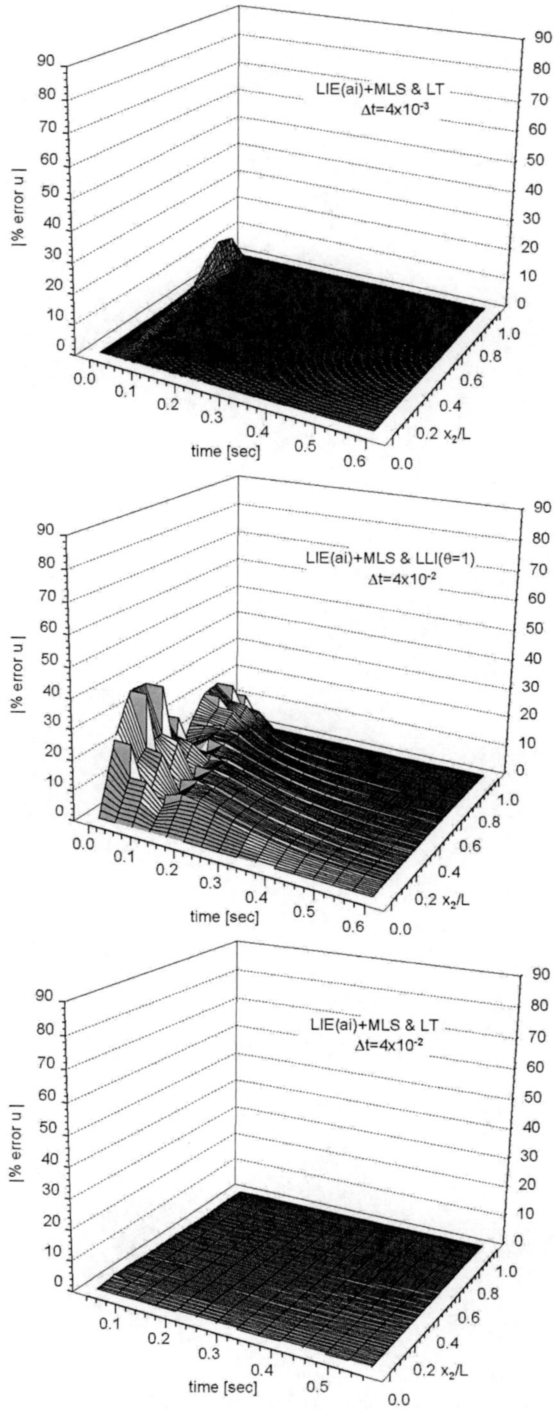

Figure 1: Continued.

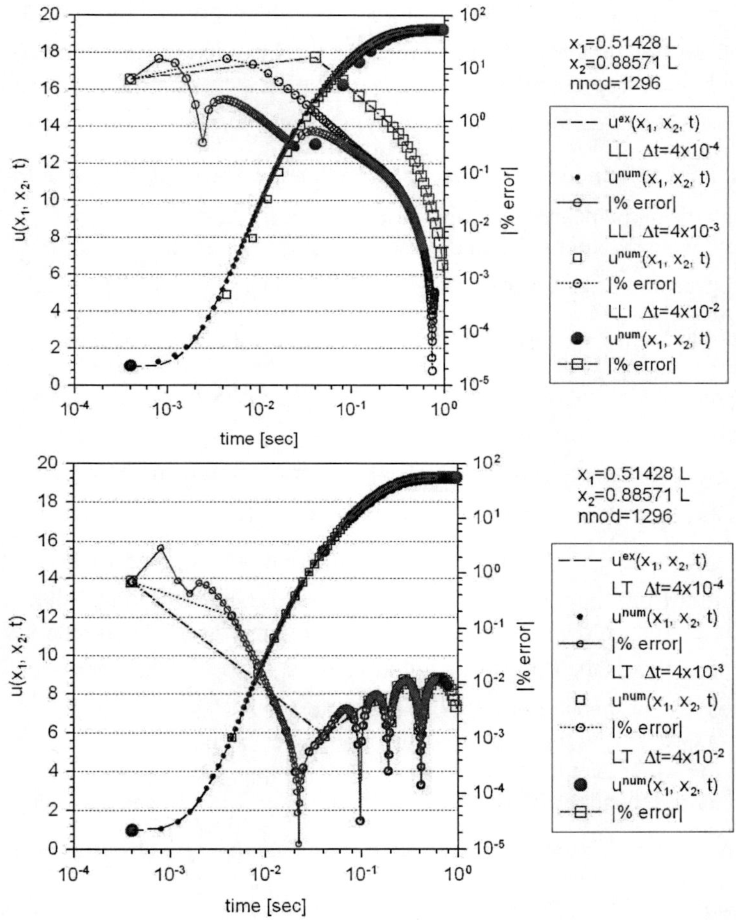

Figure 2: Time evolutions of the computed temperature and its accuracy by the LLI and LT approaches at the selected point (x_1, x_2) using 1296 nodes and three various lengths of time steps.

Recall that the prescribed boundary temperature $u_L = 20$ represents a sudden change at $t = 0$ with respect to the initial value of temperature $v = 1 = u_0$. It can be seen that better accuracy is achieved by the LT approach especially when the longer time steps are employed.

Note that three characteristic lengths play a role in this fully discretized transient field problem:

(i) $l_h = h$ represents the distance between two neighbouring nodes

(ii) $l_S = R_i h$ represents the radius of the influence domain which is used for selection of nodes contributing to the meshless approximation at certain point

(iii) $l_T = \sqrt{(\lambda / \rho c)\Delta t}$ is the characteristic length corresponding to the time step Δt and represents the horizon reached by the heat conduction during the time step with respect to certain point. In time stepping techniques, the optimal choice is $l_T \approx h$, because information at previous time instant t_{i-1} cannot reach the nearest neighbour node at t_i, if $l_T < h$; on the other hand, if $l_T > h$, information at a node from the nearest neighbour node is not fresh.

Now, we can explain the inaccuracy of numerical results by both the LLI and LT approaches at several early time instants $t_i = i\Delta t$ $(i = 1, 2, ..., 5)$ if $\Delta t = 4 \times 10^{-4}$. In this case $l_T = 2 \times 10^{-2}$, hence the choice of 1296 uniformly distributed nodes is almost optimal. Then, $h = 1/(\sqrt{1296} - 1) = 0.02857$ and at early time instants, the prescribed boundary value $u_L = 20$ (which is different from the initial value $v = 1$) affects the approximation in the boundary layer $L - (l_T + l_S) < x_2 < L$ and hence also in the boundary layer $L - (l_T + l_S) < x_2 < L - l_T$, i.e. at points which lie behind the horizon of the heat conduction from the top of the analyzed domain. Quite different is the situation near the bottom of the analyzed domain, where the prescribed boundary value of the temperature is the same as its initial value $u_0 = v = 1$. With increasing Δt, l_T is increased and hence not only the time interval but also the boundary layer of inaccurate results by the time stepping approaches become wider.

On the other hand, in the case of the LT approach the solution at a time instant is independent of the time step and therefore the accuracy at the time instants $t_i = 4 \times 10^{-4} + (i-1)\Delta t_2$ (which occurs after $(i-1)$ steps $\Delta t_2 = 4 \times 10^{-2}$) is the same as the accuracy after $10 \times (i-1)$ steps $\Delta t_3 = 4 \times 10^{-3}$ and/or after $100 \times (i-1)$ steps $\Delta t_4 = 4 \times 10^{-4}$. These conclusions are confirmed by the numerical results presented in Fig.2 and also in Fig.3.

The results in Fig.3 confirm the spatial delocalization of the error with increasing the size of the time step. In the case of the LT approach, this delocalization is suppressed by decrease of the maximum value of the error.

6 Conclusions

The local weak formulation is proposed for solution of transient heat conduction in FGM. The spatial variation of temperature is approximated by using the MLS approximation, while the time dependence is treated either by the Laplace transform or by the linear Lagrange interpolation in the time stepping method. Both the LT and LLI approaches can give results with reasonable accuracy except very early time instants after sudden change of initial values by different

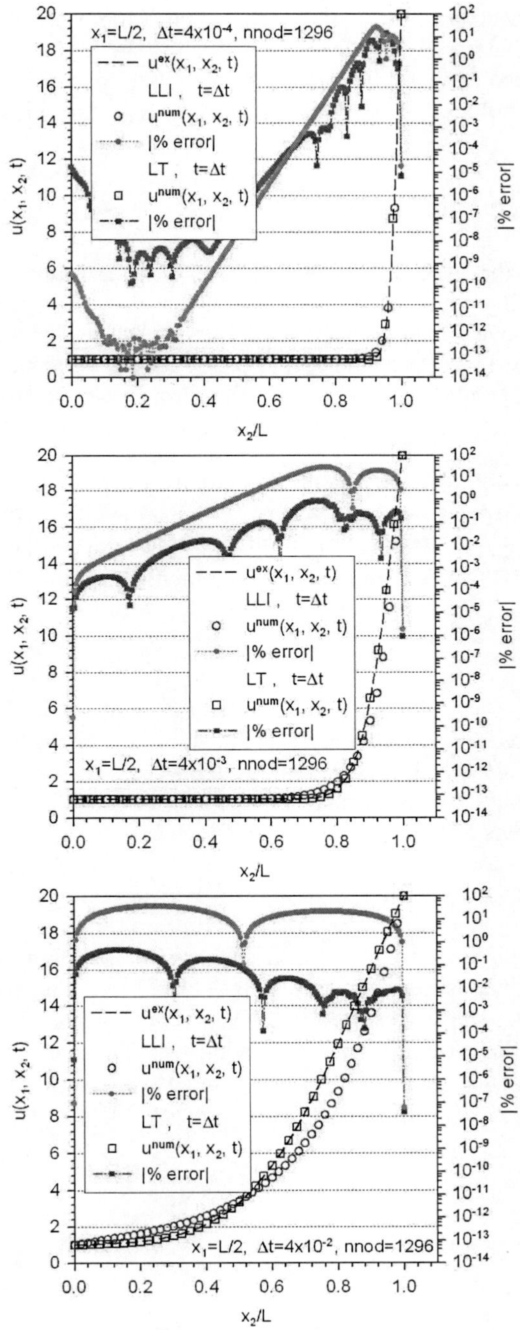

Figure 3: Spatial distributions of the temperatures and their accuracies by the LLI and LT approaches at the time instant $t = \Delta t$ for three values of the time steps Δt.

prescribed boundary values. The time stepping approach is sensitive to the choice of the time step, while the LT approach is stable. On the other hand, the computational economy is better in time stepping approaches than in the LT approach if the solution is required not only at few time instants but within a time interval.

Acknowledgements

This article has been produced with the financial assistance of the European Regional Development Fund (ERDF) under the Operational Programme Research and Development/Measure 4.1 Support of networks of excellence in research and development as the pillars of regional development and support to international cooperation in Bratislava region/Project No. 26240120020 Building the centre of excellence for research and development of structural composite materials – 2nd stage.

This work has been partially supported by the Slovak Science and Technology Assistance Agency registered under number APVV-0032-10, the Slovak Grant Agency VEGA-2/0039/09 and the German Research Foundation (DFG, ZH 15/14-1), which are gratefully acknowledged

References

[1] Sladek V., Sladek J., Local integral equations implemented by MLS-approximation and analytical integrations. *Engineering Analysis with Boundary Elements* **34**, pp. 904-913, 2010.

[2] Wrobel L.C, The Boundary Element Method, Vol1: Applications in Thermo-Fluids and Acoustics, Wiley: Chichester, 2002.

[3] Stehfest H., Algorithm 368: numerical inversion of Laplace transform. *Communication of the Association for Computing Machinery*, **13**, pp. 47-49; 624, 1970.

[4] Sladek V., Sladek J., Zhang Ch., Computation of stresses in non-homogeneous elastic solids by local integral equation method: a comparative study. *Computational Mechanics* 41, pp. 827-845, 2008.

[5] Sladek V., Sladek J., Tanaka M., Zhang Ch., Transient heat conduction in anisotropic and functionally graded media by local integral equations. *Engineering Analysis with Boundary Elements* **29**, pp. 1047-1065, 2005.

The MFS for the detection of inner boundaries in linear elasticity

A. Karageorghis[1], D. Lesnic[2] & L. Marin[3]
[1] *Department of Mathematics and Statistics, University of Cyprus, Nicosia, Cyprus*
[2] *Department of Applied Mathematics, University of Leeds, UK*
[3] *Institute of Solid Mechanics, Romanian Academy, Bucharest, Romania*

Abstract

We propose a nonlinear minimization method of fundamental solutions for the detection (shape, size and location) of unknown inner boundaries corresponding to either a rigid inclusion or a cavity inside a linear elastic body from nondestructive boundary measurements of displacement and traction. The stability of the numerical method is investigated by inverting measurements contaminated with noise.
Keywords: Cauchy–Navier equations, method of fundamental solutions, regularization.

1 Introduction

The method of fundamental solutions (MFS) [1, 2] is a meshless boundary collocation method [3] which may be used for the numerical solution of certain boundary value problems. The method has become increasingly popular over the last three decades primarily because of the ease with which it can be implemented. A comparison between the MFS and the boundary element method (BEM), as applied to direct problems, has been performed in [4]. In recent years, the MFS has been used extensively for the numerical solution of inverse problems primarily. An extensive survey of the applications of the MFS to inverse problems is provided in [5]. The most difficult class of inverse problems are the so-called *inverse geometric problems* in which the location and shape of part of the boundary of the domain of the problem in question are unknown and need to be calculated as part of the solution. The MFS was used for the first time for the solution of inverse geometric

WIT Transactions on Modelling and Simulation, Vol 52,
www.witpress.com, ISSN 1743-355X (on-line)
doi:10.2495/BE110201

problems in linear elasticity in [6], while more recent applications may be found in [7, 8].

2 Mathematical formulation

In practical nondestructive evaluation (testing) of materials the following inverse problem naturally arises: Given an elastic body Ω, detect an unknown inclusion $D \subset \Omega$ from measurements of the traction and displacement taken on the boundary $\partial\Omega$. This situation commonly arises, for instance, in fracture mechanics when some defects stem from the manufacturing process, or when the elastic properties of the material deteriorate due to the occurrence of possible damage [9].

In mathematical terms, and considering, for simplicity, a two-dimensional isotropic and homogeneous elastic simply-connected bounded body $\Omega \subset \mathbb{R}^2$, our goal is to determine the displacement $\mathbf{u} = (u_1, u_2)$ and an inclusion D compactly contained in Ω, i.e. $\overline{D} \subset \Omega$, such that $\Omega \backslash D$ is connected, satisfying the Cauchy–Navier equations (Lamé system) of elasticity

$$\begin{cases} G\left(\dfrac{\partial^2 u_1}{\partial x^2} + \dfrac{\partial^2 u_1}{\partial y^2}\right) + \dfrac{G}{1-2\overline{\nu}}\left(\dfrac{\partial^2 u_1}{\partial x^2} + \dfrac{\partial^2 u_2}{\partial x \partial y}\right) = 0, & \text{in } \Omega\backslash\overline{D}, \\ G\left(\dfrac{\partial^2 u_2}{\partial x^2} + \dfrac{\partial^2 u_2}{\partial y^2}\right) + \dfrac{G}{1-2\overline{\nu}}\left(\dfrac{\partial^2 u_1}{\partial x \partial y} + \dfrac{\partial^2 u_2}{\partial y^2}\right) = 0, & \text{in } \Omega\backslash\overline{D}, \end{cases} \tag{1a}$$

subject to the Cauchy boundary conditions on the outer boundary $\partial\Omega$

$$u_i = f_i, \quad i = 1, 2 \text{ on } \partial\Omega, \tag{1b}$$

$$t_i = g_i, \quad i = 1, 2 \text{ on } \partial\Omega, \tag{1c}$$

and either homogeneous Dirichlet conditions

$$u_i = 0, i = 1, 2 \quad \text{on} \quad \partial D, \tag{1d}$$

or homogeneous Neumann conditions

$$t_i = 0, \quad i = 1, 2 \text{ on } \partial D, \tag{1e}$$

on the inner boundary ∂D. Here f_i and $g_i, i = 1, 2$ are known displacements and tractions, respectively, G is the shear modulus, $\overline{\nu} = \nu$ in the plane strain state and $\overline{\nu} = \nu/(1+\nu)$ in the plane stress state, where ν is Poisson's ratio. According to [6], the homogeneous Dirichlet conditions (1d) on the inner boundary ∂D physically describe a rigid inclusion, while the homogeneous Neumann conditions (1e) on ∂D characterise a cavity.

www.witpress.com, ISSN 1743-355X (on-line)

The strain tensor $\boldsymbol{\varepsilon} = (\varepsilon_{ij})_{i,j=1,2}$ is related to the displacements by the kinematic relations, i.e.

$$\boldsymbol{\varepsilon} = \begin{bmatrix} \dfrac{\partial u_1}{\partial x} & \dfrac{1}{2}\left(\dfrac{\partial u_1}{\partial y} + \dfrac{\partial u_2}{\partial x}\right) \\ \dfrac{1}{2}\left(\dfrac{\partial u_1}{\partial y} + \dfrac{\partial u_2}{\partial x}\right) & \dfrac{\partial u_2}{\partial y} \end{bmatrix}, \quad (2)$$

while the stress tensor $\boldsymbol{\sigma} = (\sigma_{ij})_{i,j=1,2}$ is related to the elements of the strain tensor according to Hooke's law by

$$\boldsymbol{\sigma} = 2G \begin{bmatrix} \varepsilon_{11} + \dfrac{\overline{\nu}}{1-2\overline{\nu}}(\varepsilon_{11}+\varepsilon_{22}) & \varepsilon_{12} \\ \varepsilon_{21} & \varepsilon_{22} + \dfrac{\overline{\nu}}{1-2\overline{\nu}}(\varepsilon_{11}+\varepsilon_{22}) \end{bmatrix}. \quad (3)$$

Finally, the tractions $t_i, i = 1,2$ in (1c) are defined by $[t_1, t_2]^T = \boldsymbol{\sigma n}$, where $\boldsymbol{n} = [n_1, n_2]^T$ denotes the outward normal vector to the boundary. Equations (1a) may be written more compactly as $\nabla \cdot \boldsymbol{\sigma} = \mathbf{0}$.

Note that the boundary $\partial\Omega$ is overspecified since both the displacements and tractions are prescribed on it through equations (1b) and (1c). Consequently, we cannot expect that a solution to the above inverse problem exists for arbitrary Cauchy data $\mathbf{f}$ and $\mathbf{g}$. However, the following uniqueness result holds.

Theorem. [6] *Let $\overline{D} \subset \Omega \subset \mathbb{R}^2$ be open, bounded and simply connected domains with smooth boundaries such that the domain of elastic propagation $\Omega \backslash D$ is connected. Let also the Dirichlet and Neumann data $\mathbf{f}$ and $\mathbf{g}$ in* (1b) *and* (1c) *be such that $\mathbf{f} \in [H^{1/2}(\partial\Omega)]^2$ and $\mathbf{g} \in [H^{-1/2}(\partial\Omega)]^2$.*

(i) If $\mathbf{f} \not\equiv \mathbf{0}$ then a single pair of Cauchy data $(\mathbf{f}, \mathbf{g})$ determines (identifies) uniquely the displacement $\mathbf{u} \in [H^1(\Omega\backslash D)]^2$ and the rigid inclusion D satisfying the inverse Dirichlet problem given by equations (1a)-(1d).

(ii) If $\mathbf{f} \notin span\{(1,0),(0,1),(-y,x)\}_{(x,y)\in\partial\Omega}$ then a single pair of Cauchy data $(\mathbf{f}, \mathbf{g})$ determines (identifies) uniquely the displacement $\mathbf{u} \in [H^1(\Omega\backslash D)]^2$ and the cavity D satisfying the inverse Neumann problem given by equations (1a)-(1c) *and* (1e).

Remarks:

(a) The condition in (ii) above says that $\mathbf{f}$ does not belong to the linear space of rigid displacements on $\partial\Omega$ and it can be replaced by the condition $\mathbf{g} \not\equiv \mathbf{0}$.

(b) The inverse inclusion problems under investigations are further ill-posed because they are unstable, i.e. small noisy errors in the input data (1b) and/or (1c) cause large errors in the solution $(\mathbf{u}, D)$.

3 The method of fundamental solutions (MFS)

We approximate the displacements $u_i, i = 1, 2$ by the MFS approximations [6–8]

$$u_i^N(\boldsymbol{x}, \boldsymbol{\xi}; \boldsymbol{\alpha}, \boldsymbol{\beta}) = \sum_{j=1}^{2N} \alpha_j U_{i1}(\boldsymbol{x}, \boldsymbol{\xi}_j) + \sum_{j=1}^{2N} \beta_j U_{i2}(\boldsymbol{x}, \boldsymbol{\xi}_j), \quad \boldsymbol{x} \in \overline{\Omega} \backslash D, \tag{4}$$

where $\boldsymbol{\xi} = (\boldsymbol{\xi}_j)_{j=\overline{1,N}}$ are the singularities located in D while $\boldsymbol{\xi} = (\boldsymbol{\xi}_j)_{j=\overline{N+1,2N}}$ are the singularities located outside $\overline{\Omega}$. The vectors $\boldsymbol{\alpha} = (\alpha_j)_{j=1,\dots,2N}$, $\boldsymbol{\beta} = (\beta_j)_{j=1,\dots,2N}$ contain unknown real coefficients to be determined. The displacement fundamental solution matrix $U = (U_{ij})_{i,j=1,2}$ associated with the points $\boldsymbol{x} = (x, y)$ and $\boldsymbol{\xi} = (\xi_x, \xi_y)$ is given in [10].

On combining the kinematic relations (2), Hooke's law (3) and the MFS approximations for the displacements (4), the following MFS approximations for the tractions t_i, $i = 1, 2$ are obtained

$$t_i^N(\boldsymbol{x}, \boldsymbol{\xi}; \boldsymbol{\alpha}, \boldsymbol{\beta}) = \sum_{j=1}^{2N} \alpha_j T_{i1}(\boldsymbol{x}, \boldsymbol{\xi}_j) + \sum_{j=1}^{2N} \beta_j T_{i2}(\boldsymbol{x}, \boldsymbol{\xi}_j), \quad \boldsymbol{x} \in \partial\Omega \cup \partial D, \tag{5}$$

where the traction fundamental solution matrix $T = (T_{ij})_{i,j=1,2}$ is given in, e.g. [10].

3.1 Parametrization of the unknown boundary and choice of the boundary collocation and source points

Without loss of generality, we assume that the known outer boundary $\partial\Omega$ is a circle of radius r_o. Then, the outer boundary collocation and source points can be chosen as

$$\boldsymbol{x}_{N+k} = r_o(\cos(\vartheta_k), \sin(\vartheta_k)), \ \boldsymbol{\xi}_{N+k} = \eta_o r_o(\cos(\vartheta_k), \sin(\vartheta_k)), \ k = \overline{1, N}, \tag{6}$$

where $\vartheta_k = 2\pi(k-1)/N$, $k = \overline{1, N}$, and $\eta_o > 1$ is fixed.

We further assume that the unknown rigid inclusion or cavity D is a star-shaped domain with respect to the origin. The more general case in which the center of the star-shaped domain D is unknown can also be investigated with no major modifications, see [11]. Thus we can parameterize the boundary ∂D as

$$x = r(\vartheta)\cos\vartheta, \quad y = r(\vartheta)\sin\vartheta, \quad \vartheta \in [0, 2\pi), \tag{7}$$

where r is a $2\pi-$periodic function. The collocation form of (7) in two dimensions becomes

$$r_k = r(\vartheta_k), \quad k = \overline{1, N}, \tag{8}$$

and we choose the inner boundary and source points as

$$\boldsymbol{x}_k = r_k(\cos\vartheta_k, \sin\vartheta_k), \quad \boldsymbol{\xi}_{\boldsymbol{k}} = \eta_{int}\boldsymbol{x}_k, \quad k = \overline{1, N}, \tag{9}$$

where $\eta_{int} \in (0, 1)$ is fixed.

3.2 Penalized least-squares minimization

The coefficients $\boldsymbol{\alpha} = (\alpha_j)_{j=\overline{1,2N}}$, $\boldsymbol{\beta} = (\beta_j)_{j=\overline{1,2N}}$ and the radii $\boldsymbol{r} = (r_j)_{j=\overline{1,N}}$ can be determined by imposing the boundary conditions (1b)-(1d), or (1b), (1c) and (1e). We thus have a total of $6N$ equations in $5N$ unknowns. Note that the inverse inclusion problem under investigation is linear in the coefficients $\boldsymbol{\alpha}$ and $\boldsymbol{\beta}$, but it is nonlinear in the radii $\boldsymbol{r}$.

In the case of the inverse geometric problem (1a)-(1d) associated with the detection of the unknown rigid inclusion D, the penalized least-squares functional to be minimized is given by the sum of the residual and the regularization terms, namely,

$$\begin{aligned} S(\boldsymbol{\alpha},\boldsymbol{\beta},\boldsymbol{r}) &= Res(\boldsymbol{\alpha},\boldsymbol{\beta},\boldsymbol{r}) + Reg(\boldsymbol{\alpha},\boldsymbol{\beta},\boldsymbol{r}) \\ &= \sum_{i=1}^{2}\sum_{j=1}^{2N}\left[u_i^N(\boldsymbol{x}_j,\boldsymbol{\xi};\boldsymbol{\alpha},\boldsymbol{\beta}) - f_i(\boldsymbol{x}_j)\right]^2 \\ &+ \sum_{i=1}^{2}\sum_{j=N+1}^{2N}\left[t_i^N(\boldsymbol{x}_j,\boldsymbol{\xi};\boldsymbol{\alpha},\boldsymbol{\beta}) - g_i(\boldsymbol{x}_j)\right]^2 \\ &+ \mu_1\left(|\boldsymbol{\alpha}|^2 + |\boldsymbol{\beta}|^2\right) + \mu_2\sum_{\ell=2}^{N}\left(r_\ell - r_{\ell-1}\right)^2, \end{aligned} \tag{10}$$

where $f_i(\boldsymbol{x}_j) \equiv 0$ for $i = 1, 2$ and $j = \overline{1, N}$, and $\mu_1, \mu_2 > 0$ are regularization parameters. The functional (10) is minimized subject to the simple bounds on the variables

$$0 < r_m < r_o, \quad m = \overline{1,N}. \tag{11}$$

In (10), the last two terms, included in order to achieve stability, correspond to penalising the ℓ^2-norm of the coefficients $\boldsymbol{\alpha}$ and $\boldsymbol{\beta}$, and the H^1-discretised norm of the smooth obstacle radii $\boldsymbol{r}$.

The Cauchy data given on $\partial\Omega$ in the boundary conditions (1b) and (1c) come from practical measurements which are inherently contaminated with noisy errors, and therefore we replace f_i and g_i by f_i^p and g_i^p, $i = 1, 2$, respectively, generated as

$$f_i^p(\boldsymbol{x}_j) = (1 + \rho_j^f p_{\boldsymbol{u}}) f_i(\boldsymbol{x}_j), \quad i = 1, 2,\ j = \overline{N+1, 2N}, \tag{12}$$

$$g_i^p(\boldsymbol{x}_j) = (1 + \rho_j^g p_{\boldsymbol{t}}) g_i(\boldsymbol{x}_j), \quad i = 1, 2,\ j = \overline{N+1, 2N}, \tag{13}$$

where $p_{\boldsymbol{u}}$ and $p_{\boldsymbol{t}}$ represent the percentage of noise added into the displacements and tractions on $\partial\Omega$, respectively, and $\left(\rho_j^f\right)_{j=\overline{N+1,2N}}$ and $\left(\rho_j^g\right)_{j=\overline{N+1,2N}}$ are pseudo-random noisy variables drawn from a uniform distribution on $[-1, 1]$ using the NAG [12] routine `G05DAF`. Since the inverse problem under investigation is ill-posed being unstable, i.e. small errors $p_{\boldsymbol{u}}$ and/or $p_{\boldsymbol{t}}$ in the data (12) and/or (12) cause large errors in the solution for ∂D, the C^1-smoothness regularization term involving μ_2 was added in order to achieve stability. In addition, the regularization term involving μ_1 was added in order to deal with the ill-conditioned MFS system of equations.

The minimization of (10) is carried out using the `MINPACK` routine `lmdif` which minimizes the unconstrained sum of squares of nonlinear functions. The simple bound constraints (11) are imposed during the iterative procedure by adjustment at each iteration. The initial guess for the unknowns has been taken arbitrarily to be $\boldsymbol{\alpha}^{(0)} = \boldsymbol{\beta}^{(0)} = \mathbf{0}$ and $\boldsymbol{r}^{(0)} = 1$. The Jacobian is calculated internally using forward finite differences.

3.3 Regularization

In order to obtain a stable solution, regularization of the numerical solution can be accomplished by one of the following two approaches. In a first instance, if no regularization terms are included in the objective functional (10), i.e. $\mu_1 = \mu_2 = 0$, then, according to the discrepancy principle, we stop the iterations involved in the process of minimization once the residual $Res(\boldsymbol{\alpha}, \boldsymbol{\beta}, \boldsymbol{r})$ becomes less than the amount of noise, i.e.

$$Res(\boldsymbol{\alpha}, \boldsymbol{\beta}, \boldsymbol{r}) \leq \epsilon := \sum_{i=1}^{2} \sum_{j=N+1}^{2N} \left\{ [f_i^p(\boldsymbol{x}_j) - f_i(\boldsymbol{x}_j)]^2 + [g_i^p(\boldsymbol{x}_j) - g_i(\boldsymbol{x}_j)]^2 \right\} . \tag{14}$$

However, we do not know whether the minimization routine `lmdif` which is employed has a regularization character which justifies this stopping criterion. In order to add some rigour into the stability of the numerical solution, we consider next including some positive regularization parameters μ_1 and μ_2 in (10). Then the iteration process does not need to be stopped, i.e. it can be left to run until a user specified tolerance, say of 10^{-10}, is achieved or a maximum number of function evaluations, `maxfev`, is reached. However, one still has to choose appropriately the regularization parameters μ_1 and μ_2 and this can be done either based on the above discrepancy principle, or on the more heuristic L-curve or L-surface criterion, see [13, 14].

4 Numerical example

We consider an example for which an analytical solution is available (see [8]) in order to assess the accuracy and stability of the proposed method. In particular, we consider an isotropic linear elastic medium, e.g. copper alloy, characterised by the material constants $\bar{\nu} = \nu = 0.34$ and $G = 3.35 \times 10^{10}\,\mathrm{N/m^2}$, and occupying the two-dimensional annular domain $\Omega \setminus \overline{D}$, where

$$\Omega = \left\{ (x, y) \in \mathbb{R}^2 | x^2 + y^2 < r_o^2 \right\}, \; D = \left\{ (x, y) \in \mathbb{R}^2 | x^2 + y^2 < r_{int}^2 \right\}, \tag{15}$$

where $0 < r_{int} < r_o$. We also consider the following exact solution for the displacements

$$u_1(x,y) = \frac{1}{2G(1+\nu)}\left[V(1-\nu) - W(1+\nu)\frac{1}{(x^2+y^2)}\right]x, \tag{16}$$

$$u_2(x,y) = \frac{1}{2G(1+\nu)}\left[V(1-\nu) - W(1+\nu)\frac{1}{(x^2+y^2)}\right]y, \tag{17}$$

where $(x,y) \in \overline{\Omega}\backslash D$ and

$$V = -\frac{\sigma_o r_o^2 - \sigma_{int} r_{int}^2}{r_o^2 - r_{int}^2}, \quad W = \frac{(\sigma_o - \sigma_{int}) r_o^2 r_{int}^2}{r_o^2 - r_{int}^2}, \quad \sigma_o, \sigma_{int} \in \mathbb{R}. \tag{18}$$

The corresponding stress tensor is given by

$$\sigma_{ij}(\boldsymbol{x}) = \left[V + (-1)^{i+1} W \frac{x^2 - y^2}{(x^2+y^2)^2}\right]\delta_{ij} + 2W\frac{xy}{(x^2+y^2)^2}(1-\delta_{ij}), i, j = 1, 2. \tag{19}$$

By choosing

$$\sigma_o = \left[(1+\nu) + (1-\nu)\frac{r_{int}^2}{r_o^2}\right]\frac{\sigma_{int}}{2}, \quad \sigma_{int} = 1.0 \times 10^{10}\,\mathrm{N/m^2},$$

in expressions (16)-(19), we have that $u_1 = u_2 = 0$ on the inner boundary of the circular rigid inclusion ∂D. For these dimensional quantities, in order to have the

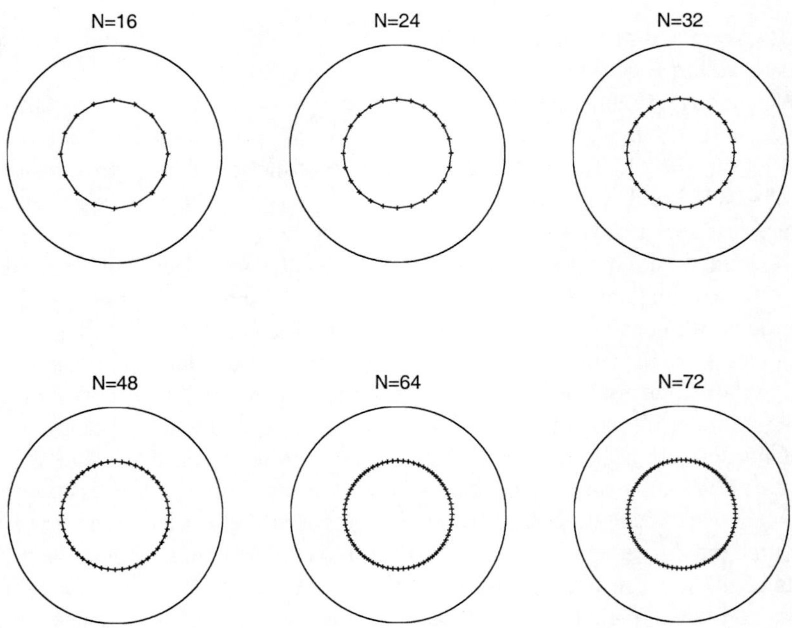

Figure 1: Results for various N with no noise and no regularization.

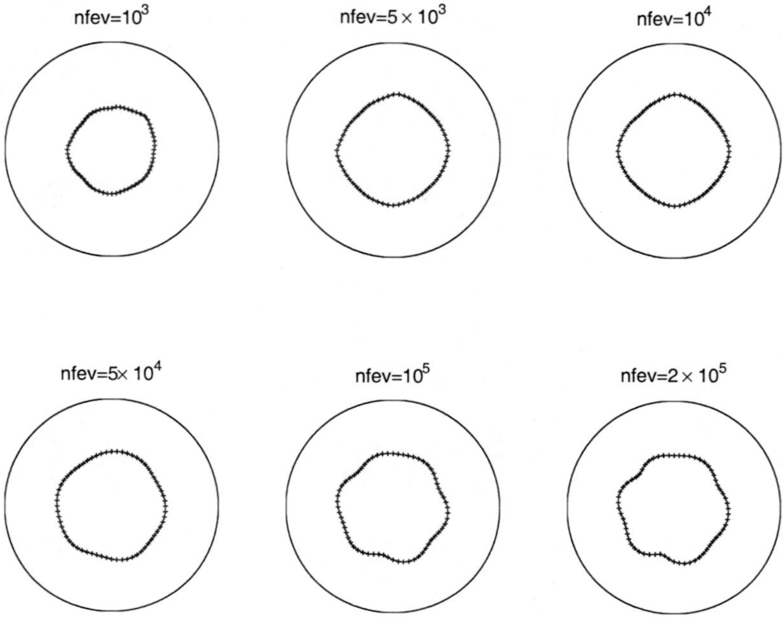

Figure 2: Results for various values of `nfev` with $p_t = 5\%$ noise and no regularization.

first two terms to be minimized in (10) of the same order of magnitude, we weigh the second term accordingly.

We conducted numerical experiments with $r_{int} = 2$ and $r_o = 4$, for different values of N in the case of no noise ($p_u = p_t = 0$) and no regularization ($\mu_1 = \mu_2 = 0$). The maximum number of function evaluations was set to `maxfev`$=10^5$ and we fixed $\eta_{int} = 0.2, \eta_o = 2$. From the results presented in Figure 1, it appears that the MFS is highly accurate.

Next, in order to investigate the stability of the numerical solution, we fix $N = 64$ and include $p_t = 5\%$ random noise in the input traction data (12). For simplicity, we consider no noise in the input displacement data (12), i.e. $p_u = 0$. In Figure 2, we present the plots of the reconstructed boundary ∂D obtained with no regularization for various numbers of function evaluations `nfev`. Since the problem under investigation is ill-posed, when no regularization is employed, an unstable solution is expected and, from Figure 2, it can be observed that as `nfev` increases beyond a certain threshold so does the instability. In this case, in order to obtain a stable solution one needs to stop the iterative process at the first `nfev` at which the discrepancy principle (14) is satisfied. In Figures 3 and 4 we present plots of the reconstructed boundary of the rigid inclusion when regularization is included in (10), namely, $\mu_1 > 0$, $\mu_2 = 0$, and $\mu_1 = 0$, $\mu_2 > 0$, respectively. In comparison with Figure 2 where no regularization was employed, from Figures 3 and 4 it can be seen that improved stable results are obtained if regularization is

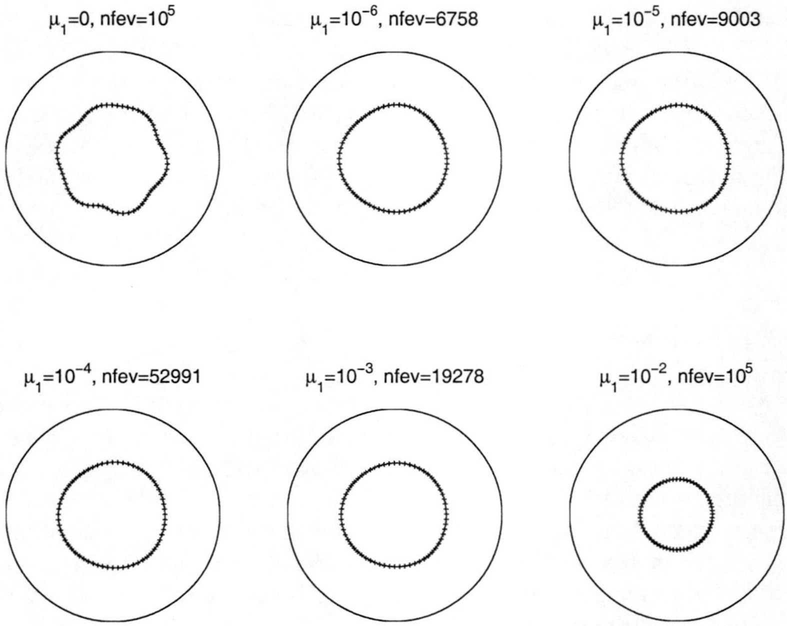

Figure 3: Results for various values of μ_1 with noise $p_t = 5\%$ noise.

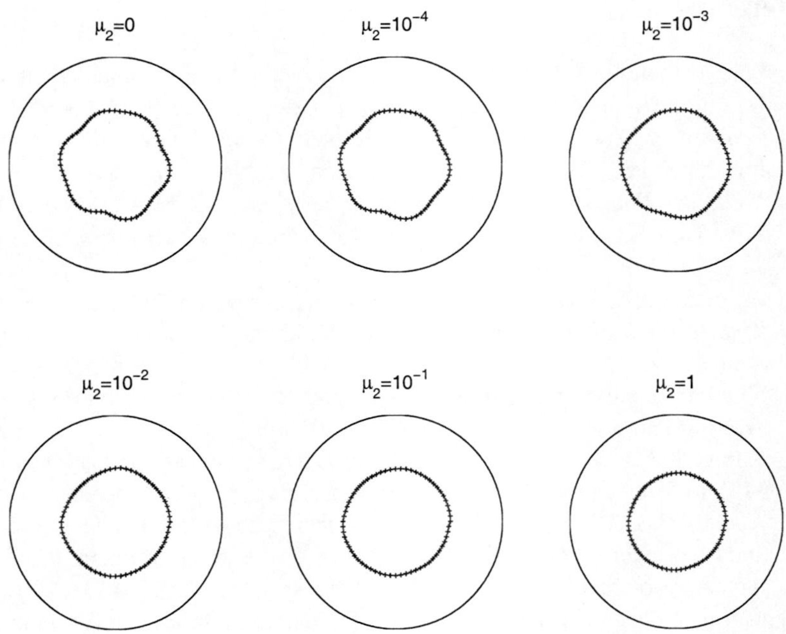

Figure 4: Results for various values of μ_2 with $p_t = 5\%$ noise.

included. Interestingly, different values of the regularization parameter $\mu_1 > 0$ in Figure 3 had the effect of producing a number nfev, for which the convergence of (10) was reached, less than the prescribed $\texttt{maxfev} = 10^5$. Improved stable and accurate results were obtained for $\mu_1 = O(10^{-5}) \div O(10^{-3})$. In contrast, for various values of μ_2, nfev reached $\texttt{maxfev} = 10^5$. Even so, the numerical results obtained with $\mu_2 = O(10^{-3}) \div O(10^1)$ in Figure 4 seem stable and reasonably accurate.

5 Conclusions

The MFS has been formulated for the solution of inverse inclusion problems arising in two-dimensional linear elasticity. The numerical experiments in the case of a rigid inclusion yield accurate results for exact data, but instabilities appear when noise is introduced into the input data. Regularization can be achieved either by appropriately limiting the number of functional evaluations, or by introducing penalty terms in the objective cost functional that is minimized. The extension of the proposed technique to inverse inclusion problems in three-dimensional linear elasticity, [15], is deferred to a future work.

References

[1] Fairweather, G. & Karageorghis, A., The method of fundamental solutions for elliptic boundary value problems. *Adv Comput Math*, **9**, pp. 69–95, 1998.

[2] Golberg, M.A. & Chen, C.S., The method of fundamental solutions for potential, Helmholtz and diffusion problems. *Boundary Integral Methods: Numerical and Mathematical Aspects*, ed. M.A. Golberg, WIT Press/Comput. Mech. Publ., Boston, MA, volume 1 of *Comput. Eng.*, pp. 103–176, 1999.

[3] Kołodziej, J.A. & Zieliński, A.P., *Boundary Collocation Techniques and their Application in Engineering*. WIT Press: Southampton, 2009.

[4] Ahmed, M.T., Lavers, J.D. & Burke, P.E., An evaluation of the direct boundary element method and the method of fundamental solutions. *IEEE Trans Magnetics*, **25**, pp. 3001–3006, 1989.

[5] Karageorghis, A., Lesnic, D. & Marin, L., A survey of applications of the MFS to inverse problems. *Inverse Probl Sci Eng*, 2011. To appear.

[6] Alves, C.J.S. & Martins, N.F.M., The direct method of fundamental solutions and the inverse Kirsch-Kress method for the reconstruction of elastic inclusions or cavities. *J Integral Equations Appl*, **21**, pp. 153–178, 2009.

[7] Marin, L., Regularized method of fundamental solutions for boundary identification in two-dimensional isotropic linear elasticity. *Int J Solids Struct*, **47**, pp. 3326–3340, 2010.

[8] Marin, L. & Johansson, B.T., Relaxation procedures for an iterative MFS algorithm for the stable reconstruction of elastic fields from Cauchy data in two-dimensional isotropic linear elasticity. *Int J Solids Struct*, **47**, pp. 3462–3479, 2010.

[9] Alessandrini, G., Bilota, A., Formica, G., Morassi, A., Rosset, E. & Turco, E., Numerical size estimates of inclusions in elastic bodies. *Inverse Problems*, **21**, pp. 133–151, 2005.

[10] Aliabadi, M.H., *The Boundary Element Method. Applications in Solids and Structures*. John Wiley and Sons: London, 2002. Volume 2.

[11] Martins, N.F.M. & Silvestre, A.L., An iterative MFS approach for the detection of immersed obstacles. *Eng Anal Bound Elem*, **32**, pp. 517–524, 2008.

[12] NAG(UK) Ltd, Wilkinson House, Jordan Hill Road, Oxford, UK, *Numerical Algorithms Group Library Mark 21*, 2007.

[13] Hansen, P.C., Regularization tools: a Matlab package for analysis and solution of discrete ill-posed problems. *Numer Algorithms*, **6**, pp. 1–35, 1994.

[14] Belge, M., Kilmer, M.E. & Miller, E.L., Efficient determination of multiple regularization parameters in a generalized L-curve framework. *Inverse Problems*, **21**, pp. 133–151, 2005.

[15] Marin, L., A meshless method for solving the Cauchy problem in three-dimensional elastostatics. *Comput Math Appl*, **50**, pp. 73–92, 2005.

IRBFN-based multiscale solution of a model 1D elliptic equation

D.-A. An-Vo, C.-D. Tran, N. Mai-Duy & T. Tran-Cong
Computational Engineering and Science Research Centre,
University of Southern Queensland, Australia

Abstract

Many engineering problems have a wide range of length scales in their solutions. Direct numerical simulations for these problems typically require extremely-large amounts of CPU time and computer memory, which may be too expensive or impossible on the present supercomputers. In this paper, we present a high-order method, based on the multiscale basis function framework and integrated radial-basis-function networks, for solving multiscale elliptic problems in one dimension.
Keywords: integrated radial basis functions, point collocation, subregion collocation, multiscale elliptic problems.

1 Introduction

In composite materials, the presence of particles/fibres in the resin gives rise to the multiscale fluctuations in the thermal or electrical conductivity. In porous media, formation properties such as permeability have a very high degree of spatial variability. These effects are typically captured at scales that are too fine for direct numerical simulation. To enable the solution of these problems, a number of advanced numerical methods have been developed. Examples include those based on the homogenisation theory (e.g. [1]), upscaling methods (e.g. [2]) and multiscale methods (e.g. [3]). The homogenisation-theory-based methods have been successfully applied for the prediction of effective properties and statistical correlation lengths for multicomponent random media. However, restrictive assumptions on the media, such as scale separation and periodicity, limit their range of application. Furthermore, when dealing with problems having many separate scales, they become very expensive because their computational cost increases exponentially with the number of scales. For upscaling methods, their design

WIT Transactions on Modelling and Simulation, Vol 52,
www.witpress.com, ISSN 1743-355X (on-line)
doi:10.2495/BE110211

principle is based on simple physical and/or mathematical motivations. A heterogeneous medium is replaced by a homogeneous medium characterised by equivalent properties, and coarse-scale equations are prescribed in explicit form. Although upscaling techniques are effective, most of their applications have been reported for the case of periodic structures. As opposed to upscaling, multiscale methods consider the full problem with the original resolution. The coarse-scale equations are formed and solved numerically, where one constructs the basis functions from the leading order homogeneous elliptic equation in coarse-scale elements. The idea of using the non-polynomial multiscale approximation space rather than the standard piecewise polynomial space was first introduced by Babuška et al. [4] for one-dimensional problems and by Hou and Wu [3] for two-dimensional elliptic problems. These methods have the ability to capture accurately the effects of fine-scale variations without the need for using global fine meshes. Multiscale methods can be categorised into multiscale finite-element (FE) methods (e.g. [5, 6]), mixed multiscale FE methods (e.g. [7, 8]) and multiscale control-volume (CV) methods (e.g. [9, 10]). Typically, there are two different meshes used: a fine mesh for computing locally the basis function space, and a coarse mesh for computing globally the solution of an elliptic partial differential equation (PDE). The multiscale bases are independent of each other and their constructions can thus be conducted in parallel. In solving the elliptic PDE, one may only need to employ a mesh that today's computing resources can afford for computing. For two-scale periodic structures, Hou et al. [11] have proved that the multiscale FE method indeed converges to the correct solution independent of the small scale in the homogenisation limit. Multiscale techniques require the solutions of elliptic PDEs which are achieved by means of discretisation schemes.

Radial-basis-function networks (RBFNs) are known as a powerful tool for the approximation of scattered data. Their application to the solution of PDEs has received a great deal of attention over the last 20 years (e.g. [12] and references therein). It is easy to implement RBF collocation methods and they can give a high-order convergence solution. A number of RBF approaches, based on local approximations, domain decompositions, preconditioning schemes and compactly-supported basis functions, have been presented, towards the solution of large-scale problems. Integrated RBFNs (IRBFNs) proposed by Mai-Duy and Tran-Cong [13, 14], in which highest-order derivatives under consideration are approximated using RBFNs and lower-order derivatives are obtained through integration, have several advantages over conventional differentiated RBFNs. The purposes of using integration (a smoothing operator) to construct the approximants are (i) to avoid the reduction in convergence rate caused by differentiation and (ii) to improve the numerical stability of a discrete solution.

This paper is concerned with the incorporation of IRBFNs and subregion collocation (i.e. control-volume (CV) formulation) into the non-polynomial approximation space approach, called msIRBFN-CV method, for solving one-dimensional multiscale elliptic problems. The remainder of the paper is organised as follows. Section 2 presents the governing equation. The proposed method is described in Section 3 and numerical results are given in Section 4. Section 5 concludes the paper.

2 Governing equation

A class of multiscale problems arising from composite material and porous media can be modelled by second-order elliptic PDEs

$$-\nabla.\left(a^{\varepsilon}\nabla u\right) = f \tag{1}$$

where the coefficient a^{ε} is an oscillatory function involving a small scale. In the case of heat conduction in composite materials, u and a represent the temperature and thermal conductivity, respectively. In the case of flows in porous media, u is the pressure and $a = k/\nu$ is the ratio of the permeability k and the fluid viscosity ν. In this work, we consider (1) in one dimension.

Conventional discretisation techniques using piecewise polynomial approximation spaces can be applied to solve (1). However, it would require the mesh size h be much smaller than the finest scale, i.e., $h \ll \varepsilon$. In contrast, the multiscale computational framework uses a coarse grid of size $h > \varepsilon$ and an adaptive basis which incorporates the small-scale features of the oscillating coefficient function $a^{\varepsilon}(\mathbf{x})$. Integrated RBFNs are designed to construct the multiscale basis function space.

3 Present msIRBFN-CV method

3.1 Integrated radial-basis-function networks

IRBFNs, which are employed with the multiquadric function, for the representation of a function y and its derivatives (e.g. up to the second order) in one dimension can be mathematically described as

$$\frac{d^2y}{dx^2}(x) = \sum_{i=1}^{m} w_i G_i(x) \tag{2}$$

$$\frac{dy}{dx}(x) = \sum_{i=1}^{m} w_i H_i(x) + C_1 \tag{3}$$

$$y(x) = \sum_{i=1}^{m} w_i \overline{H}_i(x) + C_1 x + C_2 \tag{4}$$

where C_1 and C_2 are constants of integration and

$$G_i(x) = \left((x - c_i)^2 + a_i^2\right)^{1/2} \tag{5}$$

$$H_i(x) = \int G_i(x)dx = \frac{(x-c_i)\left((x-c_i)^2+a_i^2\right)^{1/2}}{2} + \frac{a_i^2}{2}\ln\left((x-c_i)+\left((x-c_i)^2+a_i^2\right)^{1/2}\right) \quad (6)$$

$$\overline{H}_i(x) = \int H_i(x)dx = \frac{((x-c_i)^2+a_i^2)^{3/2}}{6} + \frac{a_i^2}{2}(x-c_i)\ln\left((x-c_i)+\left((x-c_i)^2+a_i^2\right)^{1/2}\right) - \frac{a_i^2}{2}\left((x-c_i)^2+a_i^2\right)^{1/2} \quad (7)$$

in which $\{c_i\}_{i=1}^m$ is a set of centres and $\{a_i\}_{i=1}^m$ is a set of RBF widths.

A set of collocation points $\{x_i\}_{i=1}^m$ is taken to be a set of centres, while the RBF width is chosen according to the following relation

$$a_i = \beta d_i,$$

where β is a factor and d_i is the minimum distance from the ith centre to its neighbouring centres. A factor β is simply chosen to be unity in the present study. Since C_1 and C_2 are to be found, we treat them like the RBF weights.

3.2 Incorporation IRBFNs into the multiscale basis function framework

The problem domain is represented using a set of N nodal points, called a global coarse-scale grid. This grid is used for solving the coarse-scale equation (1). On each interval or coarse element, $[x_{i-1}, x_i]$ with $i = \{2, 3, \ldots, N\}$, an approximation to the field variable u is sought in the form

$$u(x) = \phi_{i-1}(x)u_{i-1} + \phi_i(x)u_i, \quad (8)$$

where $x_{i-1} \leq x \leq x_i$, $u_{i-1} = u(x_{i-1})$, $u_i = u(x_i)$, and $\phi_{i-1}(x)$ and $\phi_i(x)$ the basis functions.

We employ subregion collocation to discretise (1). Each node x_i with $i = \{2, 3, \ldots, N-1\}$ is surrounded by a control volume $[x_{i-1/2}, x_{i+1/2}]$ denoted by Ω_i (Figure 1). Integrating (1) over a control volume Ω_i, one has

$$a^{\varepsilon}(x_{i+1/2})\frac{du}{dx}(x_{i+1/2}) - a^{\varepsilon}(x_{i-1/2})\frac{du}{dx}(x_{i-1/2}) + \int_{x_{i-1/2}}^{x_{i+1/2}} f dx = 0 \quad (9)$$

Taking (8) into account, one can express first derivatives in (9) in terms of nodal values of u. Opposed to traditional discretisation methods, the basis functions $\phi_{i-1}(x)$ and $\phi_i(x)$ on a coarse element $[x_{i-1}, x_i]$ are not analytical functions (e.g. not polynomials), but local numerical solutions to the following differential

www.witpress.com, ISSN 1743-355X (on-line)

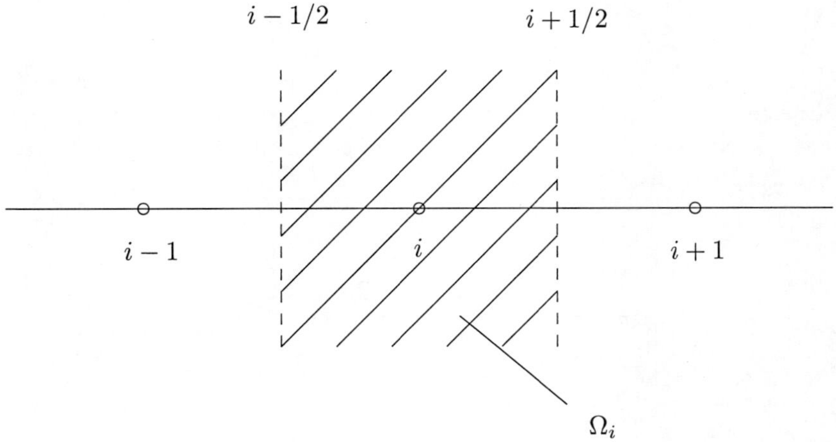

Figure 1: A CV discretisation scheme: node i and its associated control volume. The circles represent the nodes, and the vertical dash lines represent the faces of the control volume.

equation

$$\frac{d}{dx}\left(a^{\varepsilon}\frac{d\phi_k}{dx}\right)=0 \tag{10}$$

with $k=\{i-1,i\}$ and $x_{i-1}\leq x\leq x_i$. Boundary conditions for (10) are specified using the condition $\phi_k(x_j)=\delta_{kj}$ with $j=\{i-1,i\}$. Equation (10) needs be solved twice for the determination of the two ϕ_k on each element. A coarse element $[x_{i-1},x_i]$ is discretised by a set of M points, called a local fine-scale grid. Such a grid is used to capture the fine-scale structure information of the solution. Let $\{\eta_1=x_{i-1},\eta_2,\dots,\eta_M=x_i\}$ be a set of nodes of the local fine-scale grid. Collocation of (10) at a set of grid points together with imposition of Dirichlet boundary conditions at the two end points result in the following set of algebraic equations

$$\mathcal{A}\widehat{w}=\widehat{b} \tag{11}$$

where

$$\mathcal{A}=\begin{bmatrix} G_1(\eta_1), & G_2(\eta_1), & \dots, & G_M(\eta_1), & 0, & 0\\ G_1(\eta_2), & G_2(\eta_2), & \dots, & G_M(\eta_2), & 0, & 0\\ \vdots & \vdots & \ddots & \vdots & \vdots & \vdots\\ G_1(\eta_M), & G_2(\eta_M), & \dots, & G_M(\eta_M), & 0, & 0\\ \overline{H}_1(\eta_1), & \overline{H}_2(\eta_1), & \dots, & \overline{H}_M(\eta_1), & \eta_1, & 1\\ \overline{H}_1(\eta_M), & \overline{H}_2(\eta_M), & \dots, & \overline{H}_M(\eta_M), & \eta_M, & 1 \end{bmatrix}$$

$$\widehat{w}=(w_1,w_2,\dots,w_M,C_1,C_2)^T$$

$$\widehat{b} = (0, 0, \ldots, 0, \phi_k(x_{i-1}), \phi_k(x_i))^T$$

It can be seen from (11) that, owing to the presence of the two integration constants, equation (10) is forced to be satisfied exactly not only at the interior points but also at the two end points.

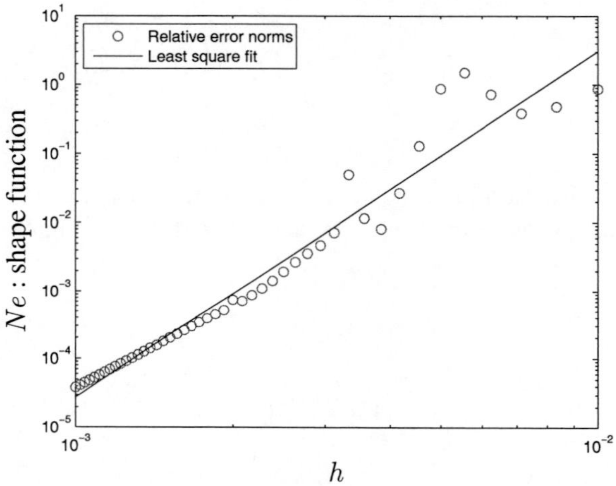

Figure 2: Fine scale convergence.

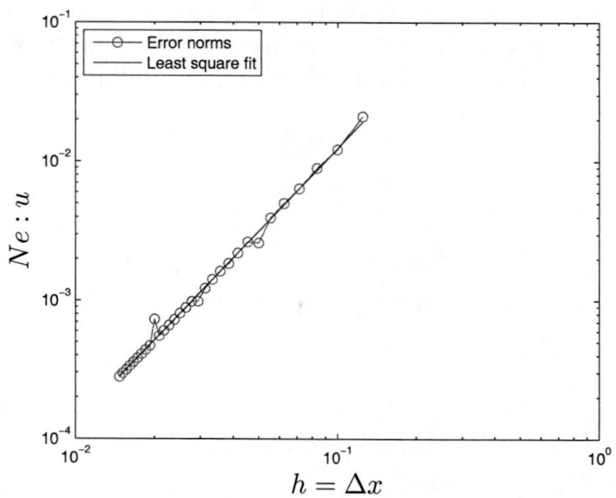

Figure 3: Coarse scale convergence.

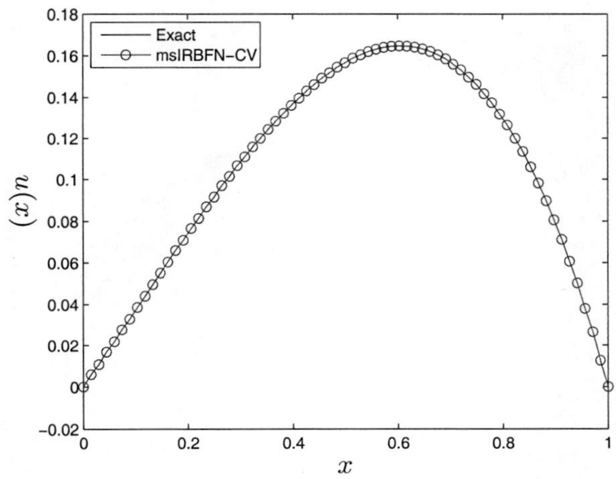

Figure 4: msIRBFN-CV solution.

4 Numerical results

Values of ϕ_k, $d\phi_k/dx$ and $d^2\phi_k/dx^2$ at an arbitrary point $x \in [x_{i-1}, x_i]$ can then be computed by

$$\phi_k(x) = \left[\overline{H}_1(x), \overline{H}_2(x), \ldots, \overline{H}_M(x), x, 1\right] \mathcal{A}^{-1}\widehat{b}, \tag{12}$$

$$\frac{d\phi_k}{dx}(x) = \left[H_1(x), H_2(x), \ldots, H_M(x), 1, 0\right] \mathcal{A}^{-1}\widehat{b}, \tag{13}$$

$$\frac{d^2\phi_k}{dx^2}(x) = \left[G_1(x), G_2(x), \ldots, G_M(x), 0, 0\right] \mathcal{A}^{-1}\widehat{b}. \tag{14}$$

Figures 2–9 summarise our multiscale modelling of $d(a^\varepsilon(x)du(x)/dx) + f(x) = 0$ where $f(x) = x$, $a^\varepsilon(x) = 1/(2 + x + \sin(2\pi x/\varepsilon))$ and $\varepsilon = 0.01$. The domain is discretised using a series of 8,10,...,68 uniform coarse elements (i.e. $N = \{9, 11, \ldots, 69\}$) and the shape functions that capture the fine scale physics in each coarse element are numerically obtained with our IRBFN collocation method using 55 uniform nodes (i.e. $M = 55$).

Figure 2 displays the behaviour of the numerical shape functions obtained by our IRBFN collocation method on the interval $0 \leq x \leq 0.1$ with a relative L_2 error norm of $Ne = 3.8451 \times 10^{-5}$ and a convergence rate of 5.0434. The coarse scale solution at the coarse grid points is obtained by a conservative control volume method. In order to have a good and consistent measure of accuracy, error norms in all cases are computed using the same 10,000 test points where the fine-scale solution is recovered via simple multiplication of the coarse results with the shape functions at the test points.

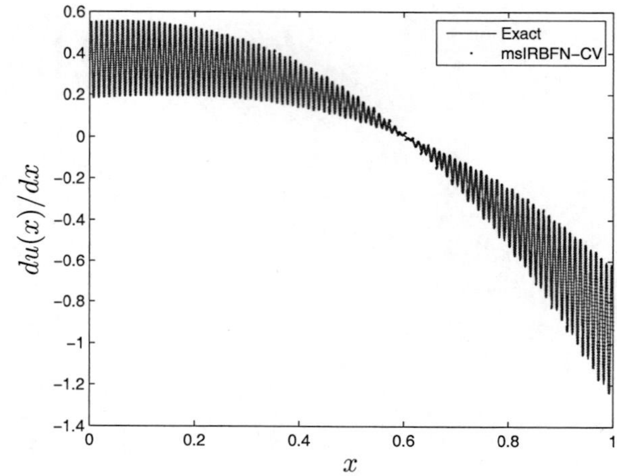

Figure 5: Fine scale recovery: $du(x)/dx$.

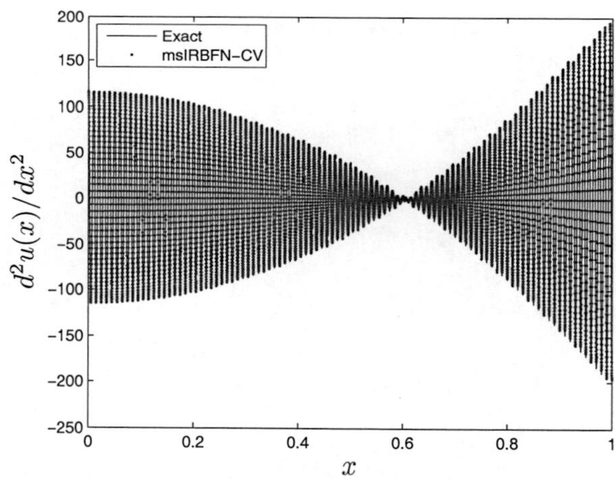

Figure 6: Fine scale recovery: $d^2u(x)/dx^2$.

Figure 3 shows that our msIRBFN-CV approach achieves a convergence rate of 1.9712 and a relative L_2 error norm of $Ne = 2.7942 \times 10^{-4}$ (69 coarse nodes) for this problem.

Figure 4 shows the coarse scale solution.

Figures 5 and 6 display the recovered fine scale results for the first and second derivatives of $u(x)$.

Figures 7, 8 and 9 display the results on a small sub-domain $0 \leq x \leq 0.1$ in order to show the fine-scale details better.

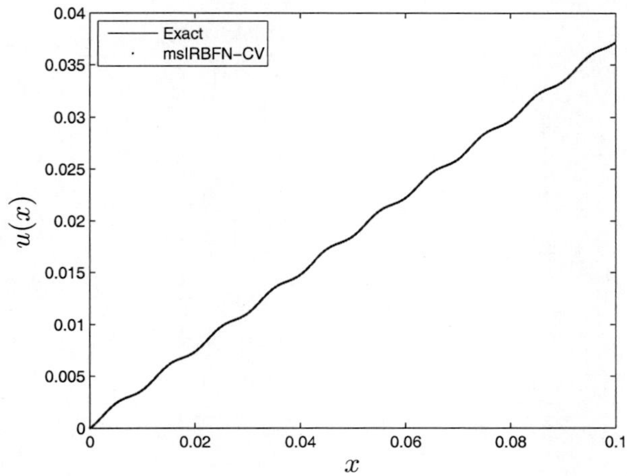

Figure 7: Fine scale: $u(x)$, $0 \leq x \leq 0.1$.

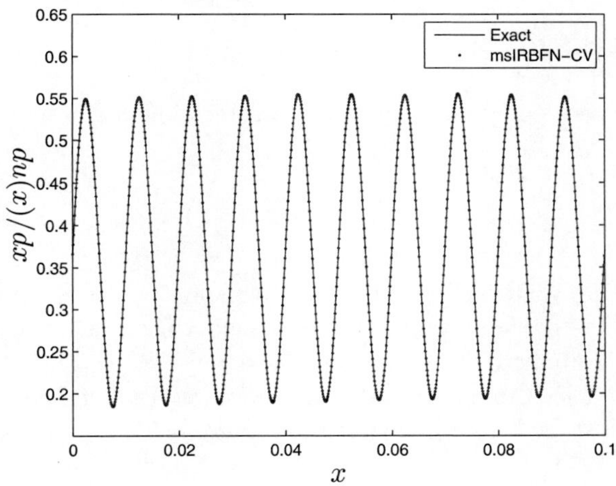

Figure 8: Fine scale: $du(x)/dx$.

5 Concluding remarks

In this paper, we have successfully introduced IRBFNs into the multiscale basis function framework. High rates of convergence for the solution of the fine scale equation are obtained. Numerical results show that our msIRBFN-CV approach captures the fine details quite well. However, we can only guarantee a C^0 solution. We are currently investigating ways to improve the choice of shape functions.

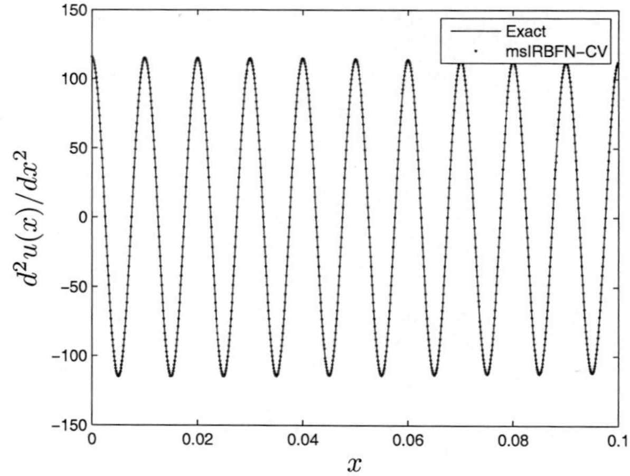

Figure 9: Fine scale: $d^2u(x)/dx^2$.

Acknowledgement

This work is supported by the Australian Research Council through a Discovery Projects grant.

References

[1] Dykaar, B.B. & Kitanidis, P.K., Determination of the effective hydraulic conductivity for heterogeneous porous media using a numerical spectral approach 1. Method. *Water Resources Research*, **28**, pp. 1155–1166, 1992.

[2] McCarthy, J.F., Comparison of fast algorithms for estimating large-scale permeabilities of heterogeneous media. *Transport in Porous Media*, **19**, pp. 123–137, 1995.

[3] Hou, T.Y. & Wu, X.-H., A multiscale finite element method for elliptic problems in composite materials and porous media. *J. of Computational Physics*, **134**, pp. 169–189, 1997.

[4] Babuška, I., Caloz, G., & Osborn, J.E., Special finite element methods for a class of second order elliptic problems with rough coefficients. *SIAM J. on Numerical Analysis*, **31**, pp. 945–981, 1994.

[5] Allaire, G. & Brizzi, R., A multiscale finite element method for numerical homogenization. *Multiscale Modeling & Simulation*, **43**, pp. 790–812, 2005.

[6] Hou, T.Y., Multiscale modelling and computation of fluid flow. *Int. J. for Numerical Methods in Fluids*, **47**, pp. 707–719, 2005.

[7] Aarnes, J.E., Kippe, V. & Lie, K.-A., Mixed multiscale finite elements and streamline methods for reservoir simulation of large geomodels. *Advances in Water Resources*, **28**, pp. 257–271, 2005.

[8] Arbogast, T., Implementation of a locally conservative numerical subgrid upscaling scheme for two-phase Darcy flow. *Computational Geosciences*, **6**, pp. 453–481, 2002.

[9] Chu, J., Efendiev, Y., Ginting, V. & Hou, T.Y., Flow based oversampling technique for multiscale finite element methods. *Advances in Water Resources*, **31**, pp. 599–608, 2008.

[10] Hajibeygi, H., Bonfigli, G., Hesse, M.A. & Jenny, P., Iterative multiscale finite-volume method. *J. of Computational Physics*, **227**, pp. 8604–8621, 2008.

[11] Hou, T.Y., Wu, X.-H. & Cai, Z., Convergence of a multiscale finite element method for elliptic problems with rapidly oscillating coefficients. *Mathematics of Computation*, **68**, pp. 913–943, 1999.

[12] Fasshauer, G.E., *Meshfree Approximation Methods With Matlab* (Interdisciplinary Mathematical Sciences - Vol. 6), World Scientific Publishers: Singapore, 2007.

[13] Mai-Duy, N. & Tran-Cong, T., Numerical solution of differential equations using multiquadric radial basis function networks. *Neural Networks*, **14**, pp. 185–199, 2001.

[14] Mai-Duy, N. & Tran-Cong, T., Numerical solution of Navier-Stokes equations using multiquadric radial basis function networks. *Int. J. for Numerical Methods in Fluids*, **37**, pp. 65–86, 2001.

A meshless solution of two dimensional density-driven groundwater flow

K. Kovarik
Faculty of Civil Engineering, University of Zilina, Slovakia

Abstract

The boundary element method (BEM) is a very useful method used for numerical models of groundwater flow. However, this method was aimed to solve problems in homogeneous domains and it presents even greater difficulties than the other numerical methods when coping with the nonhomogeneities which are so characteristic in the groundwater hydraulics. The meshless LBIE method is a very promising meshless scheme. This method is characterized as meshless since distributed nodal points, covering the domain, are employed. Nodal points can be randomly spread over the domain. Every node is surrounded by a simple surface (circle) centered at the collocation point and the boundary integral equation is written on this local boundary. The unknown variables, in the local sub-domains, are approximated by some of the interpolation methods to obtain a system of linear equations. Solving this system of equations leads to the numerical solution for the main problem. Several authors used the moving least squares (MLS) method as an interpolation method but nowadays the radial basis functions (RBF) interpolation is used instead. In this paper the combination of RBF and the dual reciprocity method is used to solve the time-dependent groundwater flow in heterogeneous domain combined with temperature transport which also influences the density and viscosity of groundwater.
Keywords: heat transfer, density driven flow, dual reciprocity, radial basis functions.

1 Introduction

The density driven groundwater flow occurs mainly in some environmental problems such as saltwater intrusion or leakage from landfills. We must solve this problem also when we study the geothermal problems. Although density driven

WIT Transactions on Modelling and Simulation, Vol 52,
www.witpress.com, ISSN 1743-355X (on-line)
doi:10.2495/BE110221

flow problems are mainly three-dimensional, they are often simulated by vertical 2D numerical models due to their very high computational costs.

The boundary element method and dual reciprocity method (DRM) are very useful methods used for numerical models of groundwater flow (see e.g. [1]). However, these method was mainly aimed to solve problems in homogeneous domains and it presents some difficulties when coping with the non homogeneities which are so characteristic of the groundwater hydraulics. The disadvantage can be overcome using some meshless scheme combined with BEM or DRM. The meshless LBIE method, introduced by Zhu et al. [2], is a very promising meshless scheme. This method is characterized as meshless since distributed nodal points, covering the domain, are employed. Nodal points are randomly spread over the domain. Every node is surrounded by a simple surface (circle) centered at the collocation point and the boundary integral equation is written on this local boundary. The unknown variables, in the local sub-domains, are approximated by some of the interpolation method to obtain a system of linear equations. Solving this system of equations leads to the numerical solution for the main problem. Zhu et al. [2] used the moving least squares (MLS) method as an interpolation method but nowadays the RBF interpolation is used instead (see e.g. [3]). In this paper the solution of coupled groundwater flow-heat transfer problem based on this RBIEM method is presented.

2 Governing equations

A density-driven groundwater flow can be described by the following equation (see also [4])

$$\rho S\frac{\partial h}{\partial t} + \epsilon\frac{\partial \rho}{\partial t} + \nabla(\rho\vec{q}) = 0 \tag{1}$$

where h is the groundwater potential, ϵ is the porosity of porous medium, ρ is the density of solution and $\vec{q}$ is the flux defined by

$$\mathbf{q} = -\mathbf{K}\left(\nabla h + \frac{\rho - \rho_0}{\rho_0}\nabla\vec{x}\right) \tag{2}$$

where $\mathbf{K}$ is matrix of hydraulic conductivities and ρ_0 initial density of water. The differential equation of 2D groundwater flow with variable density is now expressed as

$$\frac{\partial}{\partial x}\left(\rho K_x\frac{\partial h}{\partial x}\right) + \frac{\partial}{\partial y}\left[\rho K_y\left(\frac{\partial h}{\partial y} + \frac{\rho - \rho_0}{\rho_0}\right)\right] = \epsilon\frac{\partial \rho}{\partial t} + \rho S\frac{\partial h}{\partial t} \tag{3}$$

where we denote K_x, K_y hydraulic conductivities $[LT^{-1}]$ in direction x and y, respectively. Heat transfer in porous media can be described by the following equation (see e.g. [5])

$$\frac{\partial}{\partial x_i}\left(K_{TB}\frac{\partial T}{\partial x_j} - q_i\rho_L c_L T\right) = \rho_B c_B \frac{\partial T}{\partial t} \tag{4}$$

where T is temperature, K_T is thermal conductivity, c is specific heat capacity. The subscripts L and B refer to liquid and bulk phases, respectively.

3 Dual reciprocity formulation

The Eq. (3) can be transformed to the following shape

$$\begin{aligned}\frac{\partial^2 h}{\partial x^2} + \frac{K_y}{K_x}\frac{\partial^2 h}{\partial y^2} = \frac{\epsilon}{K_x\rho}\frac{\partial \rho}{\partial t} + \frac{S}{K_x}\frac{\partial h}{\partial t} - \frac{1}{\rho}\frac{\partial \rho}{\partial x}\frac{\partial h}{\partial x} - \\ -\frac{K_y}{K_x\rho}\frac{\partial \rho}{\partial y}\frac{\partial h}{\partial y} - \frac{K_y}{K_x\rho\rho_0}\frac{\partial}{\partial y}\left(\rho^2 - \rho\rho_0\right)\end{aligned} \tag{5}$$

To solve Eq. (5) in the domain Ω with boundary Γ using the DRM we should use the following transformation of co-ordinates

$$\tilde{x} = x \quad \tilde{y} = y\sqrt{\frac{K_x}{K_y}} \tag{6}$$

and we get

$$\begin{aligned}\frac{\partial^2 h}{\partial \tilde{x}^2} + \frac{\partial^2 h}{\partial \tilde{y}^2} = \frac{\epsilon}{K_x\rho}\frac{\partial \rho}{\partial t} + \frac{S}{K_x}\frac{\partial h}{\partial t} - \frac{1}{\rho}\frac{\partial \rho}{\partial \tilde{x}}\frac{\partial h}{\partial \tilde{x}} - \\ -\frac{1}{\rho}\sqrt{\frac{K_y}{K_x}}\frac{\partial \rho}{\partial \tilde{y}}\frac{\partial h}{\partial \tilde{y}} - \frac{1}{\rho\rho_0}\sqrt{\frac{K_y}{K_x}}\frac{\partial}{\partial \tilde{y}}\left(\rho^2 - \rho\rho_0\right)\end{aligned} \tag{7}$$

Applying the Green integral formula to the left side of Eq. (7) we get the following integral form

$$c_k h_k + \int_\Gamma \frac{\partial u^*_{kj}}{\partial n} h_j d\Gamma - \int_\Gamma u^*_{kj}\frac{\partial h_j}{\partial n} d\Gamma - \int_\Omega u^*_{kj} b_j d\Omega = 0 \tag{8}$$

Here b_j is the value of the right side of Eq. (7) in point j and u^*_{kj} is the fundamental solution of Laplace equation. The constant c_k has value from 0 to 1 being 0.5 if the point k is placed on the smooth boundary. The DRM is used to transform the domain integral in Eq. (8) to equivalent boundary integrals (see [6]). The basic idea is to expand the term b_j using the approximation

$$b_j \cong \sum_{i=1}^{N_b+N_i} \alpha_i f_{ij} \tag{9}$$

where α_i is a set of initially unknown coefficients and f_{ij} are approximating functions. The approximation employs N_b nodes on the boundary Γ and N_i points inside the domain Ω. The functions f_{ij} in Eq. (9) belongs to the family of radial basis function. The most often used functions are

$$f_{ij} = 1 + \sum_{m=1}^{n} r_{ij}^{m} \tag{10}$$

Now we can define the series of particular solutions $\hat{h}_{ij}$. These solutions and the approximating functions f_{ij} are linked through the equation

$$\frac{\partial^2 \hat{h}_{ij}}{\partial \tilde{x}^2} + \frac{\partial^2 \hat{h}_{ij}}{\partial \tilde{y}^2} = f_{ij} \tag{11}$$

Applying the Green's formula to the Eq. (11) and using Eq. (9) we can transform Eq. (8) and get the boundary only integral formula

$$c_k h_k + \int_\Gamma \frac{\partial u_{kj}^*}{\partial n} h_j d\Gamma - \int_\Gamma u_{kj}^* \frac{\partial h_j}{\partial n} d\Gamma = \tag{12}$$
$$= \sum_{i=1}^{N_b+N_i} \alpha_i \left(c_k \hat{h}_{ki} + \int_\Gamma \frac{\partial u_{ki}^*}{\partial n} \hat{h}_{ij} d\Gamma - \int_\Gamma u_{ki}^* \frac{\partial \hat{h}_{ij}}{\partial n} d\Gamma \right)$$

The discretized form of Eq. (12) for source point k can be written as

$$c_k h_k + \sum_{j=1}^{N_b} H_{kj} h_j - \sum_{j=1}^{N_b} G_{kj} q_j = \tag{13}$$
$$= \sum_{i=1}^{N_b+N_i} \alpha_i \left(c_k \hat{h}_{ki} + \sum_{j=1}^{N_b} H_{ki} \hat{h}_{ij} - \sum_{j=1}^{N_b} G_{ki} \hat{q}_{ij} \right)$$

where H_{kj} and G_{kj} are matrices obtained from the integrations of q^* and u^* at each boundary element. The unknown coefficients α_i can be determined from Eq. (9) as

$$\alpha_i = \sum_{j=1}^{N_b+N_i} F_{ij}^{-1} b_j \tag{14}$$

where F_{ij}^{-1} are members of the inverse matrix of values of function f_{ij}.

4 Radial basis functions interpolation

Radial basis functions (RBF) are initially known as a powerful tool for approximating multivariate functions on a scattered data. Due to their mesh-free nature RBF have received an increasing attention for solving partial differential equations (PDE) of different kinds. The first trial of such exploration was made by Kansa [7]. Full exploitation of the RBF method was constrained by the progressive ill-conditioned coefficient matrix as the number of nodes increases. To remove this difficulty, Shu et al. (see [8]) suggested using the local RBF method in which the approximation is formed by using only several local supporting points. The unknown function U is approximated in a sub-domain which forms the neighborhood or support of a reference point i by weighted sum of multiquadric functions and polynomials

$$U(x_i, y_i) = \sum_{j=1}^{n} \lambda_j R(r_{ij}) + \sum_{j=1}^{m} \chi_j p_j(x_i, y_i) \tag{15}$$

where λ_j and χ_j are weights, $R(r_{ij})$ are the RBF basis functions, and p_j is a basis for polynomial space with degree $m-1$, m is the order of R and n is a number of field nodes in the neighborhood of a reference point. Multiquadric functions are one of the most popular radial functions used for this purpose and they are defined as

$$R(r_{ij}) = \sqrt{r_{ij}^2 + \epsilon^2} \tag{16}$$

where r_{ij} is a distance between points i and j and ϵ is a so-called shape factor of multiquadric function. Coefficients λ_j, χ_j in Eq. (15) can be determined by enforcing Eq. (15) to be satisfied at these N_i nodes surrounding the point of interest. This leads to n linear equations, one for each node. The matrix form of these equations can be expressed as

$$\mathbf{U} = \mathbf{R}\lambda + \mathbf{P}\chi \tag{17}$$

where $\mathbf{U} = [U_1, U_2, \ldots U_{N_i}]$ is a vector of function values in all supporting nodes. However there are $n+m$ variables in this equation. The additional m equations can be added using the following m constraint conditions

$$\sum_{i=1}^{n} p_j(x_i, y_i)\lambda_i = 0, \quad j = 1 \ldots m \tag{18}$$

Combining Eqs.(17) and (18) yields the following set of equations in the matrix form

$$\mathbf{u} = \mathbf{A}\mathbf{a} \tag{19}$$

where we denoted

$$\mathbf{A} = \begin{bmatrix} \mathbf{R} & \mathbf{p} \\ \mathbf{p}^T & \mathbf{0} \end{bmatrix}, \quad \mathbf{a} = \begin{Bmatrix} \lambda \\ \chi \end{Bmatrix}, \quad \mathbf{u} = \begin{Bmatrix} \mathbf{U} \\ \mathbf{0} \end{Bmatrix} \tag{20}$$

The order m of multiquadric functions is equal to one and, therefore, we need one additional condition to make the interpolation problem well-posed. This condition is

$$\sum_{j=1}^{n} \lambda_j = 0 \tag{21}$$

Solving Eq. (19) we get

$$\mathbf{a} = \mathbf{A}^{-1}\mathbf{u} \tag{22}$$

Eq. (15) can be now written as (see e.g. [9])

$$U(x_i, y_i) = [\mathbf{R}^T(x_i, y_i)\ \mathbf{p}^T(x_i, y_i)]\mathbf{a} = [\mathbf{R}^T(x_i, y_i)\ \mathbf{p}^T(x_i, y_i)]\mathbf{A}^{-1}\mathbf{u} \tag{23}$$

The first n members of the product $[\mathbf{R}^T(x_i, y_i)\ \mathbf{p}^T(x_i, y_i)]\mathbf{A}^{-1}$ create set of RBF shape functions ϕ_j and Eq. (23) can be rewritten as

$$U(x_i, y_i) = \sum_{j=1}^{n} \phi_j U_j \tag{24}$$

This formula is used to develop the meshless local DRM in the next section.

5 Meshless local DRM formulation

In this section we follow the principle of RBIEM (see [3]). The area of interest Ω is covered by single nodes on the global boundary Γ and also inside the area (see Fig. 1). The local network of 16 elements was created around every node. This simplifies the generation of local boundaries and evaluation of integrals. Next four virtual internal points was added to every network (see Fig. 2) because the solution is usually more accurate (see [6]). The values of potential and its derivatives in

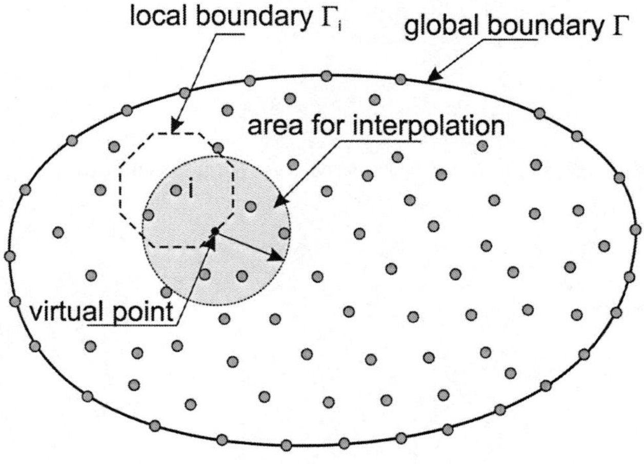

Figure 1: Nodes in the global area.

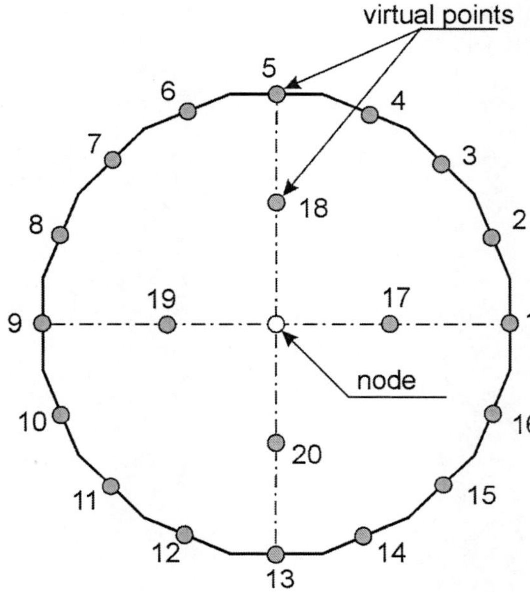

Figure 2: Local network.

x and y directions in node k can be computed from values of potentials and derivatives in virtual points using Eq. (9) as

$$h_k = \sum_{j=1}^{N_{vb}} G_{ij} q_{xj} n_x + \sum_{j=1}^{N_{vb}} G_{ij} q_{yj} n_y - \sum_{j=1}^{N_{vb}} H_{ij} h_j + \tag{25}$$

$$+ \sum_{k=1}^{N_{vb}+N_{vi}} \left\{ \alpha_k \left(c_i \hat{h}_{ik} + \sum_{j=1}^{N_{vb}} H_{ik} \hat{h}_{kj} - \sum_{j=1}^{N_{vb}} G_{ik} \hat{q}_{kj} \right) \right\}$$

$$\frac{\partial h_k}{\partial x} = \sum_{j=1}^{N_{vb}} \frac{\partial G_{ij}}{\partial x} q_{xj} n_x + \sum_{j=1}^{N_{vb}} \frac{\partial G_{ij}}{\partial x} q_{yj} n_y - \sum_{j=1}^{N_{vb}} \frac{\partial H_{ij}}{\partial x} h_j + \tag{26}$$

$$+ \sum_{k=1}^{N_{vb}+N_{vi}} \left\{ \alpha_k \left(\frac{\partial \hat{h}_{ik}}{\partial x} + \sum_{j=1}^{N_{vb}} \frac{\partial H_{ik}}{\partial x} \hat{h}_{kj} - \sum_{j=1}^{N_{vb}} \frac{\partial G_{ik}}{\partial x} \hat{q}_{kj} \right) \right\}$$

where N_{vb} is the number of virtual boundary points, N_{vi} is the number of virtual internal points, h_j are the values of potential in virtual points, q_{xj}, q_{yj} are the derivatives in boundary virtual points in x and y directions and n_x, n_y are the directional cosines of outer normal to virtual boundary, respectively. The derivative in direction y is computed according similar formula to Eq. (26). Values of potential and its derivatives can be approximated using RBF interpolation according Eq. (24) and we get

$$h_k = \sum_{j=1}^{N_{vb}} G_{ij} n_x \sum_{m=1}^{n} \phi_m q_{xm} + \sum_{j=1}^{N_{vb}} G_{ij} n_y \sum_{m=1}^{n} \phi_m q_{ym} - \qquad (27)$$

$$- \sum_{j=1}^{N_{vb}} H_{ij} \sum_{m=1}^{n} \phi_m h_m +$$

$$+ \sum_{k=1}^{N_{vb}+N_{vi}} \left\{ \alpha_k \left(c_i \hat{\Phi}_{ik} + \sum_{j=1}^{N_{vb}} H_{ik} \hat{\Phi}_{kj} - \sum_{j=1}^{N_{vb}} G_{ik} \hat{q}_{kj} \right) \right\}$$

$$\frac{\partial h_k}{\partial x} = \sum_{j=1}^{N_{vb}} \frac{\partial G_{ij}}{\partial x} n_x \sum_{m=1}^{n} \phi_m q_{xm} + \sum_{j=1}^{N_{vb}} \frac{\partial G_{ij}}{\partial x} n_y \sum_{m=1}^{n} \phi_m q_{ym} - \qquad (28)$$

$$- \sum_{j=1}^{N_{vb}} \frac{\partial H_{ij}}{\partial x} \sum_{m=1}^{n} \phi_m h_m +$$

$$+ \sum_{k=1}^{N_{vb}+N_{vi}} \left\{ \alpha_k \left(\frac{\partial \hat{h}_{ik}}{\partial x} + \sum_{j=1}^{N_{vb}} \frac{\partial H_{ik}}{\partial x} \hat{h}_{kj} - \sum_{j=1}^{N_{vb}} \frac{\partial G_{ik}}{\partial x} \hat{q}_{kj} \right) \right\}$$

where h_m, q_{xm}, and q_{ym} are values of potential and derivatives in directions x and y in supporting nodes in the neighbourhood of corresponding virtual points.

Coefficients α_k is defined by Eq. (14). The backward difference is used to approximate the time derivatives of potential on the right side of Eq. (4) and we get the resulting recurrent system of equations to solve potential and its derivatives in every node. Equations of heat transfer are solved using the same RBIE method and also the similar algorithms as the potential flow.

Equations of flow and heat transfer are then coupled by the equations of state which gives the fluid density and viscosity as functions of temperature. The coupling scheme was realized by the sequential-iterative approach using the modified Pickard algorithm:

- Step1: Solution of the transfer equations
- Step2: Update fluid flow properties ρ, μ, K_x, K_y
- Step3: Solution of potential flow
- Step4: Compute velocities of flow
- Step5: Test the convergence of the process

This modified scheme converges much faster then the classical Pickard algorithm (see [10]).

www.witpress.com, ISSN 1743-355X (on-line)

6 Numerical example

The RBIEM model has been verified with the Elder problem (see [5]) of free geothermal convection. The 2D domain is vertically oriented rectangular area filled with a homogeneous isotropic porous medium (see Fig. 3). It is a free convection problem where fluid flow is driven purely by fluid density differences. The elevated temperature of 20°C decreases water density and creates a potentially unstable situation where denser fluid overlies less dense fluid which leads to upwelling of warm water and to the formation of thermal fingering. In homogeneous isotropic media, the onset of free geothermal convection can be determined by the value of Rayleigh number *Ra* (see e.g. [5]). This number is the ratio between buoyancy forces driving free convection and conductive forces. The dimensionless thermal Rayleigh number can be defined as

$$Ra = \frac{KHc_L\Delta\rho_L}{K_T} \tag{29}$$

where K is the hydraulic conductivity, H is the height of the model domain, c_L is the specific heat of the liquid, K_T is the thermal conductivity, and $\Delta\rho$ is the fluid density difference between the top and the bottom of the domain. The minimum value of the critical Rayleigh number $Ra_{cr} = 4\pi^2$. It is valid if and only if the aspect ratio of the domain $a = L/H$ is an integer value (see [11]). According to theory, simulations with $Ra < Ra_{cr}$ are conductive, whereas systems with $Ra > Ra_{cr}$ exhibit convective and unstable flow.

The solved problem has the aspect ratio $a = L/H = 2$ and therefore the critical Rayleigh number is $Ra_{cr} = 4\pi^2 = 39.478$. The coefficients, used in the solution

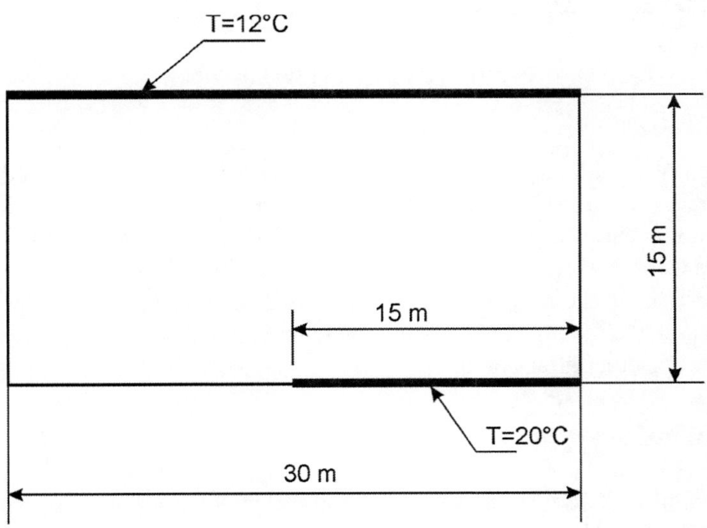

Figure 3: Elder problem.

Table 1: Parameters of the solved problem.

Quantity	Value	Unit
Porosity	0.1	—
Hydraulic conductivity	1.21E-04	ms^{-1}
Bulk thermal conductivity	1.49	$Wm^{-1}K^{-1}$
Specific heat capacity	4186	$Jkg^{-1}K^{-1}$
Initial temperature	12	^oC

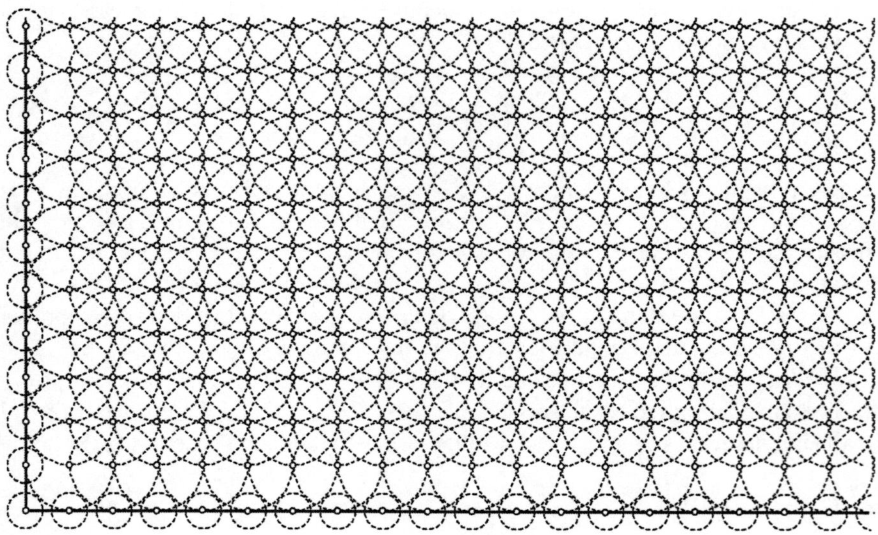

Figure 4: Detail of virtual network.

of the problem, are presented in Tab.1. The Rayleigh number computed for this problem is $Ra = 6.612 < Ra_{cr}$ and the problem is conductive and unstable flow should not occur. The domain is covered by 496 regularly distributed nodes and virtual networks were created around all node (see Fig. 4). The time step was 0.1 day. Fig. 5 shows the computed velocity field in 20 days and the increasing roll can be seen at the left part of the domain. The temperature distribution is presented in the form of isotherms at Fig. 6.

7 Conclusions

The possibility of RBIEM meshless method for modeling the density driven flow is presented in this paper. The research is at the beginning and the following study should be focused on the convective unstable situations models and on

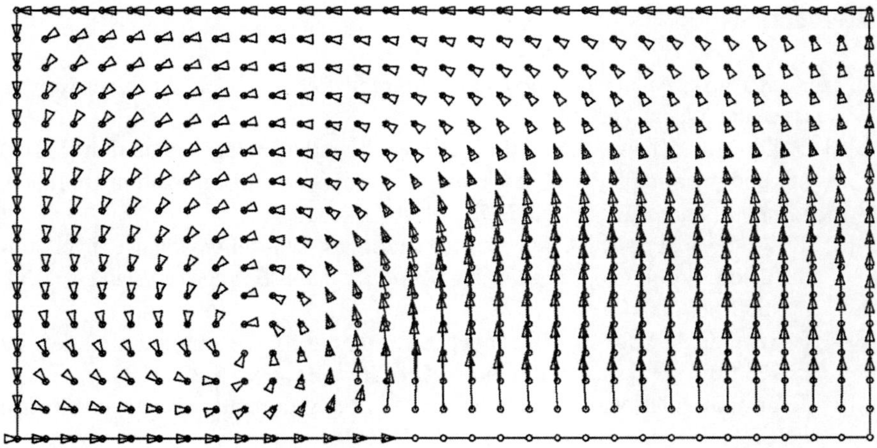

Figure 5: Resulting velocity field.

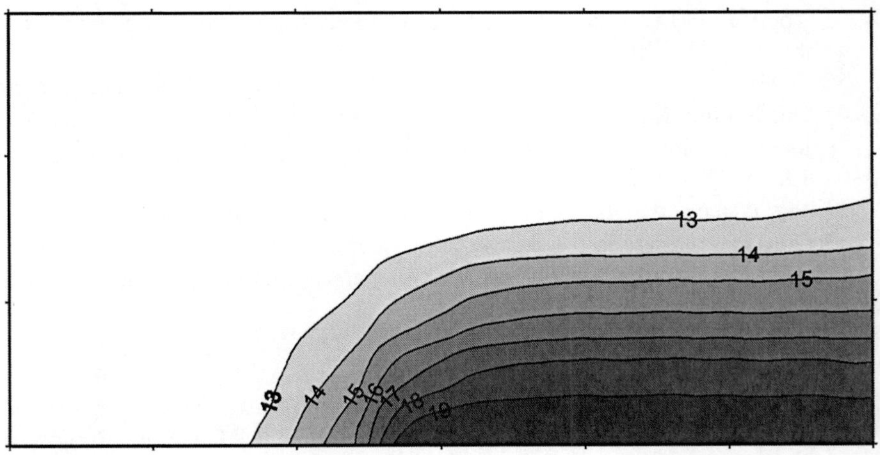

Figure 6: Isotherms.

the modification of the existing algorithms to enable distributed processing. It is necessary to choose suitable tools which give us ability to parallel solving of very large network systems which usually exist in practical solutions.

Acknowledgements

This contribution is the result of the project implementation: "Support of the Research and Development for Centre of Excellence in Transport Engineering" (ITMS: 26220120031) supported by the Research and Development Operational Programme funded by the ERDF.

References

[1] K. Kovarik, *Numerical models of groundwater pollution*, Springer-Verlag, Berlin and New York, 2000.

[2] T. Zhu, J.D. Zhang, S.N. Atluri, A local boundary integral equation (LBIE) method in computational mechanics and a meshless discretization approach, *Comput.Mech.* **21**, pp. 223-235, 1998.

[3] T.T. Bui, V. Popov, A meshless solution to two-dimensional convection-diffusion problems, *Engineering Analysis with Boundary Elements*, **34**, pp. 680-689, 2010.

[4] P. Ackerer, A. Younes, Efficient approximations for the simulation of density driven flow in porous media, *Adv. Water Res.*, **31**, pp. 15-27, 2008.

[5] T. Graf, Simulation of geothermal flow in deep sedimentary basins in Alberta, *ERCB/AGS Open File Report 2009-11*, 2009.

[6] P.W. Partridge, C. A. Brebbia, L.C. Wrobel, *The Dual Reciprocity Boundary Element Method*, CM Publications, Southampton, 1992.

[7] E.J. Kansa, Multiquadrics - A Scattered Data Approximation Scheme with Application to Computational Fluid Dynamics, *Comput. Math. Appl.*, **19**, pp. 127-145, 1990.

[8] C. Shu, H. Ding, K.S. Yeo, Local Radial Basis Function-based Differential Quadrature Method and its Application to Solve Two-dimensional Incompressible Navier-Stokes Equations, *Comput. Meth. Appl. Mech. Eng.*, **192**, pp. 941-954, 2003.

[9] M. Degham, A. Ghesmati, Numerical simulation of two-dimensional sine-Gordon solitons via a local weak meshless technique based on the radial point interpolation method (RPIM), *Computer Physics Communication*, **181**, pp. 772-786, 2009.

[10] P. Ackerer, A. Younes, M. Mancip, A new coupling algorithm for density-driven flow in porous media, *Geophys. Res. Lett.*, **31**, L12506, 2004.

[11] J.P. Caltagirone, Convection in a porous medium, *Convective transport and instability phenomena*, pp. 199-232, ed. J. Zierep, H. Oertel, Braunsche Hofbuchdruckerei und Verlag, Karlsruhe, 1982. *Comput. Math. Appl.*, **19**, pp. 127-145, 1990.

Evenly spaced data points and radial basis functions

L. T. Luh
Department of Mathematics, Providence University, Taiwan

Abstract

The purpose of this article is to introduce a kind of data setting to handle radial basis functions. Traditionally the meshless method RBF uses scattered data setting to do interpolations. This approach faces two hard problems. First, the optimal choice of the shape parameters contained in smooth radial functions are not easy to find. Second, the crucial constant ω in the exponential-type error bound, which is $O(\omega^{\frac{1}{d}})$, is too large, making this error bound meaningful only when the fill distance d is extremely small. However, in the evenly spaced data setting, an error bound of the form $O(\sqrt{d}\omega^{\frac{1}{d}})$ is established where ω is much sharper than that of the former one. What's important is that whenever this error bound is adopted, the optimal choice of the shape parameter can always be found with the fill distance d of reasonable size.

We express the effect of the shape parameter c by explicitly defined functions and present concrete criteria of the optimal choice of c, which do not require too many data points.
Keywords: radial basis function, shifted surface spline, shape parameter, interpolation.

1 Introduction

This article is intended only to open a window for the topic of choosing the shape parameter optimally. We try to avoid mentioning complicated theory and heavy techniques. Rather, we present concrete criteria of choosing the shape parameter.

The smooth RBFs, multiquadric, inverse multiquadric, gaussian, and shifted surface spline, all contain a shape parameter, usually denoted by c. Recently Cheng made a comprehensive study of its choice for MQ, IMQ and GA in Cheng [1], while the choice of c in the shifted surface spline was missing. We therefore focus

WIT Transactions on Modelling and Simulation, Vol 52,
www.witpress.com, ISSN 1743-355X (on-line)
doi:10.2495/BE110231

on this kind of RBF whose theoretical properties are equally complicated. This function is defined as follows.

$$h(x) := (-1)^m(|x|^2 + c^2)^{\frac{\lambda}{2}} \log{(|x|^2 + c^2)^{\frac{1}{2}}},\ \lambda \in Z_+,\ m = 1 + \frac{\lambda}{2},\ c > 0,$$
$$x \in R^n,\ \lambda, n\ even, \qquad (1)$$

where $|x|$ is the Euclidean norm of x, log denotes the natural logarithm, and $\lambda,\ c$ are constants. The constant c is called shape parameter whose optimal choice is a big problem. In order to provide a simple and concrete insight into our approach, we will deal with the case $\lambda = 2$ only.

Then, for any scattered set of data points $(x_1, f(x_1)), \dots, (x_N, f(x_N))$, there is a unique function

$$s(x) := \sum_{j=1}^{N} c_j h(x - x_j) + p(x) \qquad (2)$$

interpolating these data points, where $c_1, \dots, c_N$ are constants to be determined and $p(x)$ is a polynomial of degree $\leq m - 1$. The only requirement for the data points is that $x_1, \dots, x_N$ should be polynomially nondegenerate.

The interpolated functions belong to two function spaces.

Definition 1.1. *For any $\sigma > 0$, the class of band-limited functions f in $L^2(R^n)$ is defined by*

$$B_\sigma := \{f \in L^2(R^n) :\ \hat{f}(\xi) = 0\, if\ |\xi| > \sigma\},$$

where $\hat{f}$ denotes the Fourier transform of f.

A larger function space is defined as follows.

Definition 1.2. *For any $\sigma > 0$, the class of educated functions is*

$$E_\sigma := \{f \in L^2(R^n) :\ \int |\hat{f}(\xi)|^2 e^{\frac{|\xi|^2}{\sigma}} d\xi < \infty\},$$

where $\hat{f}$ denotes the Fourier transform of f. For each $f \in E_\sigma$, its norm is

$$\|f\|_{E_\sigma} := \left\{\int |\hat{f}(\xi)|^2 e^{\frac{|\xi|^2}{\sigma}} d\xi\right\}^{\frac{1}{2}}.$$

Although we restrict our interpolated functions to the two classes, other functions can be handled by triangle inequality. For example, as pointed out in Wendland [2], any function in the Sobolev space can be interpolated well by a band-limited function. Then we use $s(x)$ to interpolate that band-limited function at the same data points.

Different from Cheng's approach, we require that the data points be evenly spaced in the following sense.

Let E be an n-dimensional simplex in R^n with vertices $v_1, \ldots, v_{n+1}$. For any point $x \in E$, its barycentric coordinates are the numbers $\lambda_1, \ldots, \lambda_{n+1}$ satisfying

$$x = \sum_{i=1}^{n+1} \lambda_i v_i, \quad \sum_{i=1}^{n+1} \lambda_i = 1, \quad \lambda_i \geq 0 \text{ for all } i.$$

The definition of simplex can be found in Fleming [3].

For any n-dimensional simplex, the evenly spaced points of degree k are the points whose barycentric coordinates are of the form

$$(k_1/k, k_2/k, \ldots, k_{n+1}/k),\ k_i \text{ nonnegative integers and } k_1 + \cdots, k_{n+1} = k.$$

The interpolation occurs in a simplex and the centers(interpolation points) are evenly spaced points of that simplex. A simplex is just a line segment, triangle and tetrahedron in dimensions 1, 2, and 3, respectively.

2 The optimal choice of c

Before introducing our criteria we need some basic definitions.

Definition 2.1. *The constant ρ is defined as follows.*
(a) *Suppose $n > 5$. Let $s = \lceil \frac{n-5}{2} \rceil$. Then*

$$\rho = 1 + \frac{s}{7}.$$

(b) *Suppose $n \leq 5$. Then $\rho = 1$.*
For any $b_0 > 0$, let $\delta_0 = \frac{b_0}{6}$. Then for any $0 < \delta \leq \delta_0$, let $c_0 = 72\rho\delta$ and $c_1 = 12\rho b_0$.

Obviously $c_0 \leq c_1$. This article explores the optimal choice of c in $[c_0, \infty)$.

Based on the complicated theory developed in Luh [4–6], we know that there is a function called MN function which forms the essential part of the upper bound of $|f(x) - s(x)|$. It's defined by

$$MN(c) = \begin{cases} \sqrt{8\rho}c^{\frac{1-n}{4}} e^{c\left[\frac{\sigma}{2} + \frac{\ln\frac{2}{3}}{24\rho\delta}\right]} & \text{if } c_0 \leq c \leq c_1 \\ \sqrt{\frac{2}{3b_0}} c^{\frac{3-n}{4}} e^{\frac{c\sigma}{2}} \left(\frac{2}{3}\right)^{\frac{b_0}{2\delta}} & \text{if } c_1 \leq c < \infty \end{cases} \tag{3}$$

for $f \in B_\sigma$, and

$$MN(c) = \begin{cases} \sqrt{8\rho}c^{\frac{1-n}{4}} \sup_{\xi \in R^n} \left\{ |\xi|^{\frac{3+n}{4}} e^{\frac{c|\xi|}{2} - \frac{|\xi|^2}{2\sigma}} \right\} \left(\frac{2}{3}\right)^{\frac{c}{24\rho\delta}} & \text{if } c_0 \leq c \leq c_1 \\ \sqrt{\frac{2}{3b_0}} c^{\frac{3-n}{4}} \sup_{\xi \in R^n} \left\{ |\xi|^{\frac{3+n}{4}} e^{\frac{c|\xi|}{2} - \frac{|\xi|^2}{2\sigma}} \right\} \left(\frac{2}{3}\right)^{\frac{b_0}{2\delta}} & \text{if } c_1 \leq c < \infty \end{cases} \tag{4}$$

for $f \in E_\sigma$.

The optimal choice of the shape parameter is then the value minimizing $MN(c)$.

Theorem 2.3 of Luh [6] tells us that for any $b_0 > 0$ and simplex Q_0 of diameter b_0, there is a good bound for $|f(x) - s(x)|$ with domain a simplex $Q \subseteq Q_0$ of diameter $r \leq b_0$. The map $s(x)$ interpolates $f(x)$ at $x_1, \cdots, x_N$ which are evenly spaced points of degree $k-1$ in Q satisfying $k = \frac{r}{\delta}$ where $\delta \leq \delta_0$ can be any positive number.

We divide the criteria into two classes.

2.1 Band-limited functions

For $f \in B_\sigma$, we have the following cases, where $T := \frac{\sigma}{2} + \frac{\ln \frac{2}{3}}{24\rho\delta}$.

Case1. $\boxed{n=1}$ and $\boxed{T \geq 0}$ For any $b_0 > 0$ and positive $\delta < \frac{b_0}{6}$, if $n = 1$ and $T \geq 0$, the optimal choice of c for $c \in [c_0, \infty)$ is to let $c = c_0 := 72\delta$.

Reason: In this case $MN(c)$ in eqn.(3) is increasing on $[c_0, \infty)$.

Numerical Example:

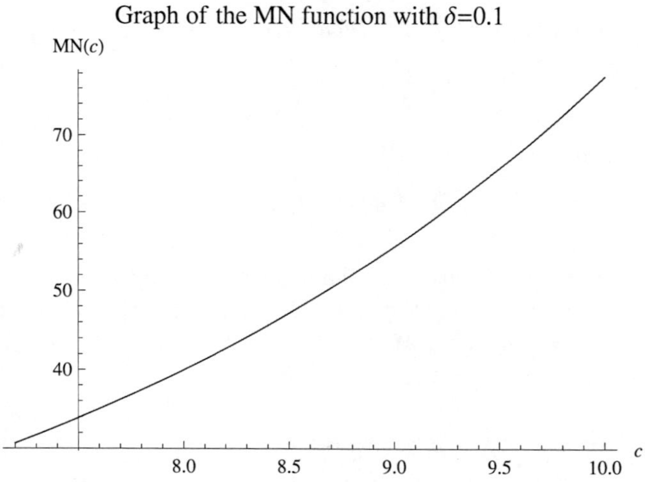

Figure 1: Here $n = 1, \sigma = 1$ and $b_0 = 1$.

Case2. $\boxed{n=1}$ and $\boxed{T < 0}$ For any $b_0 > 0$ and positive $\delta < \frac{b_0}{6}$, if $n = 1$ and $T < 0$, the optimal choice of c for $c \in [c_0, \infty)$ is $c^* \in [c_0, c_1]$ which minimizes $MN(c)$ of eqn.(3) on $[c_0, c_1]$.

Reason: In this case $MN(c)$ in eqn.(3) is increasing on $[c_1, \infty)$.

Numerical Example:

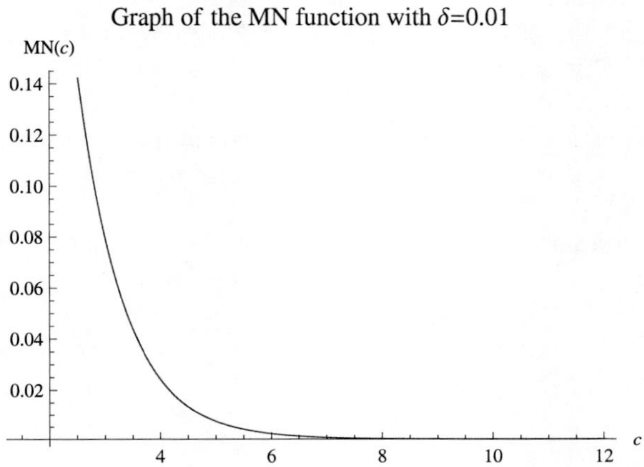

Figure 2: Here $n = 1, \sigma = 1$ and $b_0 = 1$.

Case3. $\boxed{n > 1}$ and $\boxed{T < 0}$ For any $b_0 > 0$ and positive $\delta < \frac{b_0}{6}$, if $n > 1$ and $T < 0$, the optimal choice of $c \in [c_0, \infty)$ is the value $c^* \in [c_1, \infty)$ which minimizes $MN(c)$ in eqn.(3) on $[c_1, \infty)$.

Reason: In this case $MN(c)$ in eqn.(3) decreases on $[c_0, c_1]$.

Remark: In Case3 if $n \leq 3$, $MN(c)$ will be increasing on $[c_1, \infty)$ and $c^* = c_1$.

Numerical Example:

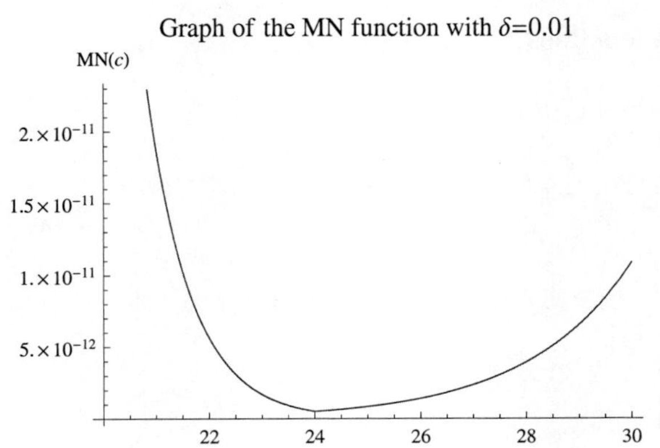

Figure 3: Here $n = 2, \sigma = 1$ and $b_0 = 2$.

Now we introduce the last case of band-limited functions.
Case4. $\boxed{n > 1}$ and $\boxed{T \geq 0}$ For any $b_0 > 0$ and positive $\delta < \frac{b_0}{6}$, if $n > 1$ and $T \geq 0$, the optimal choice of $c \in [c_0, \infty)$ is either $c^* \in [c_0, c_1]$ or $c^{**} \in [c_1, \infty)$, depending on $MN(c^*) \leq MN(c^{**})$ or $MN(c^{**}) \leq MN(c^*)$, where c^* and c^{**} minimize $MN(c)$ in eqn.(3) on $[c_0, c_1]$ and $[c_1, \infty)$, respectively.

Reason: In this case $MN(c)$ in eqn.(3) may not be monotonic on both $[c_0, c_1]$ and $[c_1, \infty)$. However, if $n \leq 3$, $MN(c)$ will be increasing on $[c_1, \infty)$ and $c^{**} = c_1$.

Numerical Example:

Graph of the MN function with b_0=10^3

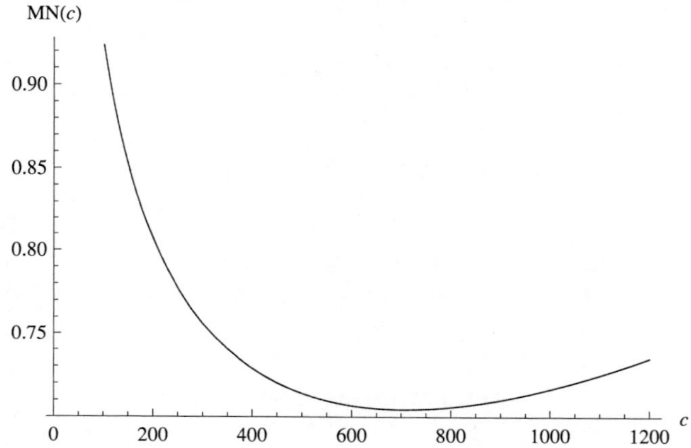

Figure 4: Here $n = 2, \sigma = 1.127$ and $\delta = 0.03$.

2.2 Educated functions

We now deal with functions in E_σ.

Case1. $\boxed{n \leq 3}$ For any $b_0 > 0$ and positive $\delta < \frac{b_0}{6}$, if $n \leq 3$, the optimal choice of $c \in [c_0, \infty)$ is $c^* \in [c_0, c_1]$ which minimizes $MN(c)$ in eqn.(4) on $[c_0, c_1]$.

Reason: In this case $MN(c)$ is increasing on $[c_1, \infty)$. Hence its minimum value happens in $[c_0, c_1]$.

Numerical Example:

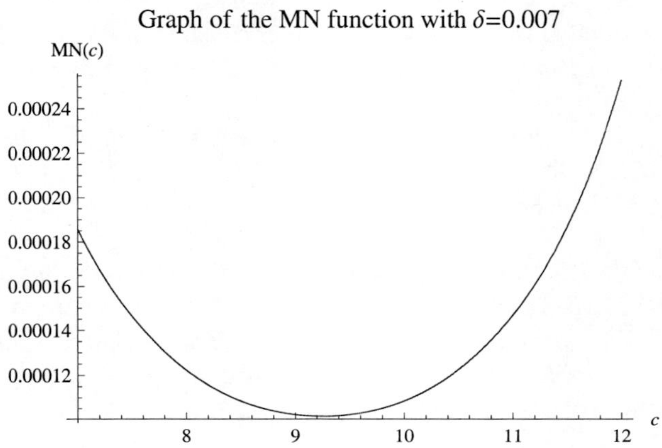

Figure 5: Here $n = 2, \sigma = 1$ and $b_0 = 1$.

Case2. $\boxed{n > 3}$ For any $b_0 > 0$ and positive $\delta < \frac{b_0}{6}$, if $n > 3$, the optimal choice of $c \in [c_0, \infty)$ is either $c^* \in [c_0, c_1]$ or $c^{**} \in [c_1, \infty)$, depending on $MN(c^*) \leq MN(c^{**})$ or $MN(c^{**}) \leq MN(c^*)$, where c^* and c^{**} minimize $MN(c)$ in eqn.(4) on $[c_0, c_1]$ and $[c_1, \infty)$, respectively.

Reason: In this case $MN(c)$ may not be monotonic on both $[c_0, c_1]$ and $[c_1, \infty)$.

Numerical Example:

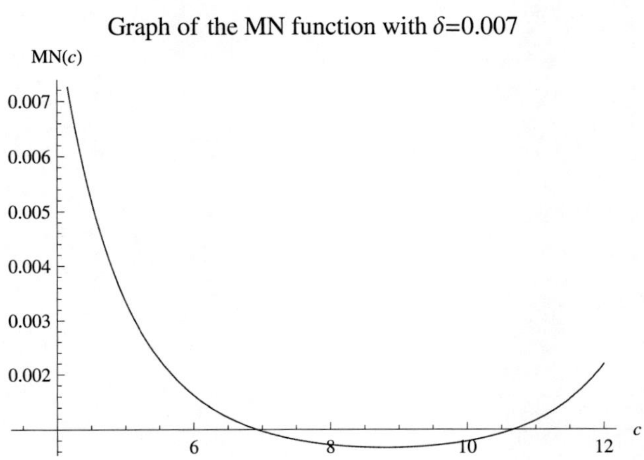

Figure 6: Here $n = 4, \sigma = 1$ and $b_0 = 1$.

References

[1] Cheng A.H.-D., *Multiquadric and Its Shape Parameter-A Numerical Investigation of Error Estimate, Condition Number, and Round-Off Error by Arbitrary Precision Computation,* submitted to Engineering Analysis with Boundary Elements.

[2] Wendland H., *Multiscale Analysis in Sobolev Spaces on Bounded Domains,* Numerische Mathematik 116, 493-517, 2010.

[3] Fleming W., *Functions of Several Variables, Second Edition,* Springer-Verlag, 1977.

[4] Luh L-T., *The Shape Parameter in the Shifted Surface Spline,* arXiv:1011.0124,2010.

[5] Luh L-T., *The Shape Parameter in the Shifted Surface Spline II,* in review.

[6] Luh L-T., *A New Error Bound for Shifted Surface Spline Interpolation,* Studies in Mathematical Sciences, Vol.1, No.1, 1-12, 2010.

www.witpress.com, ISSN 1743-355X (on-line)

An efficient implementation of the radial basis integral equation method

E. H. Ooi & V. Popov
Environmental and Fluid Mechanics, Wessex Institute of Technology, Southampton, UK

Abstract

In this paper, we propose an efficient implementation of the radial basis integral equation method (RBIEM) that does not involve discretization of the circular subdomains. By avoiding discretization on the boundaries of the subdomains, a major source of error in the numerical scheme can be eliminated. The proposed implementation is tested on the Helmholtz equation with higher gradients in the exact solution. Three different radial basis functions are investigated, namely the augmented thin plate spline, r^3 and $r^4\log(r)$. The latter two functions are augmented with the second order global polynomial. Numerical results show that the new implementation of the RBIEM produces more accurate results and is more robust in handling problems with highly variable solutions. By avoiding the boundary discretization, the tasks of keeping track of the boundary elements and the boundary nodes are not needed, which can be a daunting task especially in three-dimensional problems with complicated geometries. The proposed implementation of the RBIEM is promising and the feasibility of the approach in three-dimensional problems is currently being investigated.
Keywords: meshless, radial basis function, discretization, Helmholtz, efficiency.

1 Introduction

In the past decade, research in numerical methods has been moving towards meshless approaches for solving partial differential equations. In particular, meshless methods based on integral equations have been gaining wide attention due to the accuracy of the integral equation based methods and the reduced requirements for meshing. Meshless methods based on integral equations such as the local boundary integral equation (LBIE) method developed by Zhu et

WIT Transactions on Modelling and Simulation, Vol 52,
www.witpress.com, ISSN 1743-355X (on-line)
doi:10.2495/BE110241

al. [1, 2] have been successfully used to solve elasticity [3] and computational fluid dynamics [4] problems. While the LBIE may be considered to be a meshless approach, its implementation requires integration over part of the global boundary when the collocation nodes are located on the boundary of the solution domain. From a conceptual point of view, the approach is not truly meshless.

Popov and Bui [5] introduced the radial basis integral equation method (RBIEM), which is a meshless method based on the idea of the dual reciprocity method-multi domain (DRM-MD) with overlapping subdomains. Similar to the LBIE, the RBIEM generates around each node, a circular subdomain. The generation of the circular subdomains are by no means, the same as the domain discretization involved in the multi-domain method since there are no restrictions on how the subdomains are created. Generally, the subdomains may overlap one another and may even extend beyond the global boundary. Unlike the LBIE however, the RBIEM does not involve integration over the global boundary. All boundary integrations are performed only on the boundaries of the circular subdomains, making the RBIEM a truly meshless approach. Essentially, the RBIEM may be viewed as a meshless implementation of the DRM-MD with overlapping subdomains.

In the RBIEM [5], the boundary of each subdomain is discretized into quadratic elements, where the boundary integrals are evaluated numerically for each element. Unknown field variables on the nodes of each boundary element are interpolated using neighbouring nodes based on the radial basis functions (RBFs). Since the governing integral equations in the RBIEM are defined over each circular subdomain, it is perhaps more practical to transform the integration domain into polar coordinates and to carry out the integration numerically without involving any boundary discretization. By avoiding the discretization on the boundary of each subdomain, it is possible to improve on the accuracy of the numerical scheme.

The objective of this paper is to compare the performance of the proposed implementation of the RBIEM, whereby the boundary discretization of each subdomain is avoided, with the conventional RBIEM. The numerical schemes are tested using three different RBFs, namely $r^2\log(r)$, r^3 and $r^4\log(r)$. The first function is augmented with a global first order polynomial, while the latter two functions are augmented with second order polynomial. The choice of using r^3 and $r^4\log(r)$ is motivated by the study carried out by Partridge [8].

For simplicity, the proposed efficient implementation of the RBIEM in this paper is hereafter, referred to as RBIEM-I.

2 Mathematical formulations

2.1 The radial basis integral equation method

Consider the following partial differential equation in the two-dimensional region R enclosed by the surface S:

$$\nabla^2 u(r) = f, \quad \text{for } r \in R \cup S \tag{1}$$

where $u(r)$ is the potential variable, r is the coordinates in the Cartesian coordinate system and f is the non-homogeneous term. Equation (1) is subjected to the following boundary conditions:

$$\begin{aligned} u(r) &= u_o \quad \text{at } S_1 \\ \frac{\partial}{\partial n}u(r) &= q_o \quad \text{at } S_2 \end{aligned} \tag{2}$$

where $S_1 \cup S_2 = S$ and u_o and q_o are suitably prescribed functions. To obtain a solution of (1) subjected to the boundary conditions in (2) using the RBIEM, a set of N_t nodes denoted by ξ_i, for $i = 1, 2, \ldots, N_t - 1, N_t$, distributed across the boundary and interior of the solution domain, is selected. This is illustrated in Figure 1a.

A circular subdomain, R_i enclosed by the boundary S_i, centred about the point ξ_i is generated for each node. This is shown in Figure 1b. In principle, the subdomains can be of any shape, although the use of a circular one simplifies the generation of boundaries and the evaluation of the integrals.

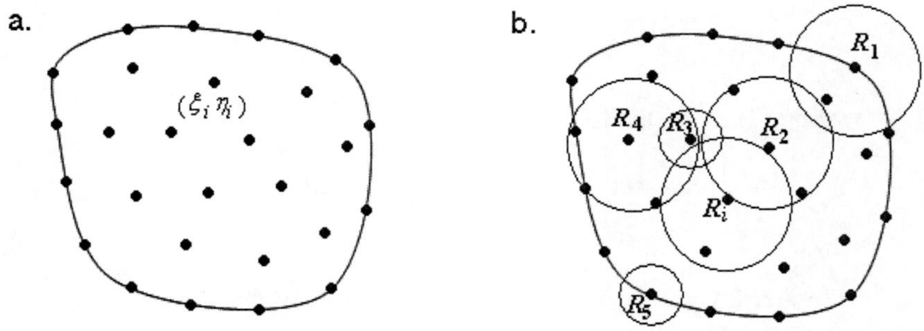

Figure 1: a) Distribution of nodes on the boundary and interior, and b) representation of some of the circular subdomains with different radii generated around the collocation nodes.

Once the collocation nodes are selected, the integral equations describing the potential and its derivatives in the x- and y- directions, respectively over each subdomain are derived, which hold in each subdomain. Using the Green's second identity, we obtain

$$\begin{aligned} u(r)\big|_{r=\xi} &= \int_{S_i} u(r)\frac{\partial}{\partial n}\Phi(r,\xi)ds(r) - \int_{S_i} n_x(r)\Phi(r,\xi)\frac{\partial}{\partial x}u(r)ds(r) \\ &\quad - \int_{S_i} n_y(r)\Phi(r,\xi)\frac{\partial}{\partial y}u(r)ds(r) + \iint_{R_i} f_i \cdot \Phi(r,\xi)dR(r), \end{aligned} \tag{3}$$

and differentiating (3) with respect to x and y, yields

$$\left.\frac{\partial u(r)}{\partial x}\right|_{r=\xi} = \int_{S_i} u(r)\frac{\partial^2}{\partial x \partial n}\Phi(r,\xi)ds(r) - \int_{S_i} n_x(r)\frac{\partial}{\partial x}\Phi(r,\xi)\frac{\partial}{\partial x}u(r)ds(r)$$
$$- \int_{S_i} n_y(r)\frac{\partial}{\partial x}\Phi(r,\xi)\frac{\partial}{\partial y}u(r)ds(r) + \frac{\partial}{\partial x}\iint_{R_i} f_i \cdot \Phi(r,\xi)dR(r), \tag{4}$$

and

$$\left.\frac{\partial u(r)}{\partial y}\right|_{r=\xi} = \int_{S_i} u(r)\frac{\partial^2}{\partial y \partial n}\Phi(r,\xi)ds(r) - \int_{S_i} n_x(r)\frac{\partial}{\partial y}\Phi(r,\xi)\frac{\partial}{\partial x}u(r)ds(r)$$
$$- \int_{S_i} n_y(r)\frac{\partial}{\partial y}\Phi(r,\xi)\frac{\partial}{\partial y}u(r)ds(r) + \frac{\partial}{\partial y}\iint_{R_i} f_i \cdot \Phi(r,\xi)dR(r), \tag{5}$$

where ds is the length of an infinitesimal part of S, dR is the area of an infinitesimal region of R and $\Phi(r, \xi)$ is the fundamental solution of the Laplace equation, which is given by

$$\Phi(r,\xi) = \frac{1}{2\pi}\ln|r-\xi|. \tag{6}$$

In writing (3) to (5), the following relation

$$\frac{\partial u(r)}{\partial n} = n_x(r)\frac{\partial u(r)}{\partial x} + n_y(r)\frac{\partial u(r)}{\partial y}, \tag{7}$$

has been utilized, where n_x and n_y are the components of the outward unit normal vector in the x- and y- directions, respectively.

The RBIEM employs discretization over the boundaries of each subdomain. For more details one may refer to Popov and Bui [5]. In our proposed formulation, no discretization of the boundaries of the subdomains is involved. Instead, we approximate the unknown variables with RBFs such that

$$\phi(r) = \sum_{j=1}^{N} \rho(r, r_j)\alpha_j, \tag{8}$$

where $\phi(r)$ is the unknown variable represented by $u(r)$, $\partial u(r)/\partial x$ and $\partial u(r)/\partial y$, $\rho(r, r_j)$ are the RBFs, α_j are unknown coefficients and N is the number of approximation points r_j in the neighbourhood of the integration node. The unknown coefficients may be expressed in terms of nodal values such that

$$\alpha_j = \sum_{j=1}^{N} W_{mj}\phi_j, \quad \text{for } m = 1, 2, \ldots, N-1, N. \tag{9}$$

where W_{mj} are explicitly given by

$$\sum_{j=1}^{N} W_{kj}\rho(r_j,r_m)=\begin{cases}0, \text{if } k\neq m\\ 1, \text{if } k=m\end{cases}; \quad \text{for } j,m=1,2,\ldots,N-1,N. \tag{10}$$

Substituting (8) and (9) into (3) to (5), we obtain

$$\begin{aligned} u(r)\big|_{r=\xi} &= \sum_{j=1}^{N} u_j \sum_{k=1}^{N} W_{kj}F_j - \sum_{j=1}^{N}\frac{\partial u_j}{\partial x}\sum_{k=1}^{N} W_{kj}G_j - \sum_{j=1}^{N}\frac{\partial u_j}{\partial y}\sum_{k=1}^{N} W_{kj}H_j \\ &+ \iint_{R_i} f_i \cdot \Phi(r,\xi)dR(r), \end{aligned} \tag{11}$$

$$\begin{aligned} \frac{\partial u(r)}{\partial x}\bigg|_{r=\xi} &= \sum_{j=1}^{N} u_j \sum_{k=1}^{N} W_{kj}\frac{\partial F_j}{\partial x} - \sum_{j=1}^{N}\frac{\partial u_j}{\partial x}\sum_{k=1}^{N} W_{kj}\frac{\partial G_j}{\partial x} - \sum_{j=1}^{N}\frac{\partial u_j}{\partial y}\sum_{k=1}^{N} W_{kj}\frac{\partial H_j}{\partial x} \\ &+ \frac{\partial}{\partial x}\iint_{R_i} f_i \cdot \Phi(r,\xi)dR(r), \end{aligned} \tag{12}$$

and

$$\begin{aligned} \frac{\partial u(r)}{\partial y}\bigg|_{r=\xi} &= \sum_{j=1}^{N} u_j \sum_{k=1}^{N} W_{kj}\frac{\partial F_j}{\partial y} - \sum_{j=1}^{N}\frac{\partial u_j}{\partial x}\sum_{k=1}^{N} W_{kj}\frac{\partial G_j}{\partial y} - \sum_{j=1}^{N}\frac{\partial u_j}{\partial y}\sum_{k=1}^{N} W_{kj}\frac{\partial H_j}{\partial y} \\ &+ \frac{\partial}{\partial y}\iint_{R_i} f_i \cdot \Phi(r,\xi)dR(r), \end{aligned} \tag{13}$$

where F_j, G_j and H_j are coefficients associated with the boundary integrals.

2.2 The dual reciprocity method

The domain integrals in (11) to (13) may be treated using, e.g., direct evaluation of the domain integral or the DRM [6]. In this case the DRM is employed, which converts the domain integral into an equivalent boundary integral using approximation of the non-homogeneous term f, see (1), based on the RBFs and the Green's second identity. By applying the DRM, (11) to (13) can be reformulated as:

$$\begin{aligned} u(r)\big|_{r=\xi} &= \sum_{j=1}^{N} u_j \sum_{k=1}^{N} W_{kj}F_j - \sum_{j=1}^{N}\frac{\partial u_j}{\partial x}\sum_{k=1}^{N} W_{kj}G_j - \sum_{j=1}^{N}\frac{\partial u_j}{\partial y}\sum_{k=1}^{N} W_{kj}H_j \\ &+ \sum_{j=1}^{N} f_j \sum_{k=1}^{N} W_{kj}\Psi_j, \end{aligned} \tag{14}$$

$$\begin{aligned} \frac{\partial u(r)}{\partial x}\bigg|_{r=\xi} &= \sum_{j=1}^{N} u_j \sum_{k=1}^{N} W_{kj}\frac{\partial F_j}{\partial x} - \sum_{j=1}^{N}\frac{\partial u_j}{\partial x}\sum_{k=1}^{N} W_{kj}\frac{\partial G_j}{\partial x} - \sum_{j=1}^{N}\frac{\partial u_j}{\partial y}\sum_{k=1}^{N} W_{kj}\frac{\partial H_j}{\partial x} \\ &+ \sum_{j=1}^{N} f_j \sum_{k=1}^{N} W_{kj}\frac{\partial \Psi_j}{\partial x}, \end{aligned} \tag{15}$$

and

$$\left.\frac{\partial u(r)}{\partial y}\right|_{r=\xi} = \sum_{j=1}^{N} u_j \sum_{k=1}^{N} W_{kj} \frac{\partial F_j}{\partial y} - \sum_{j=1}^{N} \frac{\partial u_j}{\partial x} \sum_{k=1}^{N} W_{kj} \frac{\partial G_j}{\partial y} - \sum_{j=1}^{N} \frac{\partial u_j}{\partial y} \sum_{k=1}^{N} W_{kj} \frac{\partial H_j}{\partial y} + \sum_{j=1}^{N} f_j \sum_{k=1}^{N} W_{kj} \frac{\partial \Psi_j}{\partial y}, \tag{16}$$

where Ψ_j is defined as

$$\Psi_j = \hat{u}(r)\big|_{r=\xi} - \int_{S_i} \hat{u}(r, r_j) \frac{\partial}{\partial n} \Phi(r, \xi) ds(r) + \int_{S_i} \Phi(r, \xi) \frac{\partial}{\partial n} \hat{u}(r, r_j) ds(r), \tag{17}$$

where $\hat{u}$ is the particular solution associated with the RBFs. For more details on the implementation of the DRM, one may refer to Partridge et al. [7].

3 Numerical implementation

To implement the RBIEM, (14) to (16) are collocated on the N_t nodes distributed across the boundary and interior of the solution domain. When the collocation node is located at the interior, all three equations are used in constructing the system of linear algebraic equations. When the collocation node is located on the boundary where potential is specified, only (15) and (16) are required. When the collocation node is located on the boundary where the normal derivative of potential is specified, two cases apply. When the normal derivative of potential is zero, the derivative in the direction closer to the unit normal is eliminated using (7) and the remaining two equations are used to assemble the system of linear algebraic equations. When the normal derivative of potential is not zero, the same procedure applies except that the derivative that is closer to the boundary is eliminated [5].

In the present formulation, the same set of nodes is chosen for both the interpolation of field variables and for the DRM. In this case, the N points closest to the centre of the subdomain are selected. This step simplifies the search for the neighbouring nodes, saves the overall computational time of the numerical model and ensures that all the RBF approximations are performed using the same number of nodes [5].

All boundary integrals in the proposed formulation are transformed into the polar coordinate system and are calculated numerically over the sub-domains using the Gaussian quadrature, without discretisation of the boundaries of the sub-domains.

4 Numerical examples

The performance of the proposed numerical scheme is compared with the conventional implementation of the RBIEM by solving the Helmholtz equation

$$\nabla^2 u(x, y) + \kappa^2 u(x, y) = 0, \tag{18}$$

where κ is a pre-defined positive real number, which is also known as the wavenumber. The problem is defined over a square domain bounded by the limits $-0.5 \le x \le 0.5$ and $-0.5 \le y \le 0.5$ and subjected to the following boundary conditions:

$$u(\pm 0.5, y) = \sin(\pm \frac{\kappa}{2\sqrt{2}}) \sin(\frac{\kappa}{\sqrt{2}} y),$$
$$\frac{\partial}{\partial n} u(x, \pm 0.5) = \pm \frac{\kappa}{\sqrt{2}} \cos(\pm \frac{\kappa}{2\sqrt{2}}) \sin(\frac{\kappa}{\sqrt{2}} x), \qquad (19)$$

The exact solution of this problem is given by

$$u(x, y) = \sin(\frac{\kappa}{\sqrt{2}} x) \sin(\frac{\kappa}{\sqrt{2}} y). \qquad (20)$$

Two values of κ are considered, namely 5 and 10, where the distributions of the exact solution across the solution domain for each κ are illustrated in Figures 2a and 2b, respectively. In RBIEM, 16 continuous quadratic boundary elements are discretized on the boundary of each subdomain. Both numerical schemes employ 25 nodes in the RBFs approximations of the potential and derivatives over the local boundaries.

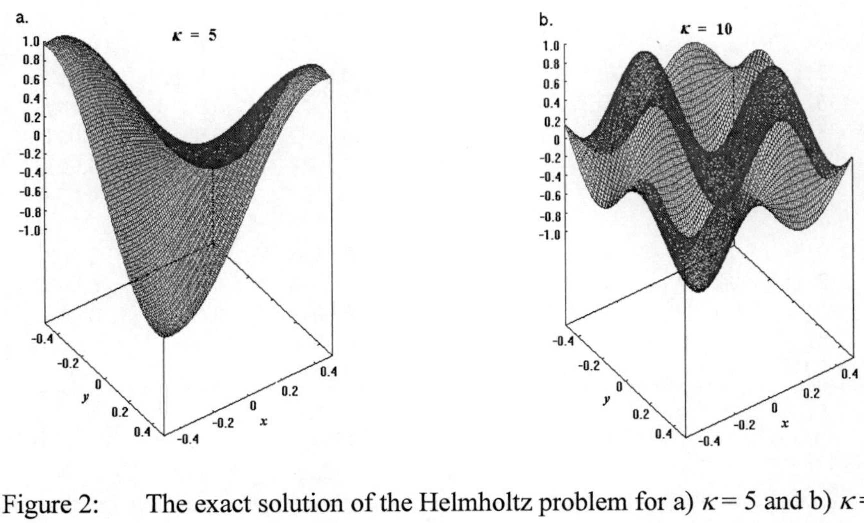

Figure 2: The exact solution of the Helmholtz problem for a) $\kappa = 5$ and b) $\kappa = 10$.

All numerical computations are coded using the programming language FORTRAN. The built-in sparse solver, PARDISO is used to solve the assembled sparse system of linear algebraic equations. A desktop computer with 2.61GHz AMD Athlon 64, X2 Dual Processor and 2.0GB of RAM is used for all computations.

Results are presented in Table 1 for five different sets of nodes, namely N_t = 2601, 3721, 6561, 8281 and 10201. Of the three RBFs tested, the worst results are obtained when the ATPS is used. The accuracies of both functions r^3 and $r^4\log(r)$ are comparable and they are by one order of magnitude more accurate than the results obtained using the ATPS. From Table 1, it is apparent that the RBIEM-I produces results that are more accurate than the RBIEM, as indicated by the lower L_2 norm errors. Significant difference in accuracy between RBIEM and RBIEM-I are obtained only when the functions r^3 and $r^4\log(r)$ are used. Unlike the RBIEM-I, no significant improvement in the accuracy of the RBIEM is observed when the number of nodes increases beyond 6561; indicating a higher convergence rate in the RBIEM-I case.

Table 1: L_2 norm errors (%) for the Helmholtz problem when $\kappa = 5$.

N_t	ATPS		r^3		$r^4\log(r)$	
	RBIEM	RBIEM-I	RBIEM	RBIEM-I	RBIEM	RBIEM-I
u						
2601	0.2235	0.2261	0.0137	0.0092	0.0140	0.0094
3721	0.1974	0.1964	0.0112	0.0054	0.0115	0.0059
6561	0.1704	0.1694	0.0099	0.0024	0.0101	0.0029
8281	0.1615	0.1605	0.0097	0.0017	0.0097	0.0022
10201	0.1552	0.1542	0.0096	0.0012	0.0097	0.0017
$\partial u/\partial x$						
2601	0.4348	0.4330	0.0326	0.0191	0.0319	0.0172
3721	0.3813	0.3790	0.0281	0.0114	0.0282	0.0108
6561	0.3298	0.3271	0.0255	0.0052	0.0258	0.0053
8281	0.3138	0.3109	0.0251	0.0037	0.0253	0.0040
10201	0.3027	0.2998	0.0248	0.0028	0.0251	0.0032
$\partial u/\partial y$						
2601	0.2318	0.2342	0.0145	0.0138	0.0111	0.0092
3721	0.2004	0.2029	0.0106	0.0085	0.0090	0.0057
6561	0.1688	0.1713	0.0086	0.0041	0.0084	0.0028
8281	0.1589	0.1614	0.0084	0.0030	0.0083	0.0020
10201	0.1519	0.1544	0.0083	0.0023	0.0083	0.0016

Figure 3 compares the distribution of percentage error between the RBIEM and RBIEM-I for $\kappa = 10$ obtained using the function $r^4\log(r)$ with 10201 nodes.

Both numerical schemes are found to produce mean percentage errors that are less than 1%. However, the RBIEM produces peak percentage errors that are greater than 10%. In the case of $\partial u/\partial x$, the maximum percentage error is 121%. Although peak errors are also observed in the RBIEM-I, the magnitudes are one order of magnitude lower than those obtained using the RBIEM.

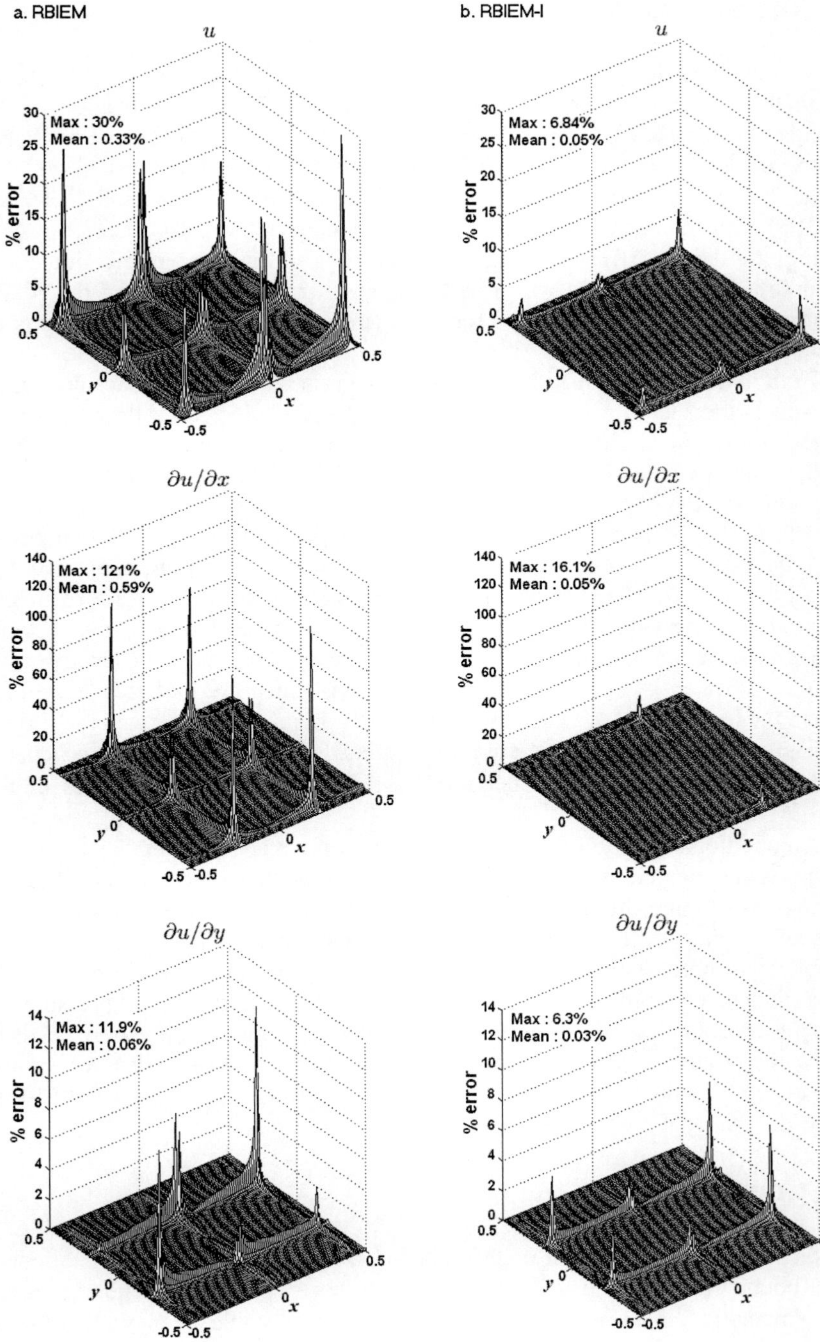

Figure 3: Distribution of percentage errors of u, $\partial u/\partial x$ and $\partial u/\partial y$ for $\kappa = 10$ obtained using a) RBIEM and b) RBIEM-I.

5 Discussions

From the numerical example presented, the RBIEM-I is found to be more accurate and more robust than the RBIEM. By avoiding discretization of the circular subdomain, a major source of error is eliminated. Significant improvement in the results are only observed when the functions r^3 and the $r^4\log(r)$ are used. When the ATPS is used, the accuracies of both the RBIEM and RBIEM-I are comparable. Since the ATPS produces the least accurate results among the three RBFs investigated, the small difference in the numerical accuracy when ATPS is used is perhaps caused by the errors from the ATPS approximation, which make any accuracy increase due to the RBIEM-I approach seem insignificant.

Both numerical schemes show convergence as the number of nodes increases. However, the increase in the number of nodes used increases the number of elements used in discretisations of the local sub-domains in the RBIEM, which further increases the numerical errors. This explains the slow convergence of the solution for RBIEM beyond 6561 nodes.

The RBIEM-I handles problems with high gradients in the solution better than the RBIEM. For problems with less variation in the solution, such as for the case of $\kappa = 5$, the RBIEM is capable of producing results that are relatively accurate. As the value of κ increases, the numerical scheme becomes unstable and fails to capture accurately the solution in areas where the gradients are large. This is indicated by the peak percentage errors, which are localized to regions where the gradients of u, $\partial u/\partial x$ and $\partial u/\partial y$ are the highest.

The same behaviour is also observed in the RBIEM-I, although this scheme is less sensitive towards the increase in κ. This indicates the robustness of the RBIEM-I.

An additional benefit of the RBIEM-I is that there is no need to keep track of the node and element numbering on the local sub-domain since no discretization is involved. While this may not be an issue in two-dimensional problems, the task of keeping track of the boundary nodes and elements in three-dimensional problems is more rigorous, especially in problems with complicated geometry. In three-dimensional problems, the RBIEM-I can be implemented by expressing the integration domain in the spherical coordinates system. Hence, one only needs to know the radius of each spherical subdomain and its centroid when performing the integration over the boundaries. The implementation of the RBIEM-I in three-dimensional problems is currently being investigated.

6 Conclusions

An efficient implementation of the RBIEM, which avoids the discretization of the boundaries of the subdomains, has been proposed. The proposed implementation is shown to be very accurate and stable even with only 20 Gaussian integration points. The new formulation is easier to implement and may be particularly useful in solving three-dimensional problems.

References

[1] Zhu T, Zhang J.D. and Atluri S.N., A local boundary integral equation (LBIE) method in computational mechanics, and a meshless discretization approach, *Computational Mechanics*, 1988; 21: 223–235.

[2] Zhu T, Zhang J.D. and Atluri S.N., A meshless local boundary integral equation (LBIE) method for solving nonlinear problems, *Computational Mechanics*, 1998; 22: 174–186.

[3] Sladek J., Sladek V. and Atluri S.N., Local boundary integral equation (LBIE) method for solving problems of elasticity with nonhomogeneous material properties, *Computational Mechanics*, 2000; 24: 456–462.

[4] Sellountos E.J. and Sequeira A., An advanced meshless LBIE/RBF method for solving two-dimensional incompressible fluid flows, *Computational Mechanics*, 2008; 41: 617–631.

[5] Popov V. and Bui T.T., A meshless solution to two-dimensional convection-diffusion problems, *Engineering Analysis with Boundary Elements*, 2010; 34: 680–689.

[6] Nardini D. and Brebbia C., A new approach in solid mechanics by an alternative boundary element procedure, *International Journal of Soil Dynamics and Earthquake Engineering*, 1983; 2: 228–233.

[7] Partridge, P.W., Brebbia, C.A. and Wrobel, L.C., *The Dual Reciprocity Boundary Element Method*. Computational Mechanics Publication, Southampton, 1992.

[8] Partridge P.W., Towards criteria for selecting approximation functions in the dual reciprocity method, *Engineering Analysis with Boundary Elements*, 2000; 24: 519–529.

Forced vibrations of tanks partially filled with liquid under seismic load

V. Gnitko, U. Marchenko, V. Naumenko & E. Strelnikova
Institute for Mechanical Engineering Problems of the Ukrainian Academy of Sciences, Ukraine

Abstract

Cylindrical tanks partially filled with liquid are the most general type of reservoirs for oil and other chemical-dangerous agent storage. Destruction of such tanks under seismic or impulsive load can lead to negative ecological consequences. The analysis method of dynamic behavior of cylindrical tanks partially filled with liquid that are under short-time impulsive load is under consideration. The method relies on reducing the problem of determining the fluid pressure to the system of singular integral equations. The coupled problem is solved using combination BEM and FEM. Differential equations of transient problem are solved numerically by Runge-Kutta method of 4th and 5th order. Numerical investigations of forced vibrations of the cylindrical tank filled with the incompressible fluid under seismic load have been carried out.
Keywords: forced vibrations, fluid-structure interaction, seismic load, boundary and finite element methods.

1 Introduction

Practicing engineers face many issues and challenges on the design and seismic evaluation of liquid storage tanks [1–3]. Liquid storage tanks are important components of lifeline and industrial facilities. Ground-supported cylindrical tanks are used to store a variety of liquids: water for drinking and firefighting, crude oil, wine, liquefied natural gas (LNG), etc. Failure of tanks, following destructive earthquakes, may lead to environmental hazard, loss of valuable contents, and disruption of fire-fighting effort. Inadequately designed or detailed tanks have suffered extensive damage in past earthquakes and have resulted in disastrous effects. Earthquake damage to steel tanks can take several forms.

WIT Transactions on Modelling and Simulation, Vol 52, © 2011 WIT Press
www.witpress.com, ISSN 1743-355X (on-line)
doi:10.2495/BE110251

Large axial compressive stresses due to beamlike bending of the tank wall can cause "elephant-foot" buckling of the wall. Sloshing liquid near the free surface can damage the roof and upper shell of tank. High stresses in the vicinity of poorly detailed base anchors can rupture the tank wall. Base shears can overcome friction causing the tank to slide.

The forces on the tanks during an earthquake depend on the dynamics of the structure, fluid stored in the tank and of the founding soil. For dynamic behavior characterization of the containers partially filled with liquid two levels of interactions need to be studied. First being the interaction between the structure and liquid stored in the container. The second interaction is that between the structure and the foundation soil. The assumption of the base of the tank being fixed is valid if it is founded on hard rock.

Procedures for the seismic analysis and design of storage tanks are generally based on the Housner multicomponent spring-mass analogy. The analogy allows the complex dynamic behaviour of a tank and its contents to be considered in simplified form. The principal modes of response include a short period impulsive mode, with a period of around 0.5 seconds or less, and a number of longer period convective (sloshing) modes with periods up to several seconds. For most tanks, it is the impulsive mode, which dominates the loading on the tank wall. The first convective mode is usually much less significant than the impulsive mode, and the higher order convective modes can be ignored.

The dynamic analysis of shell structures is often performed by use of finite element (FE) programs [4]. But, such 3-D nonlinear finite element analysis, including the contained fluid as well as the foundation soil in the system, is complex and extremely time consuming. Several simplified theoretical investigations were also conducted and a few of these studies have been used as a basis for current design standards. In [5–9] authors offer the approach based on using the boundary element method to the problem of natural vibrations of the fluid-filled elastic shells of revolution, as well as to the problem of natural liquid vibrations in the rigid vessels. This approach has the certain advantages. In the basic equations the functions and their derivatives will be defined on the domain boundaries only. That allows reducing the order of dimension of problem. This method gives new qualitative possibilities in modeling the dynamic coupled problem.

2 Problem statement

Let us consider the coupled problem of dynamic behavior of an elastic shell of revolution partially filled with liquid subjected to short-time impulsive load (Figure 1). Also free and forced vibrations of such elastic and rigid shells are under consideration.

In this study the contained liquid is assumed to be inviscid and incompressible resulting in an irrotational flow field. Let $V_x; V_y; V_z$ are the fluid velocity vector components, then the incompressibility condition can be written as following

$$divV = \frac{\partial V_x}{\partial x} + \frac{\partial V_y}{\partial y} + \frac{\partial V_z}{\partial z} = 0 \cdot \tag{1}$$

Under these suppositions, there exists a velocity potential defined as

$$V_x = \frac{\partial \varphi}{\partial x}, \quad V_y = \frac{\partial \varphi}{\partial y}, \quad V_z = \frac{\partial \varphi}{\partial z}.$$

Due to (1) potential φ satisfies the Laplace equation

$$\frac{\partial^2 \varphi}{\partial x^2} + \frac{\partial^2 \varphi}{\partial y^2} + \frac{\partial^2 \varphi}{\partial z^2} = 0 \cdot$$

The operator form of governing equations of motion for the liquid-structure system subjected to impulsive loading is given by

$$\boldsymbol{LU} + \boldsymbol{M}\ddot{\boldsymbol{U}} = P_l + Q\,, \tag{2}$$

where $\boldsymbol{L}$, $\boldsymbol{M}$ are operators of elastic and mass forces of the shell; $\boldsymbol{U} = (u_1, u_2, u_3)$ is the displacement vector; $Q(t)$ is the vector of external surface load, P_l is hydrodynamic pressure. The hydrodynamic pressure, according to the Cauchy-Lagrange integral, can be represented as follows

$$\frac{P}{\rho_l} = -\frac{\partial \varphi}{\partial t} - gz + \frac{P_0}{\rho_l}, \tag{3}$$

where φ is the velocity potential, ρ_l is the fluid density, z is coordinate of a point in liquid counted in vertical direction, g is the gravitational acceleration.

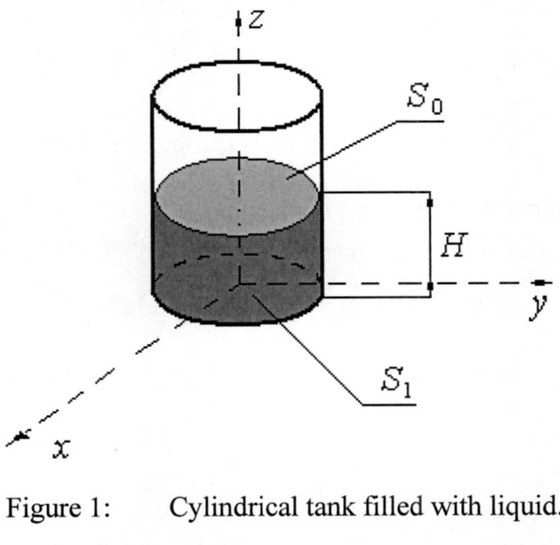

Figure 1: Cylindrical tank filled with liquid.

We denote a moistened surface of a shell through S_1 and a free surface as S_0 (Figure 1). Let refer the Cartesian coordinate system $0xyz$ connected with a shell. The free surface of the liquid S_0 coincides with the plane $x0y$ in unperturbed state. When a liquid storage tank is subjected to a dynamic load the boundary equations on the free surface are obtained by formulating dynamic and kinematics boundary conditions. The dynamic boundary condition is that the pressure on the free surface must be equal to the atmospheric one and the kinematics boundary condition is that liquid particles of the free surface remain on it during subsequent motion.

So we obtain the following boundary value problem

$$\boldsymbol{LU} + \boldsymbol{M}\ddot{\boldsymbol{U}} + \rho_l \dot{\varphi} + gz = \boldsymbol{Q},$$

$$\frac{\partial \varphi}{\partial n} = \frac{\partial w}{\partial t}, \; P \in S_1; \; \frac{\partial \varphi}{\partial n} = \dot{\zeta}, \; P \in S_0; \; \dot{\varphi} + g\zeta = 0, \; P \in S_0$$

for defining the unknown functions $\boldsymbol{U}$ and φ.

3 The mode superposition method for coupled dynamic problems

We will seek the natural modes of shell vibration in the fluid in the following form

$$U(x,y,z,t) = \sum_{k=1}^{m} u_k(x,y,z) c_k(t), \tag{4}$$

where functions $u_k(x,y,z)$ are modes of natural vibrations in vacuum, $c_k(t)$ are unknown factors.

We will seek φ as a sum of two potentials

$$\varphi = \varphi_1 + \varphi_2 .$$

To determine φ_1 we obtain the following boundary value problem:

$$\nabla^2 \varphi_1 = 0, \frac{\partial \varphi_1}{\partial n} = \frac{\partial w}{\partial t}, \; P \in S_1, \frac{\partial \varphi_1}{\partial t} = 0, \; P \in S_0 . \tag{5}$$

Here $w(x,y,z,t) = \sum_{k=1}^{m} w_k(x,y,z) c_k(t)$, and functions $w_k(x,y,z)$ are normal modes of natural vibrations in vacuum.

It would be noted, that from equation (3) and second one from (5) follows

$$\varphi_1(x,y,z,t) = \sum_{k=1}^{m} \varphi_{1k}(x,y,z) \dot{c}_k(t). \tag{6}$$

To determine φ_{1k} we have the following boundary value problem:

$$\nabla^2 \varphi_{1k} = 0,$$

$$\frac{\partial \varphi_{1k}}{\partial n} = w_k, \; P \in S_1,$$

$$\varphi_{1k} = 0, \; P \in S_0 . \tag{7}$$

To determine φ_2 we will have the following relation

$$\varphi_2(x,y,z,t) = \sum_{k=1}^{n} \varphi_{2k}(x,y,z) d_k(t),$$

where functions φ_{2k} are the natural modes of fluid vibration in rigid vessel. Let us formulate the problem of fluid vibration in rigid vessel relative to some function φ_r

$$\nabla^2\varphi_r = 0,$$
$$\frac{\partial\varphi_r}{\partial n} = 0, \; P \in S_1,$$
$$\frac{\partial\varphi_r}{\partial n} = \dot{\zeta}, \; P \in S_0,$$
$$\dot{\varphi}_r + g\zeta = 0, \; P \in S_0. \tag{8}$$

The last equation (8) follows from equation (3) and it is the dynamic condition on free surface. Differentiating this equation with respect to t we come to the following equation of fluid vibrations in the rigid vessel:

$$\ddot{\varphi}_r + g\frac{\partial\varphi_r}{\partial n} = 0, \; P \in S_0. \tag{9}$$

Let us seek for the solution of mentioned problem in the next form

$$\varphi_r(x, y, z, t) = e^{i\kappa t}\psi(x, y, z).$$

For function ψ we will have the following problem of free harmonic fluid vibrations

$$\nabla^2\psi = 0,$$
$$\frac{\partial\psi}{\partial n} = 0, \; P \in S_1,$$
$$\frac{\partial\psi}{\partial n} = \frac{\kappa^2}{g}\psi, \; P \in S_0. \tag{10}$$

Solving of this problem one can obtain the number of eigenvalues κ_k and corresponding natural modes, namely functions ψ_k. After equation (10) is solved we are looking for function φ_2 in the form

$$\varphi_2(x, y, z, t) = \sum_{k=1}^{n} d_k(t)\psi_k(x, y, z);$$
$$\varphi_{2k}(x, y, z) = \psi_k(x, y, z) \tag{11}$$

So we have $\varphi = \varphi_1 + \varphi_2$, where

$$\varphi_1(x, y, z, t) = \sum_{k=1}^{m} \varphi_{1k}(x, y, z)\dot{c}_k(t), \quad \varphi_2(x, y, z, t) = \sum_{k=1}^{n} d_k(t)\psi_k(x, y, z).$$

then

$$\nabla^2\varphi = \nabla^2\varphi_1 + \nabla^2\varphi_2 = 0,$$
$$\frac{\partial\varphi}{\partial n} = \frac{\partial\varphi_1}{\partial n} + \frac{\partial\varphi_2}{\partial n} = \frac{\partial w}{\partial t}, \; P \in S_1.$$

On free surface it is required

$$\frac{\partial\varphi}{\partial n} = \dot{\zeta}, \; P \in S_0; \; \dot{\varphi} + g\zeta = 0, \; P \in S_0.$$

www.witpress.com, ISSN 1743-355X (on-line)

Differentiating last equation with respect to t we come to the equation

$$\ddot{\varphi}+g\frac{\partial\varphi}{\partial n}=0\,,\ \ P\in S_0\,. \qquad (12)$$

We have $\dot{\varphi}_1=\ddot{\varphi}_1=0$ from relation (7). Therefore (12) leads to the system of differential equations

$$\sum_{k=1}^{n}\ddot{d}_k(t)\psi_k(x,y,z)+g\sum_{k=1}^{m}\dot{c}_k(t)\,\frac{\partial\varphi_{1k}(x,y,z)}{\partial n}+g\sum_{k=1}^{n}d_k(t)\,\frac{\partial\psi_k(x,y,z)}{\partial n}=0\,.$$

Using the relations

$$\frac{\partial\psi_k}{\partial n}=\frac{\kappa_k^2}{g}\psi_k\,,\ \ P\in S_0$$

we obtain

$$\sum_{k=1}^{n}\left[\ddot{d}_k(t)+\kappa_k^2 d_k(t)\right]\psi_k(x,y,z)+g\sum_{k=1}^{m}\dot{c}_k(t)\,\frac{\partial\varphi_{1k}(x,y,z)}{\partial n}=0\,. \qquad (13)$$

Due to orthogonality of natural modes of fluid vibrations in rigid vessel we have after dot product of equation (12) by functions ψ_l

$$\ddot{d}_l(t)+\kappa_l^2 d_l(t)+\frac{g}{(\psi_l,\psi_l)}\sum_{k=1}^{m}\dot{c}_k(t)\left(\frac{\partial\varphi_{1k}}{\partial n},\psi_l\right)=0,\quad l=1,2..,n\,. \qquad (14)$$

When functions φ_k and ψ_k are obtained we substitute them in equation (2) and obtain the following equation

$$\boldsymbol{L}\left(\sum_{k=1}^{m}c_k u_k\right)+\boldsymbol{M}\left(\sum_{k=1}^{m}\ddot{c}_k u_k\right)=-\rho_l\left(\sum_{k=1}^{m}\ddot{c}_k\varphi_{1k}+\sum_{i=1}^{n}\dot{d}_i\psi_i+gz\right)+Q\,. \qquad (15)$$

Let us ω_k, u_k are natural frequencies and free vibrations modes of the shell in vacuum. Then the following relationships are valid

$$\boldsymbol{L}u_k=\omega_k^2\boldsymbol{M}u_k\,,\qquad (\boldsymbol{M}u_k,u_j)=\delta_{kj}\,. \qquad (16)$$

Considering the result of dot product of equation (15) by u_j and taking into account relationships (14) and (16), we come to the next set of n+m second order differential equations

$$\ddot{c}_j(t)+\omega_j^2 c_j(t)+\rho_L\sum_{k=1}^{m}\ddot{c}_k(\varphi_k,u_j)+\sum_{i=1}^{n}\dot{d}_i(\psi_i,u_j)+g(z,u_j)=(Q,u_j),\ j=1,m$$

$$\ddot{d}_l(t)+\kappa_l^2 d_l(t)+\frac{g}{(\psi_l,\psi_l)}\sum_{k=1}^{m}\dot{c}_k(t)\left(\frac{\partial\varphi_{1k}}{\partial n},\psi_l\right)=0,\quad l=1,2..,n \qquad (17)$$

So we reduce the considered problem to the following. First, we have to obtain the natural frequencies and free vibrations mode shapes of the elastic shell in vacuum. The problem is solved using FEM. Second, it is necessary to obtain the frequencies and free vibrations modes of liquid in rigid shell under force of gravity. Then we define the frequencies and free vibrations modes of elastic shell without including the force of gravity. These two problems are solved using

BEM. And at the end we solve the set of second order differential equations using 4th and 5th order Runge-Kutta method.

4 System of the boundary integral equations

We use furthermore the cylindrical coordinate system and represent unknown functions as Fourier series by circumferential coordinate

$$w = w(r,z)\cos\alpha\theta, \quad \varphi = \varphi(r,z)\cos\alpha\theta. \tag{18}$$

To solve the coupled hydro-elasticity problem it is necessary to determine the potentials φ_1 and φ_2.

These problems here are reduced to the solution of the systems of singular integral equations. Determination of the potentials φ_1 was accomplished as in [10, 11]. So we obtain the following system

$$2\pi\varphi(z_0) + \int_\Gamma \varphi(z)\Theta(z,z_0)r(z)d\Gamma - \int_0^R q(\rho)\Phi(P,P_0)\rho d\rho =$$
$$= \int_\Gamma w(z)\Phi(P,P_0)r(z)d\Gamma_1, \quad P_0 \in S_1, \tag{19}$$
$$\int_\Gamma \varphi(z)\Theta(z,z_0)r(z)d\Gamma - \int_0^R q(\rho)\Phi(P,P_0)\rho d\rho = \int_\Gamma w(z)\Phi(P,P_0)r(z)d\Gamma_1, P_0 \in S_0,$$

where

$$\Theta(z,z_0) = \frac{4}{\sqrt{a+b}}\left\{\frac{1}{2r}\left[\frac{r^2 - r_0^2 + (z_0 - z)^2}{a-b}\mathsf{E}_\alpha(k) - \mathsf{F}_\alpha(k)\right]n_r + \frac{z_0 - z}{a-b}\mathsf{E}_\alpha(k)n_z\right\},$$

$$\Phi(P,P_0) = \frac{4}{\sqrt{a+b}}\mathsf{F}_\alpha(k). \tag{20}$$

Here the following notations are introduced

$$\mathsf{E}_\alpha(k) = (-1)^\alpha(1-4\alpha^2)\int_0^{\pi/2} \cos 2\alpha\theta\sqrt{1-k^2\sin^2\theta}d\theta,$$

$$\mathsf{F}_\alpha(k) = (-1)^\alpha \int_0^{\pi/2} \frac{\cos 2\alpha\theta d\theta}{\sqrt{1-k^2\sin^2\theta}}, \qquad k^2 = \frac{2b}{a+b}$$

Letting $\alpha = 0$ in the above expressions, we obtain the standard elliptic first and second kind integrals.

To determine the potential φ_2 we have to obtain functions ψ_k. Let us denote by ψ_{1k} the values of ψ_k on the wetted surface S_1 and by ψ_{0k} the values of ψ_k on the free surface S_0. Using the direct formulation of BEM to solve boundary value problem (10) and skipping for convenience an index k we can write the following system of singular integral equations

$$2\pi\psi_1 + \iint_{S_1} \psi_1 \frac{\partial}{\partial n}\left(\frac{1}{r}\right) dS_1 - \frac{\kappa^2}{g} \iint_{S_0} \psi_0 \frac{1}{r} dS_0 + \iint_{S_0} \psi_0 \frac{\partial}{\partial z}\left(\frac{1}{r}\right) dS_0 = 0$$
$$-\iint_{S_1} \psi_1 \frac{\partial}{\partial n}\left(\frac{1}{r}\right) dS_1 - 2\pi\psi_0 + \frac{\kappa^2}{g} \iint_{S_0} \psi_0 \frac{1}{r} dS_0 = 0$$
,

Suppose that

$$\psi = \psi(r,z)\cos\alpha\theta \tag{21}$$

we obtain for each harmonic term the following system of singular integral equations

$$\iint_{S_1} \psi \frac{\partial}{\partial n}\left(\frac{1}{r(P,P_0)}\right) dS_1 = \int_r \psi(z)\Theta(z,z_0) r(z) d\Gamma;$$

$$\iint_{S_0} \psi \left(\frac{1}{r(P,P_0)}\right) dS_0 = \int_0^R \psi(\rho)\Phi(P,P_0)\rho d\rho. \tag{22}$$

Here kernels $\Theta(z.z_0)$ and $\Phi(P,P_0)$ are defined in (20).

For numerical solution of system described by equations (19), (22) the boundary element method with constant approximation of unknown density on elements was used.

5 Numerical results

It would be noted that in the dynamic problem of horizontal seism we have to consider only $\alpha = 1$ in expressions (18), (21) because only first circumferential mode is excited by seismic motion.

To validate the developed numerical algorithm the comparison with FEM results was accomplished. Let us consider a cylindrical shell with a flat bottom partially filled with the fluid. The geometry of the tank is shown in Figure 2 and the parameters are following: the radius is R=1m, the thickness is h=0.01m, the length L=2m, Young's modulus $E=2\cdot10^5$ MPa, Poisson's ratio ν=0.3, the material's density is ρ=7800 kg/m^3, the fluid density ρ_l=1000 kg/m^3. The filling level of the fluid is denoted as H. Boundary conditions are following: $u_r=u_z=u_\theta=0$ to z=0 and r=R.

We analyze forced vibrations coupled problem. The radial load (Figure 3) is suddenly applied to cylindrical surface of the tank $q(r,z,t) = q_0 \cos k\varphi(r,z)\exp(-t/\tau)$, where q_0=0.1 MPa, $\tau=14.2\cdot10^{-6}$ s. Time at the end $t_n = 5\cdot10^{-3}$s.

The radial displacement response was calculated in four points that are shown in Figure 3, point 1 (node 91) is situated in the wetted part of the wall, point 2 (node 121) belongs to boundary of the liquid free surface, point 3 (node 69) is almost on the bottom whereas point 4 (node 161) is on the top of the wall. Figures 5-8 are representative of the forced motion response as calculated by the proposed method – solid lines and by the finite element complex – dash lines.

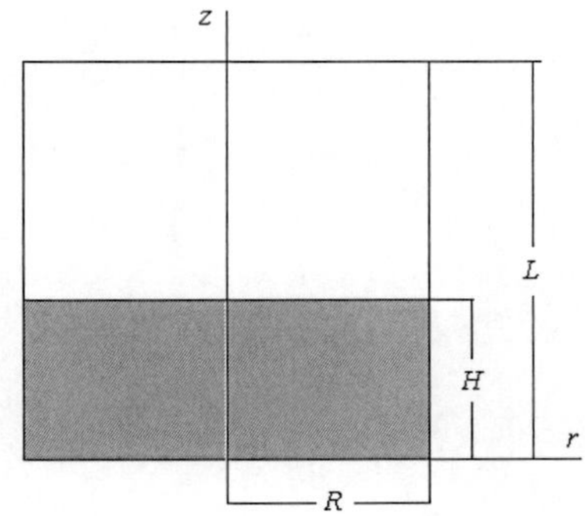

Figure 2: The scheme of the cylindrical tank.

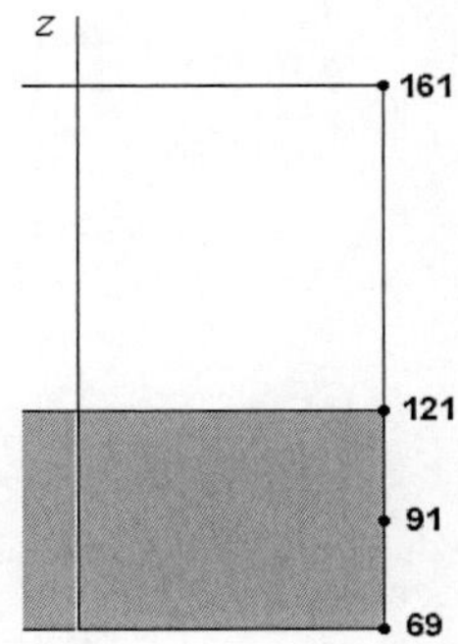

Figure 3: The scheme of nodes.

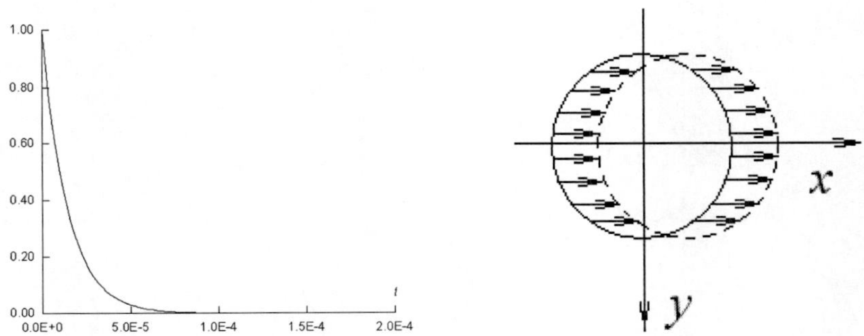

Figure 4: Impulsive load.

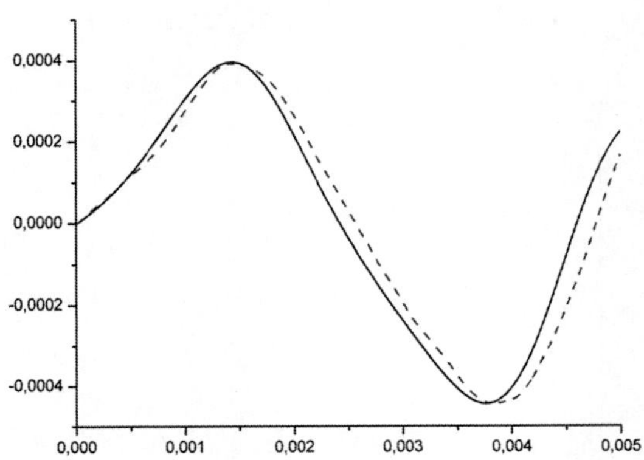

Figure 5: Time history of the radial displacement at the point 1.

www.witpress.com, ISSN 1743-355X (on-line)

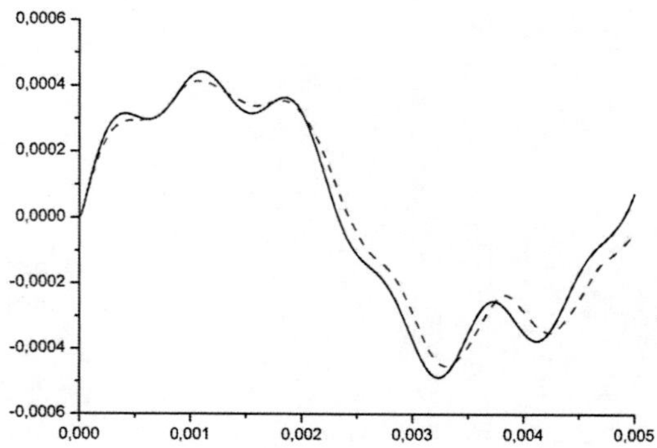

Figure 6: Time history of the radial displacement at the point 2.

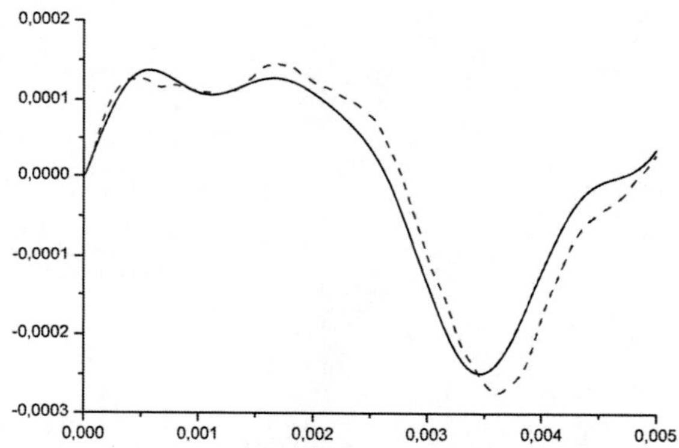

Figure 7: Time history of the radial displacement at the point 3.

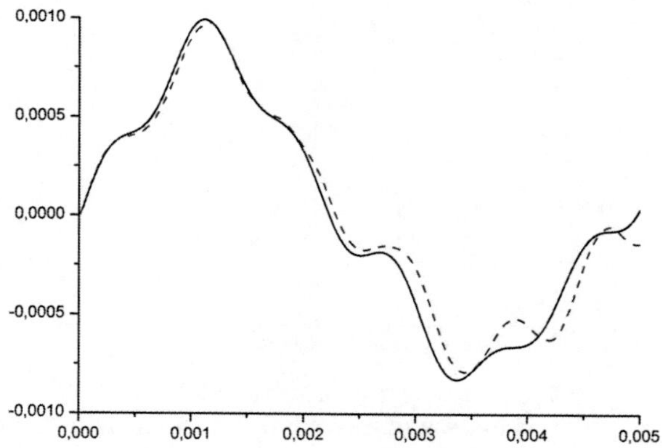

Figure 8: Time history of the radial displacement at the point 4.

The figures demonstrate good agreement of results obtained by different methods. That testifies the reliability of the method and algorithm proposed.

From previous investigation we found out that maximal radial displacement was in the point on the boundary between water and air (further "control point"). It will be interesting how the radial displacement at the control point will change subject to various filling levels of the fluid. So Figure 9 demonstrates the time history of the radial displacement at the control point when filling level of the fluid H=40 cm, H=80 cm, H=190cm.

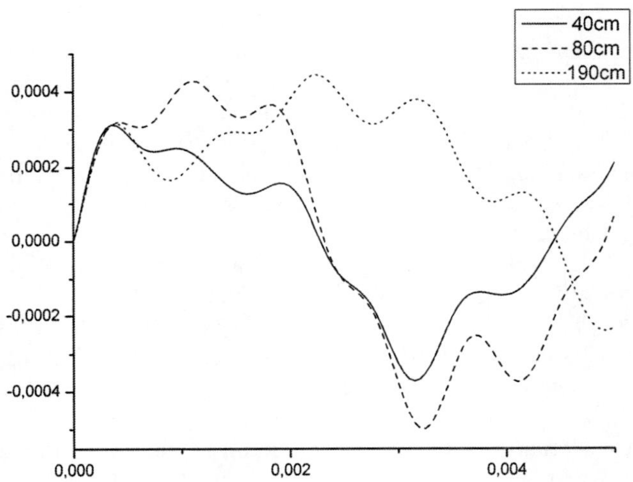

Figure 9: Time history of the radial displacement at the control point.

6 Conclusions

The numerical procedure based on a coupling the finite element formulation and the boundary element method is developed for the forced vibration analysis of shells of revolution with an arbitrary meridian partially filled with the fluid. We introduce the representation of the velocity potential as the sum of two potential corresponding to problem of the fluid free vibrations in the rigid shell and the one corresponding to problem of elastic shell with fluid without including the gravitational component. Integration by the fluid volume is reduced to integrals along the shell meridian and along the radius of the liquid free surface. It is the basic advantage of our method based on a combination of the boundary integral equations method, finite element method and expansion into Fourier series. The governing integral equations for each harmonic have been obtained. The forced vibration problem includes the liquid added masses into equations of motion. Numerical investigations of natural frequencies and forced vibrations of the cylindrical tank filled with the incompressible fluid have been carried out.

Acknowledgements

The authors gratefully acknowledge the sponsorship of this research by the Science and Technology Center in Ukraine in framework of the Project #4624. Authors also would like to acknowledge our STCU Project collaborators Professors Carlos Brebbia and Eduard Ventsel for their constant support and interest in our research.

References

[1] Sanchez-Sanchez, H., Cortes, S.C., Dominguez, A.M., Structural behaviour of liquid filled storage tanks of large capacity placed in seismic zones of high risk in Mexico// *Proc. of 13th World Conference on Earthquake Engineering,* Vancouver, B.C., Canada, 2004, Paper No. 2665.

[2] Sanchez-Sanchez, H., Cortes, S.C., Seismic response of cylindrical tanks for oil// *Proc. of 14th World Conference on Earthquake Engineering,* Beijing, China, 2008.

[3] Jhung, M.J., Jo, J.C., Jeong, S.J., Impact analysis of a water storage tank, *Nuclear Engineering and Technology*, **38(7)**, 2006.

[4] Kubenko, V.D. & Koval'chuk, P.S., Nonlinear problems of the dynamics of elastic shells partially filled with a liquid. *International Journal of Applied Mechanics*, **36(4)**, pp. 421–448, 2000.

[5] Amabili, M. & Païdoussis, M. P., Review of studies on geometrically nonlinear vibrations and dynamics of circular cylindrical shells and panels, with and without fluid-structure interaction. *Applied Mechanics Review*, **56(4)**, pp. 349–381, 2003.

[6] Kumar, V. & Ganesan, N., Dynamic analysis of conical shells conveying fluid. *Journal of Sound and Vibration*, **310(1-2)**, pp. 38–57, 2008.

[7] Malhotra, P. K., New method for seismic isolation of liquid-storage tanks. *Journal of Earthquake Engineering and Structural Dynamics*, **26(8)**, pp. 839–847, 1997.

[8] Zhang, Y.L., Gorman, D.G. & Reese, J.M., Vibration of prestressed thin cylindrical shells conveying fluid. *Thin-Walled Structures*, **41**, pp. 1103–1127, 2003.

[9] Chen, Y.H., Hwang, W.S. & Ko, C.H., Numerical simulation of the three-dimensional sloshing problem by boundary element method. *Journal of the Chinese Institute of Engineers*, **23(3)**, pp. 321-330, 2000.

[10] Brebbia, C.A., Telles, J.C.F. & Wrobel, L.C. *Boundary Element Techniques*, Springer-Verlag: Berlin and New York, 1984.

[11] Strelnikova E.,Yeseleva E., Gnitko V., Naumenko V. Free and forced vibrations of the shells of revolution interacting with the liquid// *Proc. of XXXII Conference "Boundary elements and other mesh reduction methods" WIT Press, Transaction on Modeling and Simulation*, 2010, P. 203-211.

Solving heat transfer in a timber beam exposed to fire with methods based on radial basis functions (RBFs)

L.Vrankar[1,4], G. Turk[2], E. J. Kansa[3] & F. Runovc[4]
[1] *Slovenian Nuclear Safety Administration, Slovenia*
[2] *University of Ljubljana, FGG, Slovenia*
[3] *Department of Mechanical and Aeronautical Engineering, University of California, USA*
[4] *University of Ljubljana, NTF, Slovenia*

Abstract

The objective of this paper is to represent an alternative approach to conventional numerical methods for solving heat transfer and charring behaviour of timber when exposed to fire. The model consists of differential equations for heat transfer with the corresponding boundary conditions. The char formation in the wood beam as a function of its temperature is also taken into account by the model. Picard's or Newton's methods are used for solving the second-order non-linear partial differential equations. In recent years, the RBF methods have emerged as novel computing methods in the scientific computing community. Traditionally, the most popular methods have been the finite element methods (FEM), the finite difference methods (FDM), and boundary element method (BEM). The results are tested on the one-dimensional case in standard fire conditions, for which the comparison is made with the results of one-dimensional charring rate models for wood. The same model is used to analyze a two-dimensional behaviour of a timber beam exposed to fire from three sides.
Keywords: heat transfer, charring behaviour of timber, pseudospectral mode, multiquadric, partial differential equations, Picard's method, radiation, pyrolysis, combustion.

WIT Transactions on Modelling and Simulation, Vol 52,
www.witpress.com, ISSN 1743-355X (on-line)
doi:10.2495/BE110261

1 Introduction

When wood is exposed to fire it undergoes thermal degradation. Thermal degradation or pyrolysis reduces the density by changing the wood to char and gases. The pyrolysis gases undergo flaming combustion as they leave the charred wood surface. The pyrolysis, charring, and combustion of wood have been extensively studied in [1].

The governing equations present differential equations for heat transfer with the corresponding boundary conditions. The boundary conditions prescribe the heat flow on the exposed boundaries of cross-section. In our case, different types of boundary conditions were used. The char formation in the wood beam as a function of its temperature is also taken into account by the model. The problem is solved numerically by the radial base function (RBF) methods.

Recent research on the numerical method has focused on the idea of using a meshless methodology for the numerical solution of PDEs. One of the common characteristics of all mesh-free methods is their ability to construct functional approximation or interpolation entirely from information at a set of scattered nodes, among which no relationship or connectivity is needed. In this paper, two methods will be used, Kansa's approach [2] and pseudospectral (PS) method [3,4]. The goal of this paper is to represent an alternative approach to conventional numerical methods for solving heat transfer and charring behaviour of timber when exposed to fire.

The results are tested on the one dimensional case in standard fire conditions, for which the comparison is made with the results of one-dimensional charring rate models for wood presented in the literature published by White and Nordheim [5]. The same model is used to analyse a two-dimensional behaviour of wood beam exposed to fire from three sides. The results are compared with the results obtained in literature [6]. Faster charring at the corners and typical rounding effect are observed.

2 Governing equations

In general, the heat and mass transfer is governed by the two second order non-linear partial differential equations [7]. In our case, only equation which describes heat conduction governed predominantly by temperature gradients was considered. The equation can be written as:

$$\varrho c_p \frac{\partial T}{\partial t} = k_x \frac{\partial^2 T}{\partial x^2} + k_y \frac{\partial^2 T}{\partial y^2}, \tag{1}$$

where k_x and k_y represent thermal conductivity $(\mathrm{W/mK})$ in directions x and y of a cross-section of the beam, ϱ is density $(\mathrm{kg/m^3})$, c_p specific heat $(\mathrm{J/kgK})$ and T temperature (°C). The second equation describes moisture diffusion governed by moisture potential and is not considered here.

The problem is complete when initial and boundary conditions are specified. The initial condition prescribes the temperature in the cross-section of the beam at the initial time $t = 0$

$$T(x, y, 0) = T_0(x, y). \tag{2}$$

The boundary conditions prescribe the heat flow on the exposed boundaries of cross-section. Thus, the boundary conditions at the exposed surface are given by balancing heat conduction at the surface with the radiative and convective heat flux. The boundary conditions can be written as:

$$-k_x \frac{\partial T}{\partial x} e_{nx} - k_y \frac{\partial T}{\partial y} e_{ny} = h_c(T - T_A) + \varepsilon_R \sigma (T^4 - T_R^4), \tag{3}$$

where e_{nx} and e_{ny} are components of the normal to the boundary surface and h_c is convective heat transfer coefficient $(\mathrm{W/m^2K})$. T_A is the temperature of the ambient. T_R is the temperature of the radiating surface, ε_R is the effective surface emissivity of the exterior siding and σ is the Stephan–Boltzmann constant for radiation, $(\sigma = 5.671.10^{-8}\ \mathrm{W/m^2K^4})$.

3 Solving the heat conduction equation

For the solution of eqn. (1) with the corresponding initial and boundary conditions non-symmetric RBF and PS method are used. A computer program is written in *MATLAB* environment.

3.1 Kansa's approach

A very popular non-symmetric method for the solution of PDEs with RBFs was suggested by Kansa [2]. A radial basis function is the function $\varphi_j(\mathbf{x}) := \varphi(\|\mathbf{x} - \mathbf{x}_j\|)$, which depends only on the distance between $\mathbf{x} \in \mathbb{R}^d$ and a fixed point $\mathbf{x}_j \in \mathbb{R}^d$. Here, φ_j is continuous and bounded on any bounded sub-domain $\Omega \subseteq \mathbb{R}^d$ whereas $\varphi : \mathbb{R}^d \to \mathbb{R}$. Let $r \geq 0$ denote the Euclidean distance between any pair of points in the domain Ω. The commonly used radial basis functions are linear $(\varphi(r) = r)$, cubic $(\varphi(r) = r^3)$, thin-plate spline $(\varphi(r) = r^2 \log r)$ and Gaussian $(\varphi(r) = e^{-\alpha r^2})$. The most popular globally supported C^∞ RBFs are multiquadric functions (MQ) $(\varphi(r) = (1 + (r/c)^2)^\beta), \beta = 1/2$ [8]. MQ has been already efficiently used in transport problems [9], moving-boundary problems [10], etc.

The starting point of the RBFs solution of partial differential equations is the interpolation problem. The MQ RBFs is used to interpolate the scalar level set function by using MQ basis centered at these *RBF centers*, $\Xi = \{\xi_1, \ldots, \xi_N\}$. The goal is to find an interpolant of the form

$$\Phi(\mathbf{x}) = \sum_{j=1}^{N} \alpha_j \varphi(\|\mathbf{x} - \xi_j\|), \quad \text{such that } \Phi(\mathbf{x}_i) = f_i,\ i = 1, \ldots, N, \tag{4}$$

where α_j is the weight of the radial basis function positioned at the j-th center. Knowing that the initial data values $f_1, \ldots, f_N \in \mathbb{R}$ at the data points

$\mathbf{x} = \{\mathbf{x}_1, \ldots, \mathbf{x}_N\} \subset \Omega \subset \mathbb{R}^N$, the solution of this problem leads to a linear system $\mathbf{A}\boldsymbol{\alpha} = \mathbf{f}$ with the entries of $\mathbf{A}$ given by

$$A_{ij} = \varphi(\|\mathbf{x}_i - \xi_j\|), \quad i, j = 1, \ldots, N. \tag{5}$$

We now switch to the collocation solution of partial differential equations. We consider a PDE in the general form of

$$\mathrm{L}u(\mathbf{x}) = f(\mathbf{x}), \ \text{ in } \ \Omega \subset \mathbb{R}^d, \tag{6}$$

$$\mathrm{B}u(\mathbf{x}) = g(\mathbf{x}), \ \text{ on } \ \partial\Omega, \tag{7}$$

where u is the unknown solution, d denotes the dimension, $\partial\Omega$ is the boundary of the domain Ω, L is the differential operator on the interior, and B is the operator that specifies the boundary conditions of the Dirichlet, Neumann or mixed type. $f(\mathbf{x})$ and $g(\mathbf{x})$ are given functions with sufficient smoothness mapping $\mathbb{R}^d \mapsto \mathbb{R}$, respectively.

The unknown PDE solution u is approximated by RBFs in the form:

$$u \approx U(\mathbf{x}) = \sum_{j=1}^{N} \alpha_j \varphi(\|\mathbf{x} - \xi_j\|). \tag{8}$$

The collocation matrix that arises when matching the eqn. (6) and eqn. (7) at the collocation points Ξ will be of the form

$$\mathbf{A} = \begin{bmatrix} \tilde{A}_{\mathrm{L}} \\ \tilde{A} \end{bmatrix}, \tag{9}$$

where the two blocks are generated as follows:

$$(\tilde{A}_{\mathrm{L}})_{ij} = \mathrm{L}\varphi(\|\mathbf{x} - \xi_{\mathrm{j}}\|)|_{\mathbf{x}=\mathbf{x}_{\mathrm{i}}}, \quad \mathbf{x}_{\mathrm{i}} \in \mathcal{I},\ \xi_{\mathrm{j}} \in \Xi, \tag{10}$$

$$\tilde{A}_{ij} = \mathrm{B}\varphi(\|\mathbf{x} - \xi_{\mathrm{j}}\|), \quad \mathbf{x}_{\mathrm{i}} \in \mathcal{B},\ \xi_{\mathrm{j}} \in \Xi. \tag{11}$$

The set of collocation points Ξ is split into a set of interior points $\mathcal{I}$ and a set of boundary points $\mathcal{B}$. The problem is well-posed if the linear system $\mathbf{A}\boldsymbol{\alpha} = \mathbf{b}$, with $\mathbf{b}$ as a vector consisting of entries $f(\mathbf{x}_i)$, $\mathbf{x}_i \in \mathcal{I}$, followed by $g(\mathbf{x}_i)$, $\mathbf{x}_i \in \mathcal{B}$, has a unique solution.

3.2 RBF-based pseudospectral method

The following review of non-symmetric RBF-based PS method is adapted from standard textbooks on Meshfree Approximation Methods with MATLAB [11]. An important feature of pseudospectral methods is the fact that one usually is content with obtaining an approximation to the solution on discrete set of grid points $x_i, i = 1, \ldots, N$. One way to implement the spectral method is via so-called *differentiation matrices* D. These can be found so that the following equation holds at the collocation points

$$\mathbf{u}' = D\mathbf{u}, \tag{12}$$

where $\mathbf{u} = [\hat{u}(x_1), \ldots, \hat{u}(x_N)]$ is the vector of values of the approximation solution $\hat{u}$ at the collocation points. Usually, orthogonal polynomials such as Chebyshev polynomials are used as basis functions $B_j(x)$. In our case, we used MQ RBF.

The approximate solution is expressed as:

$$\hat{u}(x) = \sum_{j=1}^{N} \beta_j B_j(x), \quad x \in \mathbb{R}, \tag{13}$$

If we evaluate (13) at the collocation points $x_i, i = 1, \ldots, N$, we get

$$\hat{u}(x_i) = \sum_{j=1}^{N} \beta_j B_j(x_i), \quad i = 1, \ldots, N, \tag{14}$$

or in matrix-vector notation

$$\mathbf{u} = \mathbf{A}\boldsymbol{\beta}, \tag{15}$$

where $A_{ij} = B_j(x_i)$.

We can do the same in the case of derivative of $\hat{u}$:

$$\frac{d}{dx}\hat{u}(x) = \sum_{j=1}^{N} \beta_j \frac{d}{dx} B_j(x). \tag{16}$$

If we evaluate again at the collocation points x_i, then we get in matrix-vector notation

$$\mathbf{u}' = \mathbf{A}_x\boldsymbol{\beta}, \tag{17}$$

We can use (15) to formally solve for the coefficient vector $\boldsymbol{\beta} = \mathbf{A}^{-1}\mathbf{u}$ and rewrite eqn. (17) as

$$\mathbf{u}' = \mathbf{A}_x\mathbf{A}^{-1}\mathbf{u}, \tag{18}$$

so that the *differentiation matrices* $\mathbf{D}$ corresponding to (12) is given by

$$\mathbf{D} = \mathbf{A}_x\mathbf{A}^{-1}, \tag{19}$$

For more complex linear differential operators $\mathcal{L}$ with constant coefficients we proceed in an analogous fashion to obtain differentiation matrix:

$$\mathbf{L} = \mathbf{A}_{\mathcal{L}}\mathbf{A}^{-1}, \tag{20}$$

where the matrix $\mathbf{A}_{\mathcal{L}}$ has entries $(A_{\mathcal{L}})_{ij} = \mathcal{L}\varphi(\|\mathbf{x} - \xi_j\|)|_{\mathbf{x}=\mathbf{x}_i}$. We can use Kansa's non-symmetric method to obtain the discretized differential operator. The RBF collocation solution is obtained by solving the linear system

$$\begin{bmatrix} \tilde{A}_{\mathcal{L}} \\ \tilde{A} \end{bmatrix} \alpha = \begin{bmatrix} \mathbf{f} \\ \mathbf{g} \end{bmatrix} \tag{21}$$

where the matrices have the same form as in Kansa's approach. If we use α from (21) and once again assume the invertibility of the system matrix, we get

$$\mathbf{u} = \mathbf{A}\boldsymbol{\alpha} = \mathbf{A} \begin{bmatrix} \tilde{A}_{\mathcal{L}} \\ \tilde{A} \end{bmatrix}^{-1} \begin{bmatrix} \mathbf{f} \\ \mathbf{g} \end{bmatrix}, \tag{22}$$

This suggests that the discretized differential operator L based on the grid points $\mathbf{x}_i, i = 1, \ldots, N$, and basis functions is given by

$$\mathrm{L}_\Gamma = \begin{bmatrix} \tilde{\mathbf{A}}_{\mathcal{L}} \\ \tilde{\mathbf{A}} \end{bmatrix} \mathbf{A}^{-1}. \tag{23}$$

3.3 Implicit discrete scheme

We consider the implicit scheme of eqns. (1) and (3):

$$\varrho c_p \frac{T^{n+1} - T^n}{\Delta t} + k_x \frac{\partial^2 T^{n+1}}{\partial x^2} + k_y \frac{\partial^2 T^{n+1}}{\partial y^2} = 0, \tag{24}$$

$$-k_x \frac{\partial T^{n+1}}{\partial x} e_{nx} - k_y \frac{\partial T^{n+1}}{\partial y} e_{ny} - h_c(T^{n+1}) \\ -\varepsilon_R \sigma((T^{n+1})^4 - T_R^4) = -h_c(T_A), \tag{25}$$

where $t_{n+1} = t_n + \Delta t$, T^{n+1} and T^n are the variable at time t_{n+1} and t_n.

The approximate solution is expressed as:

$$T(\mathbf{x}, t_{n+1}) = \sum_{j=1}^{N} \alpha_j^{n+1} \varphi_j(\mathbf{x}), \tag{26}$$

where α_j^{n+1}, $j = 1, \ldots, N$, are the unknown coefficients to be determined and $\varphi_j(\mathbf{x}) = \sqrt{(x - x_j)^2 + (y - y_j)^2 + c^2}$ are Hardy's multiquadrics functions [12].

By substituting eqn. (26) into eqns. (24) and (25) and using factorization for the radiation term $(T^4 - T_R^4)$, we obtain:

$$\sum_{j=1}^{N} \left(\varrho c_p \frac{\varphi_j(\mathbf{x}_i)}{\Delta t} + k_x \frac{\partial^2 \varphi_j(\mathbf{x}_i)}{\partial x^2} + k_y \frac{\partial^2 \varphi_j(\mathbf{x}_i)}{\partial y^2} \right) \alpha_j^{n+1} \\ = \varrho c_p \frac{T^n(\mathbf{x}_i)}{\Delta t}, \quad i = 1, \ldots, N - N_{\mathcal{B}}, \tag{27}$$

$$\sum_{j=1}^{N}\Bigg(-k_x\frac{\partial\varphi_j(\mathbf{x}_i)}{\partial x}e_{nx}-k_y\frac{\partial\varphi_j(\mathbf{x}_i)}{\partial y}e_{ny}-h_c\varphi_j(\mathbf{x}_i)$$

$$-\varepsilon_R\sigma\left(\varphi_j(\mathbf{x}_i)-T_R\right)\left(\varphi_j(\mathbf{x}_i)+T_R\right)\left({\varphi_j}^2(\mathbf{x}_i)+T_R^2\right)\Bigg)\alpha_j^{n+1}$$

$$=-h_c(T_A),\quad i=N-N_{\mathcal{B}}+1,\ldots,N, \tag{28}$$

where $N_{\mathcal{B}}$ and N present the number at boundary and all discretized points.

The system of nonlinear equations which result from the space discretization of a nonlinear PDEs can be solved by Picard's or Newton's methods. In our case, we have used the Picard's method, also known as *Successive Substitutions* [13, 14].

4 Numerical examples

4.1 One-dimensional charring

The charring rate of wood usually refers to the dimensional rate, e.g. millimetres per minute, at which wood changes to char. Many factors are involved in wood charring. Kanury and Blackshear [15] examined various physical effects, including the diffusion of condensable vapours inward, internal convection outward, properties of the partially charred wood, kinetics of pyrolysis, energetic of pyrolysis, and postdecomposition reactions. No completely satisfactorily charring model has yet been developed.

An extensive data is available for simple one-dimensional charring. Therefore, a one-dimensional case of a timber slab of spruce, with a depth *d*, exposed to the standard fire [16] is analysed in order to compare the charring rate of the wood slab with the empirical models presented in the literature.

The charring of wood may be modelled by the mass loss rate $(\mathrm{g/s})$ or by the rate of advance of the formed char front from the original surface $(\mathrm{mm/min})$. Since the material properties at elevated temperatures are difficult to obtain, constant material properties of the wood and char are used. The following data has been used:

$$T_0=20°C,\ \varrho=370\ \mathrm{kg/m^3},\ k_{wood}=0.12\ k_{char}=0.15\ \mathrm{W/mK},\ d=0.3\ \mathrm{m}$$

$$h_c=22.5\ \mathrm{W/m^2},\ \epsilon_R=0.9,\ c_{p,wood}=1530\ \mathrm{J/kgK},\ c_{p,char}=1050\ \mathrm{J/kgK}.$$

Most known models suggest constant charring rates. In our case, we used a White and Norheim non-linear empirical model for charring rate of eight different wood species. The comparison to the present model in the case of spruce is shown in Fig. 1.

In all empirical models it is assumed that the charring of woods starts instantaneously after exposure to fire. In reality, this is not the case. In our model, the charring starts when the temperature of wood reaches the temperature of pyrolysis, which is around 300 °C. This happens nearly 3 minutes after the fire starts.

www.witpress.com, ISSN 1743-355X (on-line)

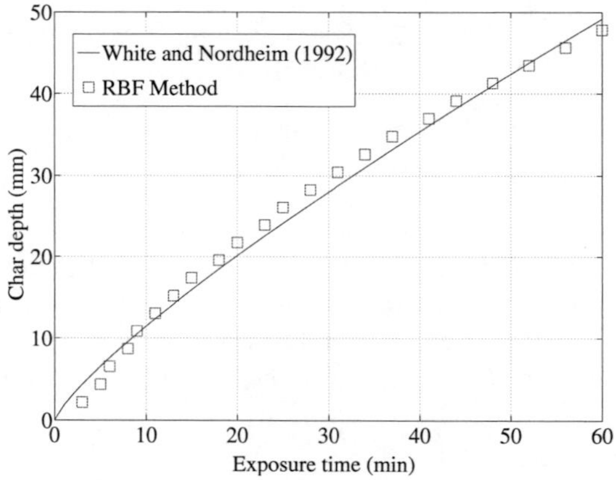

Figure 1: Comparison of White and Nordheim charring model to the present.

4.2 A two-dimensional charring

In the two-dimensional case, the formation of char in a timber beam exposed on three sides to the standard fire conditions in [16] is considered. The upper edge is thermally isolated. The original beam cross-section is rectangular with dimensions 10×15 cm. The beam cross-section is discretized by the mesh of 10×10 points. Material properties are assumed to be the same as the one-dimensional case. The results of the simulation at 10 and 30 minutes after the exposure to fire are given in figures 2 and 3.

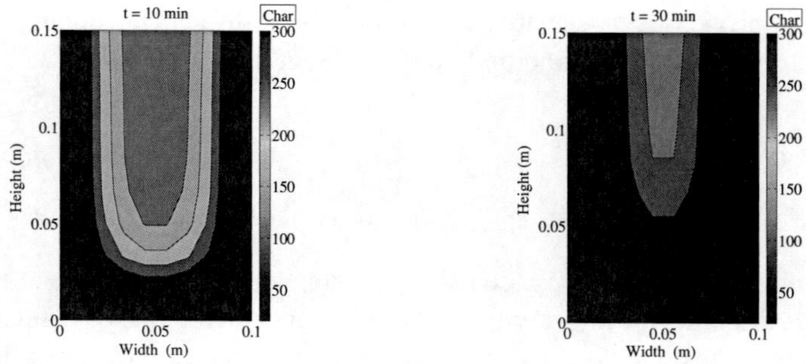

Figure 2: Temperature distribution in the cross-section of spruce beam and the transformation of wood into char at 10 and 30 minutes calculated with the Kansa approach, relaxation parameter: 0.14.

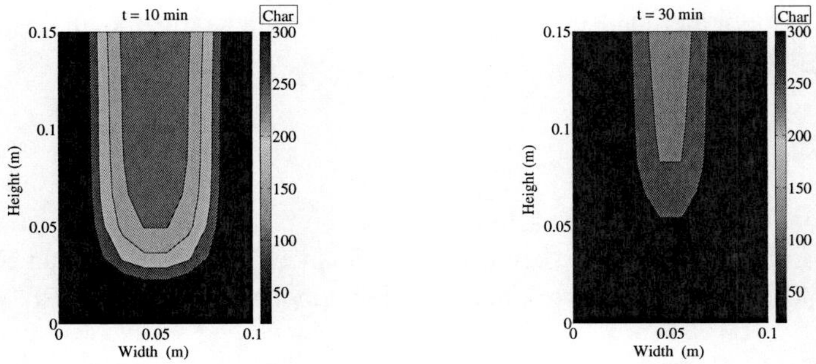

Figure 3: Temperature distribution in the cross-section calculated with the PS method, relaxation parameter: 0.14.

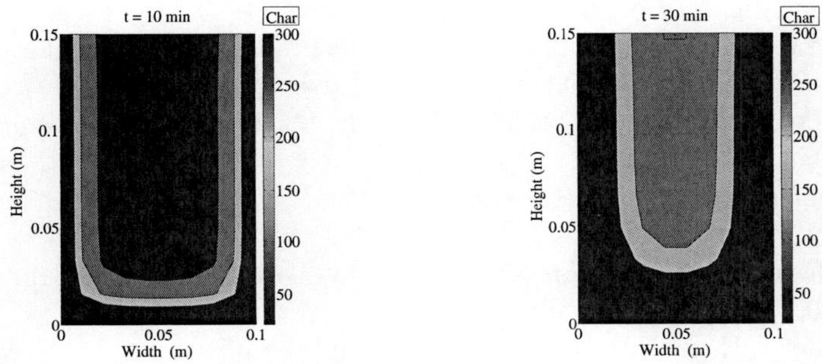

Figure 4: Temperature distribution in the cross-section calculated with the PS method, relaxation parameter: 0.1.

5 Conclusions

Since the analytical solution is seldom obtainable, the idea of this paper was to represent an alternative approach to the conventional numerical methods. Kansa's and PS methods were presented. The results were tested on the one-dimensional case for which the comparison was made with the results obtained numerically and experimentally.

In Fig. 1, we can see that our approach gives the results which are comparable to the results obtained by the model proposed by White and Nordheim. The same model was used to analyze a two-dimensional behaviour of wood beam exposed to fire from three sides. It shows that the results are comparable to the results obtained in literature [6]. Comparison of the figure 2 and 3 indicates that MQ RBFs, PS, and Picard's methods give very similar results. Comparison of the results in figures 3 and 4 also show that the results are sensitive to the relaxation parameter. The simulations also show that badly scaled or narrow basis functions (e.g. linear ($\varphi(r) = r$), cubic ($\varphi(r) = r^3$), thin-plate spline ($\varphi(r) = r^2 \log r$)) can prevent the effects of the boundary conditions from propagating inside the domain.

Therefore we can conclude that the presented methods could be an alternative to the conventional numerical methods. In our future work, moisture diffusion equations will be included and special attention should be given to the appropriate choice of RBFs.

6 Acknowledgements

The authors would like to thank the Slovenian Nuclear Safety Administration for their support. This work has been, in part, financed by the Slovenian Research Agency (ARRS) through the research program Geotechnology (PO-0268).

References

[1] Fredlund, B., Modelling of Heat and Mass Transfer in Wood Structures During Fire. *Fire Safety Journal*, **20**, pp. 39-69, 1993.

[2] Kansa, E.J., Multiquadrics – A scattered data approximation scheme with applications to computational fluid-dynamics – Solution to parabolic, hyperbolic and elliptic partial differential equations. *Comput. Math. Appl.*, **19**, pp. 147-161, 1990.

[3] Forngerg, B., *A Practical Guide to Pseudospectral Methods*, Cambridge Univ. Press, 1998.

[4] Trefethen, L.N., *Spectral Methods in MATLAB*, SIAM (Philadelphia, PA), 2000.

[5] White, R.H. & Nordheim, E.V., Charring rate of wood for ASTM E 119 exposure. *Fire Technology*, **28(1)**, pp. 5-30, 1992.

[6] Schnabl, S. & Turk, G., Coupled heat and moisture transfer beams exposed to fire. *WCTE 2006 - 9th World Conference on Timber Engineering - Portland, OR, USA*, 2006.

[7] Luikov, A.V., *Heat and Mass Transfer in Capillary-porous Bodies*, Pergamon Press, Oxford, 1966.

[8] Wang, J.G. & Liu, G.R., On the optimal shape parameters of radial basis functions used for 2-D meshless methods. *Comput. Meth. Appl. Mech. Eng.*, **30**, pp. 191-211, 2002.

[9] Vrankar, L., Turk, G. & Runovc, F., Modelling of radionuclide migration through the geosphere with radial basis function method and geostatistics. *Journal of the Chinese Institute of Engineers*, **4/27**, pp. 455-462, 2004.

[10] Vrankar, L., Kansa, E.J., Ling, L., Runovc, F. & Turk, G. Moving-boundary problems solved by adaptive radial basis functions. *Comput. Fluids*, **39**, pp. 1480-1490, 2010.

[11] Fasshauer, G.E., *Meshfree Approximation Methods with MATLAB*, Interdisciplinary Mathematical Sciences – **6**, 2007.

[12] Hardy, R.L., Multiquadric equations of topography and other irregular surfaces. *J. Geophys. Res.*, **176**, pp. 1905-1915, 1971.

[13] Stephen, W. Goode, *Differential Equations and Linear Algebra*, 3rd edition, Prentice Hall, Upper Saddle River, NJ, Appendix 4, 2005.

[14] Mai-Cao, L. Meshless Radial Basis Function Method for Unsteady Incompressible Viscous Flows. Ph.D. diss., University of Southern Queensland, Australia, 2009.

[15] Kanury, M.A. & Blackshear, P.L. Jr., Some Considerations Pertaining to the Problem of Wood-Burning. *Combustion Science and Technology*, **1**, pp. 339-356, 1970.

[16] ISO 834, Fire-resistance test-Elements of building construction-Part 1. General requirements. ISO 834-1, *International organization for standardization*, Geneva, Switzerland, 1999.

Moving least square – one dimensional integrated radial basis function networks for time dependent problems

D. Ngo-Cong[1,2], N. Mai-Duy[1], W. Karunasena[2] & T. Tran-Cong[1]
[1] *Computational Engineering & Science Research Centre, University of Southern Queensland, Australia*
[2] *Centre of Excellence in Engineered Fibre Composites, University of Southern Queensland, Australia*

Abstract

This paper presents a new numerical procedure for time-dependent problems. The partition of unity method is employed to incorporate the moving least square and one-dimensional integrated radial basis function networks (MLS-1D-IRBFN) techniques in an approach that produces a very sparse system matrix and offers as a high order of accuracy as that of global 1D-IRBFN method. Moreover, the proposed approach possesses the Kronecker-δ property which helps impose the essential boundary condition in an exact manner. Spatial derivatives are discretised using Cartesian grids and MLS-1D-IRBFN, whereas temporal derivatives are discretised using high-order time-stepping schemes, namely standard θ and fourth-order Runge–Kutta methods. Several numerical examples including two-dimensional diffusion equation, one-dimensional advection-diffusion equation and forced vibration of a beam are considered. Numerical results show that the current methods are highly accurate and efficient in comparison with other published results available in the literature.
Keywords: time-dependent problems, integrated radial basis functions, moving least square, partition of unity, Cartesian grids.

1 Introduction

In 1990, Kansa proposed a collocation scheme based on multiquadric (MQ) radial basis functions (RBF) for the numerical solution of partial differential equations

www.witpress.com, ISSN 1743-355X (on-line)
doi:10.2495/BE110271

(PDEs) [1]. Their numerical results showed that MQ scheme yielded an excellent interpolation and partial derivative estimates for a variety of two-dimensional functions over both gridded and scattered data. The main drawback of the RBF based method is the lack of mathematical theories for the choice of the RBF shape parameter. Since this original work, many meshfree methods based on RBF have been proposed. Sharan et al. [2] employed the multiquadric approximation scheme for solution of elliptic partial differential equations using the data points in arbitrary locations with an arbitrary ordering. Zerroukat et al. [3] proposed explicit and implicit meshless methods based on global RBFs for linear advection-diffusion-type partial differential equations on arbitrary collocation points system.

In recent years, a different approach for solving PDEs is the so-called Cartesian grid method where the governing equations are discretised on a Cartesian grid which does not conform to the immersed boundaries. This significantly reduces the grid generation cost and has a great potential over the conventional body-fitted methods when solving problems with moving boundaries and complicated geometry. Ye et al. [4] developed a finite-volume based Cartesian grid method for simulating two-dimensional unsteady, viscous, incompressible flows with complex immersed boundaries. In their method, the immersed boundary is represented by a series of piecewise linear segments. Based on these segments, the control volume near the immersed boundary is reformed into a body-fitted trapezoidal shape. A one-dimensional integrated radial basis function networks (1D-IRBFN) collocation method for the solution of second- and fourth-order PDEs was presented by Mai-Duy and Tanner [5]. Along grid lines, 1D-IRBFN are constructed to satisfy the governing differential equations with boundary conditions in an exact manner. The 1D-IRBFN method enjoys spectral accuracy and exponential convergence for certain problems. In this method, the Cartesian grids were used to discretise both rectangular and non-rectangular problem domains. The computational cost used for the Cartesian grid generation is negligible in comparison with that required for the body-fitted mesh. Le-Cao et al. [6] presented a numerical collocation procedure based on Cartesian grids and 1D-IRBFN for simulation of natural convection defined in 2D multiply connected domains and governed by a stream function-vorticity-temperature formulation. Ngo-Cong et al. [7] extended this method to investigate free vibration of composite laminated plates based on first-order shear deformation theory. The present paper deals with the development of MLS-1D-IRBFN methods based on two frameworks of semi-discrete and fully discrete schemes for solving several numerical examples including diffusion equation, advection diffusion equation and forced vibration of a beam. In the semi-discrete scheme, the first stage is the spatial discretisation in which the spatial derivatives of the PDEs are discretised to obtain a system of ordinary differential equations (ODEs).The system is then advanced in time using ODE solvers (e.g. fourth-order Runge–Kutta scheme) to obtain the unknowns for each time step. In the fully discrete scheme, the time derivatives of the PDEs are first discretised using high-order time-stepping schemes (standard θ-scheme) to obtain a sequence of steady problems. The spatial discretisation is then performed to obtain a full discretisation which is in the form of a system of algebraic equations. In the following, Section 2 presents the MLS-1D-IRBFN methods. Several numerical examples are then used to demonstrate the performance of the proposed methods in Section 3. Section 4 concludes the paper.

2 Moving least square – one-dimensional-integrated radial basis function networks

A schematic outline of the MLS-1D-IRBFN method is depicted in Fig. 1. For brevity, 3-node support domains are presented here. Similar explanation is applicable in the case of 5-node support domains. On an x-grid line $[l]$, a global interpolant for the field variable at a grid point x_i is sought as

$$u(x_i) = \sum_{j=1}^{n} \phi_j(x_i) u_j(x_i), \tag{1}$$

where $\{\phi_j\}_{j=1}^{n}$ is a set of the partition of unity functions constructed using MLS approximants, $u_j(x_i)$ is the nodal function value obtained from a local interpolant represented by 1D-IRBFNs, and n is the number of nodes in the support domain of x_i. In (1), MLS approximants are presently based on linear polynomials, which are defined in terms of 1 and x. Relevant derivatives of u at x_i can be obtained by differentiating (1)

$$\frac{\partial u(x_i)}{\partial x} = \sum_{j=1}^{n} \left(\frac{\partial \phi_j(x_i)}{\partial x} u_j(x_i) + \phi_j(x_i) \frac{\partial u_j(x_i)}{\partial x} \right), \tag{2}$$

$$\frac{\partial^2 u(x_i)}{\partial x^2} = \sum_{j=1}^{n} \left(\frac{\partial^2 \phi_j(x_i)}{\partial x^2} u_j(x_i) + 2 \frac{\partial \phi_j(x_i)}{\partial x} \frac{\partial u_j(x_i)}{\partial x} + \phi_j(x_i) \frac{\partial^2 u_j(x_i)}{\partial x^2} \right), \tag{3}$$

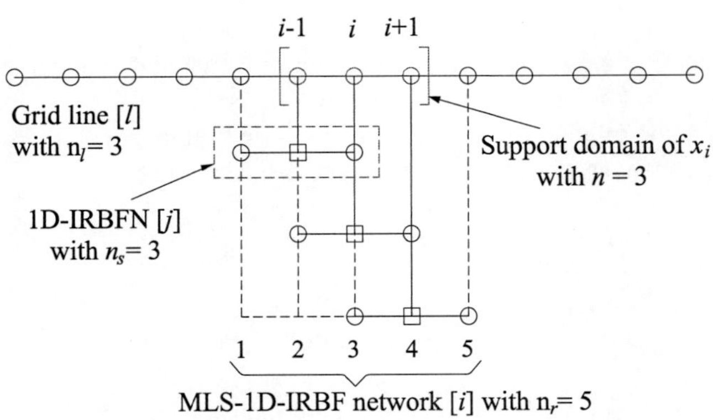

Figure 1: MLS-1D-IRBFN scheme, □ a typical $[j]$ node.

where the values $u_j(x_i)$, $\partial u_j(x_i)/\partial x$ and $\partial^2 u_j(x_i)/\partial x^2$ are calculated from 1D-IRBFNs with n_s nodes.

2.1 One-dimensional IRBFN

Consider a segment $[j]$ with n_s nodes on an x-grid line $[l]$ as shown in Fig. 1. The variation of the nodal function u_j along this segment is sought in the IRBF form. The second-order derivative of u_j is decomposed into RBFs; the RBF network is then integrated once and twice to obtain expressions for the first-order derivative of u_j and the function u_j itself as follows.

$$\frac{\partial^2 u_j(x)}{\partial x^2} = \sum_{k=1}^{n_s} w^{(k)} G^{(k)}(x), \tag{4}$$

$$\frac{\partial u_j(x)}{\partial x} = \sum_{k=1}^{n_s} w^{(k)} H_{[1]}^{(k)}(x) + c_1, \tag{5}$$

$$u_j(x) = \sum_{k=1}^{n_s} w^{(k)} H_{[0]}^{(k)}(x) + c_1 x + c_2, \tag{6}$$

where $\left\{w^{(k)}\right\}_{k=1}^{n_s}$ are RBF weights to be determined; $\left\{G^{(k)}\right\}_{k=1}^{n_s}$ are known RBFs, e.g., for the case of multiquadrics $G^{(k)}(x) = \sqrt{(x - x^{(k)})^2 + a^{(k)2}}$, $a^{(k)}$ – the RBF width; $H_{[1]}^{(k)}(x) = \int G^{(k)}(x)dx$; $H_{[0]}^{(k)}(x) = \int H_{[1]}^{(k)}(x)dx$; and c_1 and c_2 are integration constants which are also unknown.

It is more convenient to work in the physical space than in the network-weight space. The RBF coefficients including two integration constants can be related to the physically meaningful nodal variable values. The second- and first-order derivatives of the variable u are expressed in terms of nodal variable values as follows.

$$\frac{\partial^2 u(x)}{\partial x^2} = \bar{D}_{2x}\hat{u}^{[j]} + k_{2x}(x), \qquad \frac{\partial u(x)}{\partial x} = \bar{D}_{1x}\hat{u}^{[j]} + k_{1x}(x), \tag{7}$$

where k_{1x} and k_{2x} are scalars whose values depend on x and boundary values; and $\bar{D}_{1x}$ and $\bar{D}_{2x}$ are known vectors of length n_s.

By application of equation (7) to n_s nodes on the segment $[j]$, the second- and first-order derivatives of u_j at node x_i can be determined as

$$\frac{\partial^2 \hat{u}_j(x_i)}{\partial x^2} = \hat{\mathbf{M}}_{2x(idk,:)}\hat{u}^{[j]} + \hat{k}_{2x(idk)}, \tag{8}$$

$$\frac{\partial \hat{u}_j(x_i)}{\partial x} = \hat{\mathbf{M}}_{1x(idk,:)}\hat{u}^{[j]} + \hat{k}_{1x(idk)}, \tag{9}$$

where $\hat{\mathbf{M}}_{1x}$ and $\hat{\mathbf{M}}_{2x}$ are known matrices of dimension $n_s \times n_s$; $\hat{k}_{1x}$ and $\hat{k}_{2x}$ are known vectors of length n_s; and idk is the index number indicating the location of node x_i over the local network j.

2.2 Incorporation of MLS and 1D-IRBFN

By substituting equations (8) and (9) into equations (2) and (3), the second- and first-order derivatives of the variable $u(x_i)$ can be expressed as

$$\frac{\partial u(x_i)}{\partial x} = \hat{\mathbf{d}}_{1x}^{[i]}\hat{u}^{[i]} + k_{1x}^{[i]}, \quad \frac{\partial^2 u(x_i)}{\partial x^2} = \hat{\mathbf{d}}_{2x}^{[i]}\hat{u}^{[i]} + k_{2x}^{[i]}. \tag{10}$$

where $\hat{u}^{[i]} = \left(u^{(1)}, u^{(2)}, \ldots, u^{(n_r)}\right)^T$, n_r is the number of nodes in the MLS-1D-IRBF network $[i]$, $k_{1x}^{[i]}$ and $k_{2x}^{[i]}$ are known scalars, and $\hat{\mathbf{d}}_{1x}^{[i]}$ and $\hat{\mathbf{d}}_{2x}^{[i]}$ are known vectors of length n_r.

The values of first- and second-order derivatives of u with respect to x at the nodal points on the grid line $[l]$ can be given by

$$\frac{\partial \hat{u}_i}{\partial x} = \hat{\mathbf{D}}_{1x}^{[l]}\hat{u}^{[l]} + \hat{k}_{1x}^{[l]}, \quad \frac{\partial^2 \hat{u}_i}{\partial x^2} = \hat{\mathbf{D}}_{2x}^{[l]}\hat{u}^{[l]} + \hat{k}_{2x}^{[l]}, \tag{11}$$

where $\hat{u}^{[l]} = \left(u^{(1)}, u^{(2)}, \ldots, u^{(n_l)}\right)^T$, $\hat{k}_{1x(i)}^{[l]}$ and $\hat{k}_{2x(i)}^{[l]}$ are known vectors of length n_l, $\hat{\mathbf{D}}_{1x(i,idi)}^{[l]}$ and $\hat{\mathbf{D}}_{2x(i,idi)}^{[l]}$ are known matrices of dimension $n_l \times n_l$, and n_l is the number of nodes on the grid line $[l]$.

The values of first- and second-order derivatives of u with respect to x at the nodal points over the problem domain can be given by

$$\frac{\partial \tilde{u}}{\partial x} = \tilde{\mathbf{D}}_{1x}\tilde{u} + \tilde{k}_{1x}, \quad \frac{\partial^2 \tilde{u}}{\partial x^2} = \tilde{\mathbf{D}}_{2x}\tilde{u} + \tilde{k}_{2x}, \tag{12}$$

where $\tilde{u} = \left(u^{(1)}, u^{(2)}, \ldots, u^{(N_{ip})}\right)^T$, $\tilde{\mathbf{D}}_{1x}$ and $\tilde{\mathbf{D}}_{2x}$ are known matrices of dimension $N_{ip} \times N_{ip}$; $\tilde{k}_{1x}$ and $\tilde{k}_{2x}$ are known vectors of length N_{ip}; and N_{ip} is the total number of interior nodal points.

Similarly, the values of the second- and first-order derivatives of u with respect to y at the nodal points over the problem domain can be given by

$$\frac{\partial \tilde{u}}{\partial y} = \tilde{\mathbf{D}}_{1y}\tilde{u} + \tilde{k}_{1y}, \quad \frac{\partial^2 \tilde{u}}{\partial y^2} = \tilde{\mathbf{D}}_{2y}\tilde{u} + \tilde{k}_{2y}, \tag{13}$$

3 Numerical results

The MLS-1D-IRBFN based on 5-node support domains and 5-node 1D-IRBFNs are applied to solve the following examples.

3.1 Example 1: Two-dimensional diffusion equation

This example is concerned with the following 2D diffusion equation

$$\frac{\partial u}{\partial t} = \frac{\partial^2 u}{\partial x^2} + \frac{\partial^2 u}{\partial y^2} + f(x,y,t) \text{ with } f(x,y,t) = \sin x \sin y(2\sin t + \cos t), \tag{14}$$

www.witpress.com, ISSN 1743-355X (on-line)

Table 1: Two-dimensional diffusion equation: comparison of relative error and CPU time between 1D-IRBFN and MLS-1D-IRBFN methods. The fully discrete scheme with standard θ-scheme for temporal discretisation is employed. (1) 1D-IRBFN, (2) MLS-1D-IRBFN.

Grid	Δt	Exact	Numerical solution		Relative error		CPU time (s)	
			(1)	(2)	(1)	(2)	(1)	(2)
11	1.0000	−0.389450	−0.387900	−0.387891	3.98E-03	4.00E-03	0.02	0.19
21	0.2500	−0.389450	−0.389404	−0.389400	1.19E-04	1.30E-04	0.16	0.78
31	0.1111	−0.389450	−0.389441	−0.389437	2.29E-05	3.47E-05	1.61	1.92
41	0.0625	−0.389450	−0.389448	−0.389443	7.23E-06	1.93E-05	8.36	3.76
51	0.0400	−0.389450	−0.389449	−0.389444	2.95E-06	1.51E-05	40.97	6.48
61	0.0278	−0.389450	−0.389450	−0.389445	1.42E-06	1.36E-05	127.39	10.47

Table 2: Two-dimensional diffusion equation: MLS-1D-IRBFN solutions at the interior point $x = 0.8, y = 0.8$ in comparison with those of BEM [9] and 1D-IRBFN methods with the same time step $\Delta t = 0.25$. The same grid of 21×21 is employed for 1D-IRBFN and MLS-1D-IRBF methods, while a much finer mesh is used for BEM. The fully discrete scheme with standard θ-scheme for temporal discretisation is employed. (1) 1D-IRBFN, (2) MLS-1D-IRBFN.

T	Exact	Numerical solution			Relative error		
		BEM	(1)	(2)	BEM	(1)	(2)
0.25	0.127314	0.126800	0.127206	0.127205	4.04E-03	8.45E-04	8.55E-04
0.50	0.246712	0.245300	0.246679	0.246676	5.72E-03	1.34E-04	1.45E-04
0.75	0.350771	0.348500	0.350701	0.350697	6.47E-03	2.00E-04	2.12E-04
1.00	0.433021	0.430000	0.432997	0.432992	6.98E-03	5.48E-05	6.62E-05
1.25	0.488347	0.484800	0.488313	0.488308	7.26E-03	6.98E-05	8.12E-05
1.50	0.513311	0.509400	0.513313	0.513307	7.62E-03	4.23E-06	7.36E-06
1.75	0.506359	0.502400	0.506363	0.506357	7.82E-03	7.23E-06	4.44E-06
2.00	0.467924	0.464100	0.467956	0.467951	8.17E-03	6.89E-05	5.70E-05
2.25	0.400396	0.397000	0.400433	0.400429	8.48E-03	9.27E-05	8.07E-05
2.50	0.307974	0.305200	0.308031	0.308027	9.01E-03	1.85E-04	1.73E-04
2.75	0.196403	0.194400	0.196462	0.196460	1.02E-02	3.03E-04	2.90E-04
3.00	0.072620	0.071500	0.072690	0.072689	1.54E-02	9.54E-04	9.39E-04
3.25	−0.055677	−0.055800	−0.055611	−0.055611	2.21E-03	1.19E-03	1.20E-03
3.50	−0.180513	−0.179700	−0.180447	−0.180446	4.50E-03	3.63E-04	3.74E-04
3.75	−0.294125	−0.292300	−0.294070	−0.294066	6.21E-03	1.90E-04	2.01E-04
4.00	−0.389450	−0.386800	−0.389404	−0.389400	6.81E-03	1.19E-04	1.30E-04

Table 3: Two-dimensional diffusion equation: MLS-1D-IRBFN solutions at the interior point $x = 0.8, y = 0.8$ in comparison with those of BEM [9] and 1D-IRBFN methods with the same time step $\Delta t = 0.25$. The same grid of 21×21 is employed for 1D-IRBFN and MLS-1D-IRBF methods, while a much finer mesh is used for BEM. The semi-discrete scheme with fourth-order Runge–Kutta method for temporal discretisation is employed. (1) 1D-IRBFN, (2) MLS-1D-IRBFN.

T	Exact	Numerical solution			Relative error		
		BEM	(1)	(2)	BEM	(1)	(2)
0.25	0.127314	0.126800	0.127315	0.127314	4.04E-03	6.45E-06	1.64E-06
0.50	0.246712	0.245300	0.246714	0.246712	5.72E-03	7.33E-06	2.14E-06
0.75	0.350771	0.348500	0.350774	0.350770	6.47E-03	7.65E-06	2.32E-06
1.00	0.433021	0.430000	0.433024	0.433020	6.98E-03	7.84E-06	2.43E-06
1.25	0.488347	0.484800	0.488351	0.488346	7.26E-03	7.97E-06	2.50E-06
1.50	0.513311	0.509400	0.513315	0.513309	7.62E-03	8.08E-06	2.57E-06
1.75	0.506359	0.502400	0.506363	0.506358	7.82E-03	8.19E-06	2.63E-06
2.00	0.467924	0.464100	0.467928	0.467923	8.17E-03	8.31E-06	2.69E-06
2.25	0.400396	0.397000	0.400400	0.400395	8.48E-03	8.45E-06	2.78E-06
2.50	0.307974	0.305200	0.307976	0.307973	9.01E-03	8.68E-06	2.91E-06
2.75	0.196403	0.194400	0.196404	0.196402	1.02E-02	9.14E-06	3.17E-06
3.00	0.072620	0.071500	0.072621	0.072620	1.54E-02	1.11E-05	4.28E-06
3.25	−0.055677	−0.055800	−0.055677	−0.055677	2.21E-03	4.19E-06	3.58E-07
3.50	−0.180513	−0.179700	−0.180514	−0.180513	4.50E-03	6.97E-06	1.94E-06
3.75	−0.294125	−0.292300	−0.294128	−0.294125	6.21E-03	7.50E-06	2.24E-06
4.00	−0.389450	−0.386800	−0.389453	−0.389449	6.81E-03	7.74E-06	2.37E-06

defined on a square domain $0 < x, y < 1, t > 0$ and subject to Dirichlet boundary conditions. The boundary and initial conditions can be derived from the analytical solution $u_E = \sin x \sin y \sin t$.

In this example, the time step Δt is taken based on the parameter $d = \kappa \Delta t / \Delta x^2$, where κ is the diffusion coefficient, presently $\kappa = 1$. The parameter d is the ratio of time step Δt to the characteristic diffusion time $\Delta x^2/\kappa$, which is roughly the time required for a disturbance to be transmitted by diffusion over a distance Δx [8]. The parameter d is here chosen to be 100.

The grid convergence study and CPU-time requirement for both 1D-IRBFN and MLS-1D-IRBFN based on the fully discrete framework with standard θ-scheme for temporal discretisation are presented in Table 1. It can be seen that the numerical solutions for both methods are converging well from the coarse mesh to the fine mesh. MLS-1D-IRBFN yields the same order accuracy as that of 1D-IRBFN, and offers a significant improvement in terms of efficiency when dealing with fine meshes. Table 2 presents the numerical solutions of 1D-IRBFN and MLS-1D-IRBFN methods at the interior point $x = 0.8, y = 0.8$ in comparison

Table 4: One-dimensional advection-diffusion equation: MLS-1D-IRBFN solutions at time $t = 1$ in comparison with those of DRBFN [3] and 1D-IRBFN methods with the same grid of 17 and $\Delta t = 0.01$. (1) 1D-IRBFN, (2) MLS-1D-IRBFN.

x	Exact	Numerical solution			Absolute error		
		DRBFN	(1)	(2)	DRBFN	(1)	(2)
0.0625	0.998874	–	0.998880	0.998882	–	6.51E-06	8.44E-06
0.1250	0.902800	–	0.902804	0.902810	–	4.18E-06	9.95E-06
0.1875	0.815967	–	0.815970	0.815982	–	2.76E-06	1.51E-05
0.2500	0.737486	0.737486	0.737487	0.737506	9.89E-03	1.15E-06	1.98E-05
0.3125	0.666553	0.666554	0.666553	0.666576	8.45E-03	3.45E-07	2.27E-05
0.3750	0.602443	0.602444	0.602441	0.602467	6.37E-03	1.88E-06	2.41E-05
0.4375	0.544499	–	0.544495	0.544523	–	3.44E-06	2.43E-05
0.5000	0.492128	0.492129	0.492123	0.492151	1.78E-03	5.05E-06	2.36E-05
0.5625	0.444794	–	0.444787	0.444816	–	6.73E-06	2.22E-05
0.6250	0.402013	0.402014	0.402004	0.402033	1.82E-03	8.48E-06	2.02E-05
0.6875	0.363346	–	0.363336	0.363364	–	1.03E-05	1.77E-05
0.7500	0.328399	0.328400	0.328387	0.328414	3.66E-03	1.21E-05	1.48E-05
0.8125	0.296813	–	0.296799	0.296825	–	1.42E-05	1.15E-05
0.8750	0.268265	0.268266	0.268249	0.268272	4.01E-03	1.57E-05	7.34E-06
0.9375	0.242463	–	0.242444	0.242465	–	1.93E-05	1.85E-06

with those of BEM, which were published in [9] using the first-order finite difference approximation for the time derivative and boundary element method for spatial discretisation. The corresponding results of 1D-IRBFN and MLS-1D-IRBFN based on the semi-discrete framework with fourth-order Runge–Kutta method for temporal discretisation are described in Table 3. The same grid of 21×21 is used for 1D-IRBFN and MLS-1D-IRBF methods, while a much finer mesh was used for BEM. It can be seen in these tables that the results of MLS-1D-IRBFN and 1D-IRBFN methods are slightly different and both are more accurate than that of BEM.

3.2 Example 2: One-dimensional advection-diffusion equation

The present method is here applied to solve the 1D advection-diffusion equation $\partial u(x,t)/\partial t = \kappa \partial^2 u(x,t)/\partial x^2 + \nu \partial u(x,t)/\partial x$, defined on a domain $0 < x < 1$, $t > 0$ and subject to boundary and initial conditions $u(0,t) = ae^{bt}$, $t > 0$, $u(1,t) = ae^{bt-c}$, $t > 0$, and $u(x,0) = ae^{-cx}$, $0 \leq x \leq 1$. The problem has an analytical solution $u_E = ae^{bt-cx}$, where $c = (\nu + \sqrt{\nu^2 + 4\kappa b})/(2\kappa) > 0$, $\kappa = 0.1$, $b = 0.1$, $a = 1.0$, and $\nu = 0.1$.

Table 4 presents the MLS-1D-IRBFN numerical solutions of the 1D advection-diffusion equation at time $t = 1s$ in comparison with those of DRBFN [3] and 1D-IRBFN methods. The DRBFN result was obtained using implicit Crank-Nicholson

Table 5: Forced vibration of simply supported beam: comparison of deflection w at time $t = 1$s. (1) 1D-IRBFN, (2) MLS-1D-IRBFN.

x (cm)	Exact	Numerical solution		Relative error	
		(1)	(2)	(1)	(2)
5	−0.009151	−0.009140	−0.009143	1.24E-03	9.04E-04
10	−0.016306	−0.016289	−0.016293	1.05E-03	7.72E-04
15	−0.019916	−0.019891	−0.019896	1.24E-03	9.84E-04
20	−0.020138	−0.020109	−0.020114	1.45E-03	1.21E-03
25	−0.017585	−0.017560	−0.017564	1.44E-03	1.21E-03
30	−0.012916	−0.012898	−0.012901	1.44E-03	1.20E-03
35	−0.006820	−0.006809	−0.006810	1.70E-03	1.45E-03

Table 6: Forced vibration of simply supported beam: comparison of velocity v at time $t = 1$s. (1) 1D-IRBFN, (2) MLS-1D-IRBFN.

x (cm)	Exact	Numerical solution		Relative error	
		(1)	(2)	(1)	(2)
5	1.558434	1.560281	1.561015	1.19E-03	1.66E-03
10	2.776816	2.779817	2.780941	1.08E-03	1.49E-03
15	3.391601	3.394314	3.395441	8.00E-04	1.13E-03
20	3.429398	3.430837	3.431784	4.20E-04	6.96E-04
25	2.994631	2.995018	2.995757	1.29E-04	3.76E-04
30	2.199611	2.199324	2.199909	1.31E-04	1.35E-04
35	1.161464	1.160977	1.161331	4.20E-04	1.15E-04

Thin Plate Spline scheme. For comparison purposes, 1D-IRBFN and MLS-1D-IRBFN are implemented based on the fully discrete scheme with standard θ-scheme for temporal discretisation, and using the same grid and time step as those in [3]. It can be seen that the MLS-1D-IRBFN result is more accurate than that of DRBFN, but slightly less accurate than that of 1D-IRBFN.

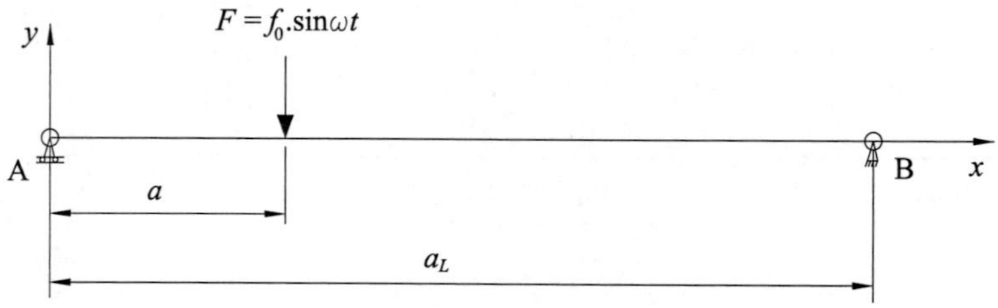

Figure 2: Forced vibration of a simply supported beam.

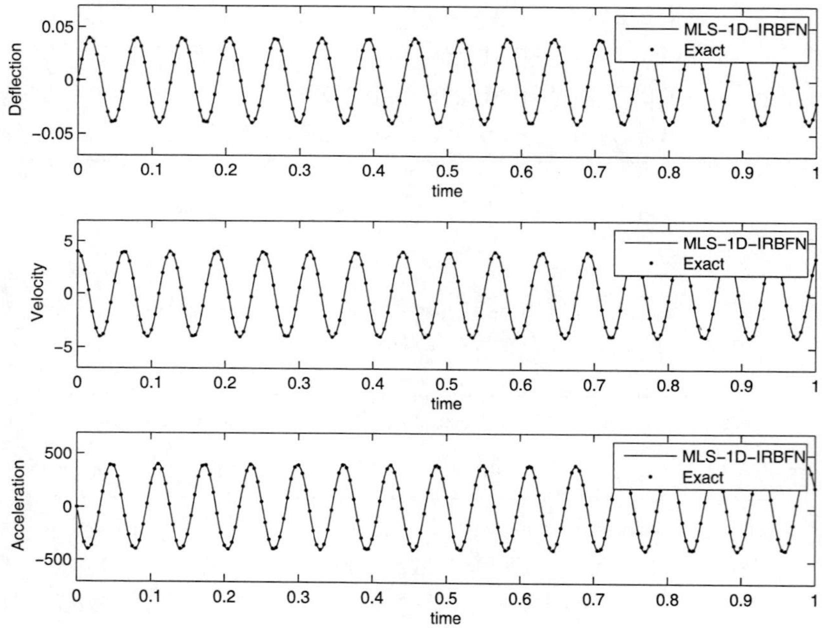

Figure 3: Steady state response of the mid-point of a simply supported beam, using a grid of 81 and $\Delta t = 10^{-3}$.

3.3 Example 3: Forced vibration of a beam

This example deals with the dynamic behaviour of a simply supported beam subject to a harmonic external force $F(t) = f_0 \sin \omega t$ applied at $x = a$, as shown in Fig. 2 (where $f_0 = 100\,\text{N}$, $\omega = 100\,\text{rad/s}$ and $a = 10\,\text{cm}$). The problem geometry and material parameters of the beam used here are: the length of the beam $a_L = 40\,\text{cm}$, the cross-section area $A = 1\,\text{cm}^2$, the moment of inertia $I = 0.0833\,\text{cm}^4$, Young's modulus $E = 300\,\text{GPa}$ and material density $\rho = 732.4\,\text{kg/m}^3$. The equation of motion for forced lateral vibration of a beam is given by $EI\partial^4 w/\partial x^4 + \rho A \partial^2 w/\partial t^2 = f(x,t)$. The boundary and initial

www.witpress.com, ISSN 1743-355X (on-line)

conditions for the simply supported beam can be described as $w = 0$, $\frac{\partial^2 w}{\partial x^2} = 0$, at $x = 0, x = a_L$; $w = 0$, $\frac{\partial w}{\partial t} = v_0$, at $t = 0$. An analytical solution to this problem can be found in [10].

Tables 5 and 6 present numerical results of the deflection and velocity of the simply supported beam at time $t = 1s$, using a grid of 81 and time step of 10^{-3}. The fully discrete scheme with Newmark method for temporal discretisation is employed here. It can be seen that MLS-1D-IRBFN yields more accurate result for deflection than 1D-IRBFN, while 1D-IRBFN produces more accurate result for velocity than MLS-1D-IRBFN. Fig. 3 shows the steady-state responses of the forced vibration system using MLS-1D-IRBFN in comparison with the analytical solutions. The numerical results are in good agreement with the analytical solutions as shown in this figure.

4 Conclusion

A moving least square – one-dimensional-integrated radial basis function networks approach based on the semi-discrete and fully discrete frameworks is developed for time-dependent problems. Spatial discretisation is carried out using MLS-1D-IRBFN and Cartesian grids, while the time derivatives are discretised using the high order schemes (e.g. standard θ, fourth-order Runge–Kutta). The numerical results obtained show that the proposed methods yield a high convergence order of accuracy as that of 1D-IRBFN, while requires less computational effort. The present methods have a promising capability to solve unsteady incompressible viscous flow problems.

Acknowledgement

D. Ngo-Cong is supported by a University of Southern Queensland Postgraduate Research Scholarship.

References

[1] Kansa, E.J., Multiquadrics - A Scattered Data Approximation Scheme with Applications to Computational Fluid-Dynamics - II: Solutions to parabolic, Hyperbolic and Elliptic Partial Differential Equations. *Computers & Mathematics with Applications*, **19(8/9)**, pp. 147–61, 1990.

[2] Sharan M., Kansa E.J. & Gupta S., Application of multiquadric method for numerical solution of elliptic partial differential equations. *Applied Mathematical Computations*, **84(2-3)**, pp. 275–302, 1997.

[3] Zerroukat M., Djidjeli K. & Charafi A., Explicit and implicit meshless methods for linear advection-diffusion-type partial differential equations. *International Journal for Numerical Methods in Engineering*, **48**, pp. 19–35, 2000.

[4] Ye T., Mittal R., Udaykumar H.S. & Shyy W., An accurate Cartesian grid method for viscous incompressible flows with complex immersed boundaries. *Journal of Computational Physics*, **156**, pp. 209–240, 1999.
[5] Mai-Duy N. & Tanner R.I., A Collocation Method based on One-Dimensional RBF Interpolation Scheme for Solving PDEs. *International Journal of Numerical Methods for Heat & Fluid Flow*, **17(2)**, pp. 165–186, 2007.
[6] Le-Cao K., Mai-Duy N. & Tran-Cong T., An effective integrated-RBFN Cartesian-grid discretization for the stream function-vorticity-temperature formulation in nonrectangular domains. *Numerical Heat Transfer, Part B*, **55**, pp. 480–502, 2009.
[7] Ngo-Cong D., Mai-Duy N., Karunasena W. & Tran-Cong T., Free vibration analysis of laminated composite plates based on FSDT using one-dimensional IRBFN method. *Computers & Structures*, **89**, pp. 1–13, 2011.
[8] Ferziger J.H. & Peric M., *Computational Methods for Fluid Dynamics*, 3rd edition, Springer: Berlin, pp. 143–144, 2002.
[9] Ingber M. & Phan-Thien N., A boundary element approach for parabolic differential equations using a class of particular solutions. *Applied Mathematical Modelling*, **16(3)**, pp. 124–132, 1992.
[10] Rao S.S., *Mechanical vibrations*, 4th edition, Pearson Prentice Hall: New Jersey, pp. 619–620, 2004.

The radial basis integral equation method for 2D Helmholtz problems

H. Dogan & V. Popov
Wessex Institute of Technology, Environmental and Fluid Mechanics, Southampton, UK

Abstract

A meshless method for the solution of 2D Helmholtz equation has been developed by using the Boundary Integral Equation (BIE) combined with Radial Basis Function (RBF) interpolations. BIE is applied by using the fundamental solution of the Helmholtz equation, therefore domain integrals are not encountered in the method. The method exploits the advantage of placing the source point always in the centre of circular sub-domains in order to avoid singular or near-singular integrals. Three equations for two-dimensional (2D) or four for three-dimensional (3D) potential problems are required at each node. The first equation is the integral equation arising from the application of the Green's identities and the remaining equations are the derivatives of the first equation in respect to space coordinates. RBF interpolation is applied in order to obtain the values of the field variable and partial derivatives at the boundary of the circular sub-domains, providing in this way the boundary conditions for solution of the integral equations at the nodes (centres of circles). The accuracy and robustness of the method has been tested on some analytical solutions of the problem. Two different RBFs are used, namely $f_1(R) = R^2 \ln(R) + x + y + 1$ and $f_2(R) = R^4 \ln(R) + x^2 + y^2 + xy + x + 1$. The latter has been found to produce more accurate results.

Keywords: meshless method, 2D Helmholtz equation, circular sub-domains, radial basis functions.

1 Introduction

The 2-D modelling of acoustics has been of significant importance for a variety of engineering problem in areas such as ultrasonics, microfluidics, aeoracoustics, etc. The numerical modelling of the problem is essential in case a complete

WIT Transactions on Modelling and Simulation, Vol 52, © 2011 WIT Press
www.witpress.com, ISSN 1743-355X (on-line)
doi:10.2495/BE110281

theoretical approach is not possible, i.e. multiphase flows or in the absence of experimental research. Hence, robust numerical algorithms are needed. The robustness of the numerical scheme plays a more important role whilst dealing with high frequencies since the numerical pollution and dispersion affect the accuracy in a significant way [1]. Meshless methods have received attention in the last a few decades due to their several advantages over usual boundary element methods hence some approaches, using Galerkin method [2] and hybrid boundary node method [3], are present for solving the Helmholtz equation. In this report, we present a truly meshless Radial Basis Integral Equation Method (RBIEM) in order to solve the governing Helmholtz equation in a 2-D setting.

In the direct BEM formulation for the Helmholtz equation, difficulties in implementation arise due to singular integrals which require a special treatment [4], and furthermore the solution loses its accuracy significantly in the interior part of the domain close to the boundary which could be handled by integral transformations as [5] performed contour integration.

Similarly to the local boundary integral equation method [6], the RBIEM [7] implements the BEM equation over circular sub-domains where the source points are placed in the centres of the circles, which eliminates the above mentioned singular integrals. Three equations at each source point for 2D problems are solved, one for the field variable using the direct BEM formulation while the other two equations are the derivatives of the original equation in respect to spatial coordinates at the source point. The field variable and gradients over the circular boundaries are interpolated by using radial basis functions (RBFs). The augmented thin plate spline (ATPS) and fourth order ATPS with second order polynomial augmentation are used in this work. The fundamental solution, Hankel function of the second kind, of the interior 2D Helmholtz problem is used. The matrix coefficients resulting from the integration over the circular boundaries will be same provided that the radius remains the same, therefore they are evaluated only once.

The RBIEM differs from the LBIE in certain aspects [7]: (i) for the solution on the boundary, the LBIE replaces the circular domain with part of the global boundary and the remaining part of the circle, whereas the RBIEM keeps the circular integration approach on the. (ii) the LBIE uses the concept of "companion solution" in order to avoid the gradients/normal derivatives inside the problem domain while the RBIEM solves for the potential and the partial derivatives at each source node including the global boundary of the domain. (iii) The boundary conditions in the RBIEM are directly imposed at the source points on the global boundary, each replacing one of the three equations.

Since the integral domains are always circular, the integrals are regular regardless of the order of the derivative; though the integral kernels for the 2D Helmholtz equation include Hankel functions up to second order, whose evaluation is a challenging task [4, 5].

The method is briefly described in Sections 2 and 3, while numerical examples are presented in Section 4 with the conclusions in Section 5. The details concerning the solution procedures in the RBIEM can be found in [7].

www.witpress.com, ISSN 1743-355X (on-line)

2 The radial basis integral equation method

In this section the radial basis integral equation method (RBIEM) is introduced.

Let us consider the following Helmholtz equation:

$$(\nabla^2 + k^2)u(r) = 0 \tag{1}$$

where $u(r)$ is a potential field, r is a position vector. Given a point r inside a domain Ω, by applying the Green integral formula equation (1) can be transformed into the following integral form:

$$\lambda(r)u(r) + \int_{\Gamma} q^*(r,\xi)u(\xi)d\Gamma_{\xi} - \int_{\Gamma} u^*(r,\xi)q(\xi)d\Gamma_{\xi} = 0 \tag{2}$$

where $u^*(r,\xi)$ is the fundamental solution of the Helmholtz problem, $q(\xi) = \partial u(\xi)/\partial n$ and $q^*(r,\xi) = \partial u^*(r,\xi)/\partial n$. For a 2D problem the fundamental solution is given by

$$u^*(\mathrm{r},\xi) = -\frac{i}{4}H_0^2(kR) \tag{3}$$

$$H_0^2(kR) = J_0(kR) - iY_0(kR) \tag{4}$$

where J_0 and Y_0 are Bessel functions of first and second kind, respectively and R is the distance from the point of application of the concentrated unit source to any other point under consideration, i.e. $R = |r - \xi|$. The derivative of the fundamental solution could be evaluated analytically;

$$q^*(\mathrm{r},\xi) = \frac{\partial R}{\partial n}\frac{ik}{4}H_1^2(kR) \tag{5}$$

The constant $\lambda(r)$ has value from 0 to 1 being equal to $1/2$ for smooth boundaries and 1 if the source point r is inside the domain.

The proposed formulation solves in each interior node three integral equations in order to obtain the potential u, and the partial derivatives $\partial u/\partial x_j$. Equation (6) is used to find the potential while the equations for derivatives $\partial u/\partial x_j$ are obtained by differentiating (2) in respect to x_j, where x_j are components of r. The derivatives of (2) are given below:

$$\frac{\partial u(r)}{\partial x_j} = -\int_{\Gamma_i} \frac{\partial q^*(r,\xi)}{\partial x_j}u(\xi)d\Gamma_{\xi} + \int_{\Gamma_i} \frac{\partial u^*(r,\xi)}{\partial x_j}q(\xi)d\Gamma_{\xi} \tag{6}$$

Note that the derivatives of the fundamental solution and its gradient appearing in the integral Kernels of Eqn. (6) can be evaluated analytically as given below:

www.witpress.com, ISSN 1743-355X (on-line)

$$\frac{\partial u^*(r,\xi)}{\partial x_j} = \frac{\partial R}{\partial x_j}\frac{\partial u^*(r,\xi)}{\partial R} = \frac{\partial R}{\partial x_j}\frac{ik}{4}H_1^2(kR) \tag{7}$$

$$\begin{aligned}\frac{\partial q^*(r,\xi)}{\partial x_j} &= \frac{\partial}{\partial x_j}\left\{\frac{\partial R}{\partial n}\frac{ik}{4}H_1^2(kR)\right\} = \frac{\partial^2 R}{\partial x_j \partial n}\frac{ik}{4}H_1^2(kR) + \frac{\partial R}{\partial n}\frac{\partial R}{\partial x_j}\frac{\partial}{\partial R}\left\{\frac{ik}{4}H_1^2(kR)\right\} \\ &= \frac{\partial^2 R}{\partial x_j \partial n}\frac{ik}{4}H_1^2(kR) + \frac{\partial R}{\partial n}\frac{\partial R}{\partial x_j}\frac{ik}{4}\left\{kH_2^2(kR) - \frac{H_1^2(kR)}{R}\right\}\end{aligned} \tag{8}$$

The discretized form of (6) for the unknown $\partial u/\partial x_j$ at node i is given as:

$$\frac{\partial u_i}{\partial x_j} = -\sum_{k=1}^{J_{bn}}\frac{\partial H_{ik}}{\partial x_j}u_k + \sum_{k=1}^{J_{bn}}\frac{\partial G_{ik}}{\partial x_j}q_k \tag{9}$$

The normal derivative q in (9) can be written as:

$$q = \frac{\partial u}{\partial n} = \nabla u \cdot \vec{n} = \sum_k \frac{\partial u}{\partial x_k}n_k \tag{10}$$

where n_k are components of the unit normal vector. According to (10) equation (2) can be discretized as:

$$u_i + \sum_{j=1}^{J_{bn}} H_{ij}u_j - \sum_m \sum_{j=1}^{J_{bn}} G_{ijm}\frac{\partial u_j}{\partial x_m} = 0 \tag{11}$$

where $G_{ijm} = G_{ij}n_m$. Substituting q from (10) into (9) yields:

$$\frac{\partial u_i}{\partial x_j} = -\sum_{n=1}^{J_{bn}}\frac{\partial H_{in}}{\partial x_j}u_n + \sum_p \sum_{n=1}^{J_{bn}}\frac{\partial G_{inp}}{\partial x_j}\frac{\partial u_n}{\partial x_p} \tag{12}$$

where $G_{inp} = G_{in}n_p$. Equation (11) can be written in matrix form as

$$\mathbf{Hu} - \sum_m \mathbf{G}_m \frac{\partial \mathbf{u}}{\partial x_m} = 0 \tag{13}$$

and equations (12) can be written as:

$$\frac{\partial \mathbf{u}}{\partial x_j} = -\frac{\partial \mathbf{H}}{\partial x_j}\mathbf{u} + \sum_m \frac{\partial \mathbf{G}_m}{\partial x_j}\mathbf{q}_m \tag{14}$$

3 Interpolation for the unknown values at the circular boundary of the sub-domain

In order to perform the integration over the local boundaries of the circular sub-domains, values of the potentials and partial derivatives must be known on the circles. In this work four quadratic continuous elements were employed to perform the integration over the circular boundaries. Eight fictitious nodes were introduced on the circular boundaries in order to define the four quadratic elements used in the integration over the circles. The values of the field variables at the eight nodes were determined through interpolation using the values of field variables at neighbouring nodes. The final system of equations solves for potentials and derivatives only at specified nodes for solution of the problem, which are located at centres of circular sub-domains, and not at the fictitious nodes on the circular boundaries. Only nodes at centres of sub-domains are used in the interpolation for obtaining the values of field variables at fictitious nodes on the circular boundaries. The unknown potential at one of the eight nodes, denoted by ω is approximated by n neighbouring nodes x_i by the following formula:

$$u(\omega)=\sum_{i=1}^{n} f(\omega,x_i)\cdot a_i \tag{15}$$

The n neighbouring nodes can be given as an input to the code, which would in part reduce the meshless nature of this approach, or can be generated automatically as was done in the present case. Two possibilities were considered: (i) to select the interpolation nodes by defining a suitable radius around ω which would contain the desired number of interpolation nodes, or (ii) to define the required number of interpolation nodes n and let the numerical scheme find the nearest n nodes to ω. In this case the option (ii) was adopted since it offered better control over the number of nodes used in the interpolation of the field variables.

Here f is the ATPS function and a_i are the unknown coefficients. The unknown coefficients a_i are determined by constructing a system of equations which is obtained by applying (15) on neighbour nodal points x_i:

$$u(x_j)=\sum_{i=1}^{n} f(x_j,x_i)\cdot a_i \tag{16}$$

The following system of equations is formed

$$\boldsymbol{u}_0=\boldsymbol{F}_0\boldsymbol{a} \tag{17}$$

where $u_0=[u(x_1), u(x_2), \ldots, u(x_n)]^{\mathrm{T}}$, $F_0=f_{ji}=f(x_j,x_i)$, $j=1,2,\ldots,n$; $i=1,2,\ldots,n$.

The unknown coefficients a_i are determined by:

$$\boldsymbol{a}=\boldsymbol{F}_0^{-1}\boldsymbol{u}_0 \tag{18}$$

The potential at point ω can be written as

$$u(\omega)=F(\omega,x_i)F_0^{-1}u_0 \tag{19}$$

where $F(\omega,x_i)=[f(\omega,x_1), f(\omega,x_2),\ldots,f(\omega,x_n)]$.

The partial derivatives at ω are obtained as:

$$\frac{\partial u(\omega)}{\partial x_l}=F(\omega,x_i)F_0^{-1}q_{0l} \tag{20}$$

where $q_{0l}=[q_l(x_1), q_l(x_2),\ldots, q_l(x_n)]^{\mathrm{T}}$ and $q_l=\partial u/\partial x_l$.

Equation (20) produces more accurate approximation of the partial derivatives.

www.witpress.com, ISSN 1743-355X (on-line)

4 Numerical results

4.1 Solution of the helmholtz equation

The Helmholtz Eqn. is used in order to verify the validity of the approach. A square domain with length *L=1* at each side is considered. The $\partial\Phi/\partial x$ and $\partial\Phi/\partial y$ approach zero on the lines x=0 and y=0, therefore the domain is shifted (0.5, 0.5) from the origin for a better visualisation of relative error. The following boundary conditions are applied:

$$\Phi\big|_{x=1.5} = J_0(kr) + i(\cos(1.5k) + \sin(ky)) \tag{21}$$

$$\Phi\big|_{y=1.5} = J_0(kr) + i(\cos(kx) + \sin(1.5k)) \tag{22}$$

$$\left.\frac{\partial\Phi}{\partial n}\right|_{x=0.5} = \frac{k}{2}J_1(kr)/r + i\,k\sin(\frac{k}{2}) \tag{23}$$

$$\left.\frac{\partial\Phi}{\partial n}\right|_{y=0.5} = \frac{k}{2}J_1(kr)/r - i\,k\cos(\frac{k}{2}) \tag{24}$$

which yield the solution $\Phi = J_0(kr) + i(\cos(kx) + \sin(ky))$.

The ATPS is used in all of the RBF computations. The built-in Fortran functions *getri* and solver *pardiso* are used to compute the inverse matrix formed in RBF interpolation and to solve the overall sparse system, respectively. The models of sizes up to 361201 source nodes in the domain with uniform distribution, equivalent to 1,083,603 degrees of freedom (DOF), have been solved on a workstation Intel Xeon 3.2 GHz.

4.1.1 The parameters that affect the efficiency of the method

In this section, we investigate the parameters that affect the efficiency and accuracy of the algorithm. Although the results obtained for $\Phi = J_0(kr) + i(\cos(kx) + \sin(ky))$ are presented here, these optimal values show the same behaviour when tested individually with the functions *Re*(Φ) and *Im*(Φ). The effect of two major components: (i) the RBF approximation and (ii) circular integration around nodes, on the numerical error is analyzed. Keeping the number of the nodes used in the computations sufficiently large, numerical experiments were performed in order to determine more optimal values for each of the parameters. As the base case values we set *Nodes*=10201, N_ω=24, N_d=16, N_b=12, $R_d = D_x$ *and* $R_b = D_x$. The results are displayed by examining the L2-norms for the components *Re*(Φ), *Re*($\partial\Phi/\partial x$), *Re*($\partial\Phi/\partial y$), *Im*(Φ), *Im*($\partial\Phi/\partial x$) and *Im*($\partial\Phi/\partial y$), where *k* is wavenumber; λ is wavelength; N_ω is the number of fictitious points used in the circular integration around each node; R_d is the radius of the circular sub-domain in the

domain; R_b is the radius of the circular sub-domain on the boundary; N_d is the number of points used in RBF interpolation in the domain; N_b is the number of points used in RBF interpolation on the boundary; D_x is the distance between nodes in x and y directions in case of uniform distribution.
The L2-norm error for Φ is defined as

$$e_{L2} = \sqrt{\frac{\sum_{i=1}^{N}(\Phi_i - \Phi_i^n)^2}{\sum_{i=1}^{N}\Phi_i^2}} 100\% \qquad (25)$$

where N is the number of nodes, Φ_i^n the potential at node i by numerical method and Φ_i the analytical solution at node i.

Firstly, we investigated the parameters related to the circular integration around nodes. Tests were carried out in order to determine the optimal integration radius R by prescribing analytical solution values of both potential and the derivatives on the circular boundary. We conclude that a ratio $R_d/\lambda \leq 0.01$ is required for accurate results in the computations. Next, the influence of ratio R_d/D_x on the accuracy of the solution has been investigated. The numerical experiments suggested a value of $R_d/D_x = 1$ as one leading to accurate results, as can be seen in Figure 1.

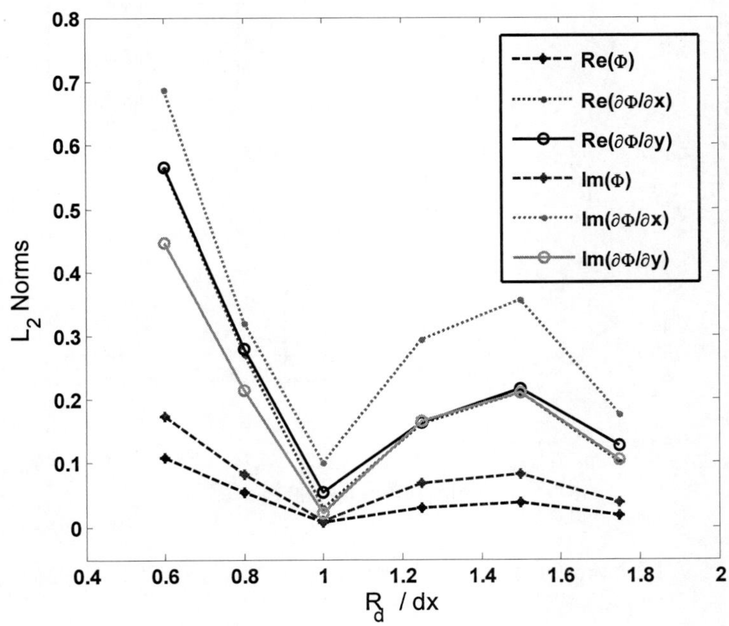

Figure 1: L2 Norm error for different R_d/D_x with *Nodes*=10201, N_ω=24, N_d=16, N_b=12 and $R_b = D_x$.

The number of neighbouring interpolation points used in the RBF has been increased up to 48, which is sufficiently large. The results show slight improvement up to this point, whereas numerical error dominates afterwards. The results reach a nearly stable behaviour for $N_d \geq 28$ (Fig. 2). Nevertheless, a choice of 16 gives the best accuracy which is also preferable to the large numbers due to the significant reduction in the sparsity of the overall matrix and computation time in that regard.

The behaviour of the method is investigated near the boundaries by varying the integration radius R_b and the number of nodes in the RBF for the nodes on the boundary, N_b, close to the boundary or far inside the domain. Note that keeping the same value would decrease the computation time since the circular integration results in the same coefficients for fixed radius and number of boundary elements used. Reducing the R_b and N_b show slightly more accurate results, hence related figures are not shown here.

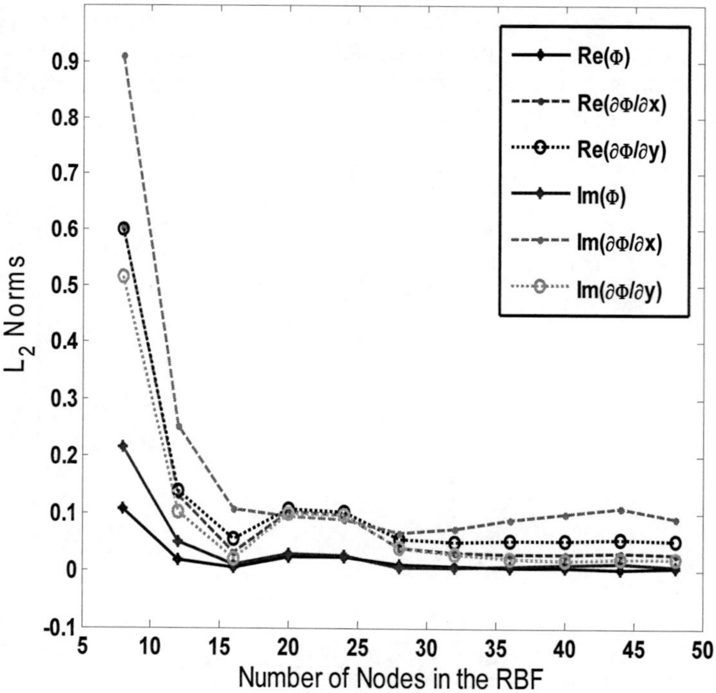

Figure 2: L2 Norm error for different N_d with *Nodes*=10201, N_ω=24, $R_d = D_x$ and $R_b = D_x$.

The overall convergence of the method is stated clearly by the Figure 3 where the computations are carried with up to 360201 nodes.

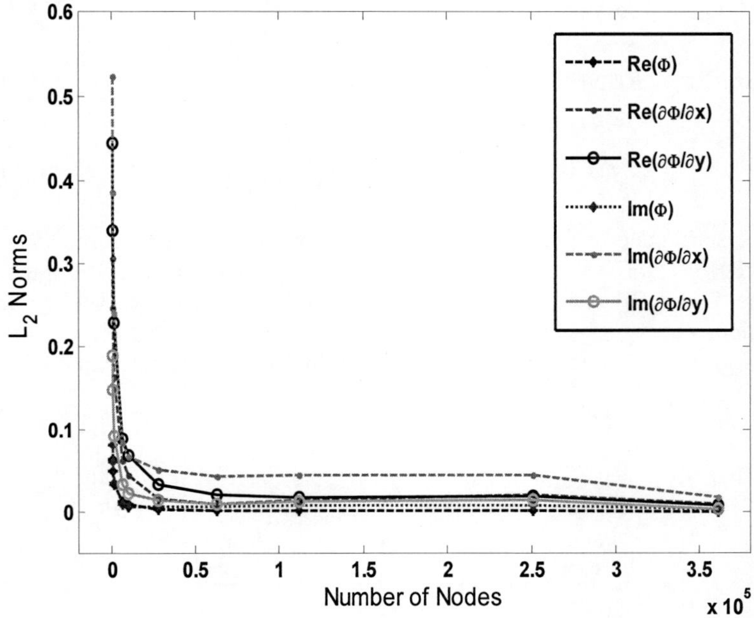

Figure 3: L2 Norm convergence with increasing the number of Nodes, $N_\omega=24$, $N_{d=}16$, $N_b=12$, $R_d= D_x$ and $R_b= D_x$.

4.2 Example 2

Another solution to the Helmholtz equation is used in order to test the effect of the RBFs. A square domain with length $L=1$ at each side is considered. The following boundary conditions are applied:

$$\Phi\Big|_{y=0.5} = Sin\left(\frac{\sqrt{2}}{2}kx\right)Sin\left(\frac{\sqrt{2}}{4}k\right) \tag{26}$$

$$\Phi\Big|_{y=-0.5} = Sin\left(\frac{\sqrt{2}}{2}kx\right)Sin\left(-\frac{\sqrt{2}}{4}k\right) \tag{27}$$

$$\frac{\partial\Phi}{\partial n}\bigg|_{x=0.5} = kCos\left(\frac{\sqrt{2}}{4}k\right)Sin\left(\frac{\sqrt{2}}{2}ky\right) \tag{28}$$

$$\frac{\partial\Phi}{\partial n}\bigg|_{x=-0.5} = kCos\left(-\frac{\sqrt{2}}{4}k\right)Sin\left(\frac{\sqrt{2}}{2}ky\right) \tag{29}$$

www.witpress.com, ISSN 1743-355X (on-line)

which yield the solution $\Phi = \text{Sin}\left(\frac{\sqrt{2}}{2}kx\right)\ \text{Sin}\left(\frac{\sqrt{2}}{2}ky\right)$.

Two different RBFs are used:

$$\text{(i)}\ f_1(R) = R^2\ln(R) + x + y + 1 \tag{30}$$

$$\text{(ii)}\ f_2(R) = R^4\ln(R) + x^2 + y^2 + xy + x + 1 \tag{31}$$

The tests related to the best accuracy of the method were performed for the second RBF, f_2, as well. The optimum parameters were found to be the same with the first RBF except the number of fictitious nodes on each circle of integration, e.g. N_ω=24, N_{d}=16, N_b=12 *and* $R_b = D_x$. For the sake of brevity, related figures are not re-plotted here. As seen in Table 1, the solution converges

Table 1: L2 Norm error comparison for f_1 and f_2 RBFs with k=6.

f_1	f_2	
***Re*(Φ)** **L2 Norm**	***Re*(Φ)** **L2 Norm**	**Number of Nodes**
2.4937	0.1624	441
0.7906	0.0348	1681
0.2887	0.0024	6561
0.1849	0.0033	10201
0.0237	0.0071	40401
0.0149	0.0075	63001
0.0280	0.0081	160801
0.0624	0.0078	251001

Table 2: L2 Norm error for f_2 RBFs with high wavenumbers.

k	***Re*(Φ)** **(L2 Norm)**	***Re*(∂Φ/∂x)** **(L2 Norm)**	***Re*(∂Φ/∂x)** **(L2 Norm)**	**Nodes**
20	0.0156	0.0242	0.0207	63001
36.2	0.0557	0.0489	0.0536	251001
43.18	0.0543	0.0515	0.0480	361201
50.02	0.0457	0.0541	0.0506	361201
61.36	0.0573	0.0554	0.0543	361201
72.05	0.0962	0.0989	0.1039	361201
78.29	0.2180	0.2163	0.2156	361201

more rapidly using f_2 in comparison to f_1. Therefore, the usage of f_2 allows us to solve the equation with higher wavenumbers. In Table 1, we can also notice the phenomena related to the optimum choice of R_d/λ, namely a further increase of the source nodes after an accurate solution is reached yields the so called pollution effect. Finally in Table 2, the results are provided with the high wavenumbers for the analytical solution $\mathrm{Sin}\left((\sqrt{2}/2)kx\right)$ $\mathrm{Sin}\left((\sqrt{2}/2)ky\right)$. The results are acceptable up to $k=78.29$. The resonant frequencies were avoided as much as possible while choosing the wavenumbers.

5 Conclusions

A meshless method for the solution of 2D Helmholtz equation has been developed by using the Boundary Integral Equation (BIE) combined with Radial Basis Function (RBF) interpolations. BIE is applied by using the fundamental solution of the Helmholtz equation, therefore domain integrals are not encountered in the method. The method exploits the advantage of placing the source point always in the centre of circular sub-domains in order to avoid singular or near-singular integrals. RBF interpolation is applied in order to obtain the values of the field variable and partial derivatives at the boundary of the circular sub-domains, providing this way the boundary conditions for solution of the integral equations at the nodes (centres of circles). The accuracy, robustness and efficiency of the method have been validated on some analytical solutions of the problem. The parameters emerging from the BIE and RBF, such as radius of circular integration, number of neighbouring nodes in the RBF interpolation, choice of RBF etc., have been tested to certain extent to determine more optimal values which yield to more accurate solutions. In the placement of source nodes, a distance of 0.01-0.02 of the wavelength between the nodes is essential to catch the wave behaviour precisely. An integration radius equal to this distance is found to give more accurate results. Among the two RBFs used, $f_2(R) = R^4 \ln(R) + x^2 + y^2 + xy + x + 1$ *has shown more rapid* convergence and accuracy than $f_1(R) = R^2 \ln(R) + x + y + 1$.

Acknowledgement

The present study was supported by the SONO project, contract number: 228730, as part of the Seventh Framework Programme (FP7-NMP-2008-Large-2).

References

[1] Wenderodt C, v. Estorff, O. Dispersion analysis of the meshfree radial point interpolation method for the Helmholtz equation. *Int. J. For Numer. Methods In Engineering* 2009; 77: 1670-1689.

[2] Bouillar Ph., Suleau S. Element-Free Galerkin solutions for Helmholtz problems: formulation and numerical assessment of the pollution effect. *Compu. Methods Appl. Mech. Eng.* 1998; 162: 317-335.
[3] Miao Y., Wang Y., Wang Y. H. A meshless hybrid boundary-node method for Helmholtz problems. */Engineering Analysis with Boundary Elements/* 2009; 33: 120–127.
[4] Tadeu A. J. B., Santos P. F. A., Kausel E. Closed-form integration of singular terms for constant, linear and quadratic boundary elements. Part 2. SV-P wave propagation. *Engineering Analysis with Boundary Elements* 1999; 23: 757-768.
[5] Elliot D., Johnston P. R. Gauss-Legendre quadrature for the evaluation of integrals involving the Hankel function. *Jour. Of Computational and Applied Mathematics* 2008; 211: 23–35.
[6] Zhu T, Zhang J-D, Atluri S.N. A local boundary integral equation (LBIE) method in computational mechanics, and a meshless discretization approach. *Computational Mechanics* 1988; 21: 223-235.
[7] Popov V., Bui T. T. A meshless solution to two-dimensional convection-diffusion problems. *Engineering Analysis with Boundary Elements*, 2010; 34; 680-689.

Meshless computation for partial differential equations of fractional order

P. H. Wen[1] & Y. C. Hon[2]
[1]*School of Engineering and Material Science, University of London, UK*
[2]*Department of Mathematics, City University of Hong Kong, Hong Kong SAR, China*

Abstract

Fractional advection-dispersion equations are used in groundwater hydrology to model the transport of passive tracers carried by fluid flow in a porous medium. In this paper, the numerical properties of partial differential equations of fractional order $1 \le \alpha \le 2$ are investigated by the use of a radial basis function interpolation scheme. The differential equations of fractional order are first analyzed in the Laplace transformed domain and the Durbin inversion method is then used to determine the solutions in the time domain. The accuracy and stability of these methods are investigated for several standard types of problems involving partial differential equations of fractional order.
Keywords: fractional derivatives, partial differential equations, radial basis function, Laplace transform, Durbin algorithm.

1 Introduction

Analysis of the diffusion-wave equation in mathematical physics has been of considerable interest in the literature. Ordinary and partial differential equations of fractional order have been the focus of many studies because of their frequent appearance in various applications in fluid mechanics, viscoelasticity, biology, physics, and engineering. Fractional calculus in mathematics is a natural extension of integer-order calculus and gives a useful mathematical tool for modeling many processes in nature. One of these processes, in which fractional derivatives have been successfully applied, is called diffusion [1]. Fractional derivatives have recently been applied to many problems in physics [2–8], finance [9, 10], and hydrology [11]. Fractional space derivatives are used to

www.witpress.com, ISSN 1743-355X (on-line)
doi:10.2495/BE110291

model anomalous diffusion or dispersion, where a particle plume spreads at a rate inconsistent with the classical Brownian motion model. When a fractional derivative replaces the second derivative in a diffusion or dispersion model, it leads to enhanced diffusion (also called super diffusion). For one dimensional advection–dispersion model with constant coefficients, analytical solutions are available using Fourier transform methods [12, 13]. Many practical problems, however, are formulated with variable coefficients [14] whose analytical solutions are still not available.

In the last decade, the radial basis functions (RBFs) have been under intensive research in the areas of multivariate function interpolation and partial differential equations solver [15, 16]. In this paper we propose a numerical scheme by the use of RBFs for solving the ordinary and partial differential equations of fractional order as follow:

Consider a simple form for transport equations

$$D^{\alpha}(u) + A(x)\frac{\partial u}{\partial t} + B(x)\frac{\partial^2 u}{\partial t^2} + C(x)u = f(x,t), \quad 0 < t, \ 0 \le x \le 1 \quad (1)$$

where the coefficients $A(x)$, $B(x)$ and the source team $f(x,t)$ are given. In (1), D^{α} denotes a fractional derivative operator defined as

$$D^{\alpha}(u) = \frac{1}{\Gamma(m-\alpha)}\frac{\partial^m}{\partial x^m}\int_0^x \frac{u(s)ds}{(x-s)^{\alpha-m+1}} \quad (2)$$

where $\alpha > 0$ and m is an integer such that $m > \alpha > m-1$. The boundary conditions and initial conditions are given respectively by

$$\begin{aligned} &u(x,0) = g_1(x), \dot{u}(x,0) = g_2(x) \\ &u(0,t) = 0 \\ &u_{,x}(0,t) = 0 \end{aligned} \quad (3)$$

or

$$\begin{aligned} &u(x,0) = g_1(x), \dot{u}(x,0) = g_2(x) \\ &u(0,t) = 0 \\ &u(1,t) = h(t). \end{aligned} \quad (4)$$

For simplicity, we only consider fractional derivatives in space and the index α is assumed to lie within the range $1 \le \alpha \le 2$ and therefore $m = 2$. By applying the Laplace transformation on equations (1) and (3, 4) with consideration of initial condition, we obtain

$$\begin{aligned} &D^{\alpha}(\tilde{u}) + A(x)[p\tilde{u} - g_1(x)] + B(x)[p^2\tilde{u} - pg_1(x) - g_2(x)] + C(x)\tilde{u} \\ &= \tilde{f}(x), 0 < t, \ 0 \le x \le 1 \end{aligned} \quad (5)$$

with the boundary condition

$$\tilde{u}(0)=0,\ \tilde{u}_{,x}(0)=0 \ \text{ or } \ \tilde{u}(0)=0,\ \tilde{u}(1)=\tilde{h}\,. \tag{6}$$

We note here that although there are several numerical methods developed to solve differential equations of fractional order, there is little information on the numerical solution of partial differential equations of fractional order.

This paper is organized as follows. In Section 2, radial basis function interpolation scheme is introduced. The numerical schemes for the solutions of (1) are then derived in Section 3. In Section 4 several numerical tests and solutions of the transport equilibrium equation are given for the verification of the accuracy and efficiency of the proposed numerical schemes. Finally, conclusion is given in Section 5.

2 Approximation scheme using radial basis functions

RBFs have been initially used for scattered data fitting and general multi-dimensional data interpolation problems, see Ref. [15], and were later applied by Kansa [17, 18] for the numerical approximations of various types PDEs. For one dimensional problem, this interpolation becomes very simple. In order to guarantee unique solution of the interpolation problem, a polynomial term should be added to the interpolation. Thus, the interpolated distribution of function u at the point x can be expressed by

$$u(x)=\sum_{k=1}^{n}R_k(x,x_k)a_k+\sum_{j=1}^{t}P_j(x)b_j=\mathbf{R}(x)\mathbf{a}+\mathbf{P}(x)\mathbf{b} \tag{7}$$

along with the constraints

$$\sum_{k=1}^{n}P_j(x_k)a_k=0,\qquad 1\le j\le q \tag{8}$$

where the vector $\mathrm{R}(x)=\{R_1(x,x_1),R_2(x,x_2),...,R_n(x,x_n)\}$ is the set of radial basis functions at point x, $\mathrm{P}(x)=\{P_1(x),P_2(x),...,P_q(x)\}$ is the set of polynomial functions, $\{x_k\}_1^n$ is the set of distributed collocation points in the range $0\le x\le 1$, $P_j=x^{j-1}$. Two unknown vectors $\mathrm{a}=\{a_1,a_2,...,a_n\}^T$ and $\mathrm{b}=\{b_1,b_2,...,b_t\}^T$ are to be determined by nodal values respectively. A set of linear equations to determine the coefficients **a** and **b** can be written, in matrix form, as

$$\mathbf{R}_0\mathbf{a}+\mathbf{P}_0\mathbf{b}=\mathbf{u},\qquad \mathbf{P}_0^T\mathbf{a}=\mathbf{0} \tag{9}$$

with nodal values $\mathrm{u}=\{u_1,u_2,...,u_n\}^T$ at collocation points $\{x_k\}_{k=1}^n$ and matrices

$$\mathbf{R}_0 = \begin{bmatrix} R_1(x_1) & R_2(x_1) & \ldots & R_n(x_1) \\ R_1(x_2) & R_2(x_2) & \ldots & R_n(x_2) \\ . & . & \ldots & . \\ . & . & \ldots & . \\ . & . & \ldots & . \\ R_1(x_n) & R_2(x_n) & \ldots & R_n(x_n) \end{bmatrix},$$

$$\mathbf{P}_0 = \begin{bmatrix} P_1(x_1) & P_2(x_1) & \ldots & P_t(x_1) \\ P_1(x_2) & P_2(x_2) & \ldots & P_t(x_2) \\ . & . & \ldots & . \\ . & . & \ldots & . \\ . & . & \ldots & . \\ P_1(x_n) & P_2(x_n) & \ldots & P_t(x_n) \end{bmatrix}. \tag{10}$$

Solving these equations given in (9) gives

$$\begin{aligned} \mathbf{b} &= \left(\mathbf{P}_0^T \mathbf{R}_0^{-1} \mathbf{P}_0\right)^{-1} \mathbf{P}_0^T \mathbf{R}_0^{-1} \mathbf{u} = \mathbf{Bu}, \\ \mathbf{a} &= \mathbf{R}_0^{-1} \left[\mathbf{I} - \mathbf{P}_0 \left(\mathbf{P}_0^T \mathbf{R}_0^{-1} \mathbf{P}_0\right)^{-1} \mathbf{P}_0^T \mathbf{R}_0^{-1} \right] \mathbf{u} = \mathbf{Au} \end{aligned} \tag{11}$$

where **I** denotes the diagonal unit matrix. In this paper, the radial basis function is selected to be the infinitely smooth multiquadrics:

$$R_k(x,x_k) = \sqrt{c^2 + (x - x_k)^2} \tag{12}$$

where c is a free parameter. Finally the approximated variable can be presented as

$$u(x) = [\mathbf{R}(x)\mathbf{A} + \mathbf{P}(x)\mathbf{B}]\mathbf{u} = \sum_{k=1}^{n} N_k(x) u_k = \mathbf{Nu} \tag{13}$$

where $N_k(x)$ is called shape function. From the expression (13), it can be seen that the partial derivatives of the unknown function can be obtained by simply evaluating the partial derivatives of the shape functions. From (13), we have

$$u' = u_{,x} = \left[\mathbf{R}_{,x}(x)\mathbf{A} + \mathbf{P}_{,x}(x)\mathbf{B}\right] \mathbf{u} = \overline{\mathbf{N}}\mathbf{u} \quad \text{and}$$

$$u'' = u_{,xx} = \left[\mathbf{R}_{,xx}(x)\mathbf{A} + \mathbf{P}_{,xx}(x)\mathbf{B}\right] \mathbf{u} = \overline{\overline{\mathbf{N}}}\mathbf{u} \tag{14}$$

where

$$R_{k,x}(x) = \frac{x - x_k}{\sqrt{c^2 + (x - x_k)^2}} \text{ and } P_{j,x}(x) = (j-1)x^{j-2} \tag{15}$$

For second order derivative of the shape function, we have

$$R_{k,xx}(x) = \frac{c^2}{\sqrt{c^2 + (x - x_k)^2}} \text{ and } P_{j,xx}(x) = (j-1)(j-2)x^{j-3}. \tag{16}$$

3 Numerical scheme of fractional derivatives

The space variable x is defined to be the unit length interval in (5). By simple integration by part, the fractional derivative can be written as

$$\int_0^x \frac{\tilde{u}(s)ds}{(x-s)^{\alpha-1}} = -\tilde{u}(0)x^{\alpha} + \frac{\tilde{u}'(0)}{\alpha(\alpha+1)}x^{\alpha+1} + \frac{1}{\alpha(\alpha+1)}\int_0^x (x-s)^{\alpha+1}\tilde{u}''(s)ds \tag{17}$$

Therefore, the differential equation with fractional order (5) becomes

$$\frac{1}{\Gamma(2-\alpha)}\int_0^x \frac{\tilde{u}''(s)ds}{(x-s)^{\alpha-1}} + p[A(x) + pB(x)]\tilde{u}$$

$$+ C(x)\tilde{u} + \frac{\tilde{u}(0)}{\Gamma(1-\alpha)}x^{-\alpha} + \frac{\tilde{u}_{,x}(0)}{\Gamma(2-\alpha)}x^{1-\alpha} = G(x) \tag{18}$$

$$G(x) = \tilde{f}(x) + A(x)g_1(x) + B(x)[pg_1(x) + g_2(x)].$$

It is worthy to notice that there are singularities in (18) at origin if $1 \le \alpha \le 2$. It may cause troubles for numerical calculation. Normally the boundary condition $\tilde{u}(0) = \tilde{u}_{,x}(0) = 0$ is assumed [19]. Substituting (13) (14) into (18) results

$$\frac{1}{\Gamma(2-\alpha)}\int_0^x \frac{\overline{\overline{\mathbf{N}}}(s)\tilde{\mathbf{u}}ds}{(x-s)^{\alpha-1}} + [pA(x) + p^2B(x) + C(x)]\mathbf{N}(x)\tilde{\mathbf{u}}$$

$$+ \frac{\mathbf{N}(0)\tilde{\mathbf{u}}}{\Gamma(1-\alpha)}x^{-\alpha} + \frac{\overline{\mathbf{N}}(0)\tilde{\mathbf{u}}}{\Gamma(2-\alpha)}x^{1-\alpha} = G(x) \tag{19a}$$

or

$$\left\{\begin{array}{l}\dfrac{1}{\Gamma(2-\alpha)}\displaystyle\int_0^x \frac{\overline{\overline{\mathbf{N}}}(s)ds}{(x-s)^{\alpha-1}}+[pA(x)+p^2B(x)+C(x)]\mathbf{N}(x) \\ +\dfrac{\mathbf{N}(0)}{\Gamma(1-\alpha)}x^{-\alpha}+\dfrac{\overline{\mathbf{N}}(0)}{\Gamma(2-\alpha)}x^{1-\alpha}\end{array}\right\}\tilde{\mathbf{u}}=G(x)\,. \quad (19b)$$

In this paper, we set up a grid of points, $x_k=(k-1)/N$, $\mathrm{k}=1,2,...,n$, $n=N+1$. For each collocation point except the end points x_1 and x_{N+1}, we have

$$\left\{\begin{array}{l}\dfrac{1}{\Gamma(2-\alpha)}\displaystyle\int_0^{x_k} \frac{\overline{\overline{\mathbf{N}}}(s)ds}{(x_k-s)^{\alpha-1}}+[pA(x_k)+p^2B(x_k)+C(x_k)]\mathbf{N}(x_k) \\ +\dfrac{\mathbf{N}(0)}{\Gamma(1-\alpha)}x_k^{-\alpha}+\dfrac{\overline{\mathbf{N}}(0)}{\Gamma(2-\alpha)}x_k^{1-\alpha}\end{array}\right\}\tilde{\mathbf{u}}=G(x_k) \quad (20)$$

for $k=2,3,...,N$, with the boundary conditions

$$\tilde{u}(0)=\tilde{u}_1=0,\ \tilde{u}_{,x}(0)=\sum_{k=1}^{n}\overline{N}_k(0)\tilde{u}_k=0 \ \text{ or } \ \tilde{u}(0)=\tilde{u}_1=0,\ \tilde{u}(1)=\tilde{u}_n=\tilde{h}\,. \quad (21)$$

Thus, we have obtained a complete system of algebraic equations for the computation of (N+1) nodal unknowns ũ in the Laplace transform domain. As there is weak singularity in the integral of (10), we can transfer the integral into a regular integral as

$$\int_0^{x_k}\frac{F(s)ds}{(x_k-s)^{\alpha-1}}=\frac{1}{2-\alpha}\int_0^{x_k^{2-\alpha}}F(x_k-\lambda^{1/(2-\alpha)})d\lambda \quad (22)$$

whose integration can be performed by any standard integral scheme.

There are many inversion methods available for numerical computing the inverse Laplace transform. Here, the method proposed by Durbin [20] is adopted. A series of samples (L+1) in the transformation space p_l, $l=0,1,2,...,L,$ are selected as complex variable. Transformed variables are evaluated for these specified transform parameters respectively. The physical variables in the time domain can then be determined by the Laplace inversion technique. Demonstration of the Durbin's inverse method was made by Wen et al [21, 22] for the elasticity wave propagations for two and three dimensional problems. The formula of inversion used is written as

$$F(t)=\frac{2e^{\omega t}}{T}\left[-\frac{1}{2}\tilde{F}(p_0)+\sum_{l=0}^{L}\mathrm{Re}\left\{\tilde{F}(p_l)e^{2l\pi it/T}\right\}\right] \quad (23)$$

where $\tilde{F}(p_l)$ denotes the transformed variable in the Laplace domain and $i=\sqrt{-1}$. The parameter of the Laplace transform is taken to be $s_l=\omega+2l\pi i/T$. The selection of two free parameters ω and T has small effect on the accuracy of inversion. By large number of numerical tests, we can conclude that Durbin inversion method is stable and convergent for large range selection of free parameters ω and T particularly for the solid dynamic mechanics.

4 Numerical verification

In the following test cases, we assume that $1 \le \alpha \le 2, m = 2$. For numerical error estimation, we define the relative average error by the following formula

$$MEU(x) = \frac{1}{Nu^*_{\max}} \sum_{i=1}^{M} \left| u(x,t_i) - u^*(x,t_i) \right| \tag{24}$$

where M is the number of time observations at collocation point x and u^* denotes the analytical solutions at time t_i in the region $[0, t_{\max}]$. The free parameters is chosen to be $\omega = 5/T$ and T=20. To test the numerical schemes, it is important to use simple analytical models. Hence, in the first example we consider the case of evolution of a density profile to an equilibrium solution [19]. The equilibrium density profile is the solution of the equation,

$$D^{\alpha}(u) = f(x), \quad 0 \le x \le 1. \tag{25}$$

In order to have a simple solution, we use a simple form for the source function

$$f(x) = 1 - x^{2-\alpha}. \tag{26}$$

Along with the boundary condition $u(0) = 0,\ u'(0) = 0$. In this case, the analytical solution for any value of $\alpha > 1$ is given by

$$u^*(x) = \frac{x^{\alpha}}{\Gamma(1+\alpha)} - \frac{\Gamma(3-\alpha)x^2}{2}. \tag{27}$$

Numerical results are listed in Table 1 when fractional order $\alpha = 1.5$. It is apparent that the accuracy depends on the selection of free parameter c. For small c, we are able to obtain high accurate results even with small number of collocation points. However, the accuracy does not increase with the increasing of collocation number. In addition, the divergence of solution will occur when $c = 10/n$ and $n > 31$ even when double precision of variable is used. This fact is well known in the use of radial basis functions for multivariate interpolation and PDE solver. Therefore, in the following tests, parameter c is chosen to be

Table 1: Relative error in (23) for the first example

$$u^*_{\max} = 1/\Gamma(1+\alpha) - \Gamma(3-\alpha)/2.$$

c	$n=11$	$n=21$	$n=31$	$n=41$
$0.1/n$	9.0454×10^{-4}	4.0698×10^{-4}	2.6690×10^{-4}	1.9813x10-4
8	6.9659×10^{-4}	2.4753×10^{-4}	1.3557×10^{-4}	8.7012x10-5
$10/n$	1.9478×10^{-4}	1.2055×10^{-4}	4.8050×10^{-1}	5.8850x10-1

$c = 1/n$. In the second example, we consider the following differential equation with non-zero boundary condition at origin

$$D^{\alpha}(u) + u(x) + \frac{\partial^2 u}{\partial x^2} = f(x), \quad 0 \le x \le 1 \tag{28}$$

along with the boundary condition $u(0) = 0$, $u(1) = 1$ and source term

$$f(x) = \frac{x^{2-\alpha}}{\Gamma(3-\alpha)} + x^2 + 2. \tag{29}$$

The analytical solution is $u(x) = x^2$. Numerical results are presented in Table 2 for different fractional order α.

Table 2: Relative error for the second example, $u^*_{\max} = 1$.

	$n=5$	$n=11$	$n=21$
1.001	9.7484×10^{-8}	2.3098×10^{-8}	8.7972×10^{-8}
1.1	1.8599×10^{-8}	2.1025×10^{-8}	1.7394×10^{-8}
1.5	6.5070×10^{-9}	1.2149×10^{-8}	3.4291×10^{-9}
1.9	2.2595×10^{-8}	2.4772×10^{-8}	9.4626×10^{-7}
1.999	1.6621×10^{-7}	2.2757×10^{-6}	3.4531×10^{-7}

For different fractional order, same order of accuracy can be achieved even with different number of collocation points. Apparently this numerical scheme is of high accuracy and convergent to solve partial differential equation of fractional order. In the final example, we consider the following differential equation of fractional order with time dependence as

$$D^{\alpha}(u) - e^{x}\left(\frac{\partial u}{\partial t} + \frac{\partial^2 u}{\partial t^2}\right) = f(x,t),$$
$$f(x,t) = (t + e^{x})e^{-t}/\Gamma(1+\alpha), \;\; 0 < t, \; 0 \le x \le 1 \tag{30}$$

$$u(x,0) = 0, \; \frac{\partial u(x,0)}{\partial t} = x^{\alpha} \tag{31a}$$

$$u(0,t) = 0, \; u'(0,t) = 0. \tag{31b}$$

The analytical solution is determined by $u^{*}(x,t) = x^{\alpha} t e^{-t}$. The numerical solutions by using the proposed scheme are plotted in Figure 1 for $0 \le t \le 5$. In this case fractional order is fixed to be $\alpha = 1.5$. Two curves are plotted which correspond to the numbers of collocation point $n = 21$ and 41 respectively.

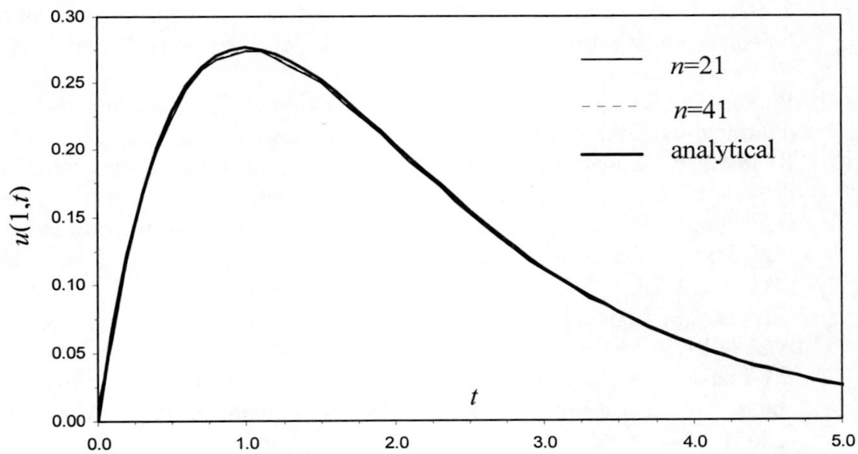

Figure 1: Variation of $u(1,t)$ against time t.

5 Conclusion

In this paper we consider a partial differential equation of fractional order α with variable coefficient and propose a new numerical scheme using radial basis function interpolation. For zero boundary conditions, $u(0,t) = u'(0,t) = 0$, the singularities are vanished in the differential equation and accurate numerical solutions can be obtained. Although this is not required, for simplicity, all collocation points are chosen to be uniformly distributed in the range $0 \le x \le 1$. The Laplace transform technique with Durbin inversion method for time variable is applied for solving the time dependent problems. Compared with analytical

www.witpress.com, ISSN 1743-355X (on-line)

solutions, excellent agreement with the proposed numerical scheme is obtained. We can conclude with following observations: (1) The radial basis function interpolation is suitable to solve partial differential equation of fractional order; (2) The time dependent problem can be dealt with in the Laplace transform domain; (3) Numerical solutions are not sensitive with the selections of collocation point number n; (4) Nonlinear problems will be investigated in separate paper.

Acknowledgement

The work described in this paper was fully supported by a grant from the Research Grants Council of the Hong Kong Special Administrative Region, China (Project No. CityU 101310).

References

[1] Ciesielski M. and Leszczynsky J., Numerical treatment of an initial-boundary value problem for fractional partial differential equations, *Signal Process* 86 (2006), 2619–2631.

[2] Barkai E., Metzler R., Klafter J., From continuous time random walks to the fractional Fokker–Planck equation, *Phys. Rev.* E 61 (2000) 132–138.

[3] Blumen A., Zumofen G., Klafter J., Transport aspects in anomalous diffusion: Levy walks, *Phys. Rev.* A 40 (1989) 3964–3973.

[4] Bouchaud J. P., Georges A., Anomalous diffusion in disordered media—statistical mechanisms, models and physical applications, *Phys. Rep.* 195 (1990) 127–293.

[5] Chaves A., Fractional diffusion equation to describe Levy lights, *Phys. Lett.* A 239 (1998) 13–16.

[6] Meerschaert M. M., Benson D., ScheXer H. P., Baeumer B., Stochastic solution of space-time fractional diffusion equations, *Phys. Rev.* E 65 (2002) 1103–1106.

[7] Meerschaert M.M., ScheXer H. P., Semistable Levy Motion, *Frac. Calc. Appl. Anal.* 5 (2002) 27–54.

[8] Zaslavsky G., Fractional kinetic equation for Hamiltonian chaos. Chaotic advection, tracer dynamics and turbulent dispersion, *Phys.* D 76 (1994) 110–122.

[9] Raberto M., Scalas E., Mainardi F., Waiting-times and returns in high-frequency financial data: an empirical study, *Physica* A 314 (2002) 749–755.

[10] Sabatelli L., Keating S., Dudley J., Richmond P., Waiting time distributions in financial markets, *Eur. Phys. J.* B 27 (2002) 273–275.

[11] Baeumer B., Meerschaert M. M., Benson D., Wheatcraft S. W., Subordinated advection–dispersion equation for contaminant transport, *Water Resour. Res.* 37 (2001) 1543–1550.

[12] Benson D., Wheatcraft S., Meerschaert M. M., Application of a fractional advection–dispersion equation, *Water Resour. Res.* 36 (2000) 1403–1412.

[13] Chaves A., Fractional diffusion equation to describe Levy lights, *Phys. Lett.* A 239 (1998) 13–16.
[14] Barkai E., Metzler R., Klafter J., From continuous time random walks to the fractional Fokker–Planck equation, *Phys. Rev.* E 61 (2000) 132–138.
[15] Hardy R. L., Multiquadric equations of topography and other irregular surfaces, *J. Geophys. Res.* 176, (1971) 1905-1915.
[16] Franke, R., Scattered data interpolation: test of some methods, *Math. Comput.* 38, (1982) 181-200.
[17] Kansa, E. J., Multiquadrics—a scattered data approximation scheme with applications to computational fluid dynamics—I, *Computers Math. Appl.* 19(8/9), (1990) 127-145.
[18] Madych, W. R., and Nelson, S. A., Multivariate interpolation and conditionally positive definite functions I, *J. Approx. Theory and Its Appl.* 4, (1988) 77-89.
[19] Lynch V. E., Carreras B. A., del-Castillo-Negrete D., Ferreira-Mejias K. M., Hicks H. R., Numerical methods for the solution of partial differential equations of fractional order, *Journal of Computational Physics* 192 (2003) 406–421.
[20] Durbin F., Numerical inversion of Laplace transforms: an efficient improvement to Dubner and Abate's method. *The Computer Journal* 1975; 17:371–376.
[21] Wen P. H., Chen C. S., The method of particular solutions for solving scalar wave equations. *Commun. Numer. Meth. Engng.* (2009).
[22] Wen P. H., Aliabadi M. H., Rooke D. P., The influence of elastic waves on dynamic stress intensity factors (three dimensional problem). *Archive of Applied Mechanics* 1996; 66(6):384–385.

Author Index